CONCRETE SHELL ROOFS

CONCRETE SHELL ROOFS

C.B. WILBY

B.Sc., Ph.D., C.Eng., F.I.C.E., F.I.Struct.E.
Professor of Civil and Structural Engineering;
Chairman of Undergraduate and Postgraduate
Schools of Civil and Structural Engineering,
University of Bradford, England

and

I. KHWAJA

B.Sc., Ph.D., C.Eng., M.I.Struct.E.
Lecturer in Civil and Structural Engineering,
University of Bradford, England

APPLIED SCIENCE PUBLISHERS LTD
LONDON

APPLIED SCIENCE PUBLISHERS LTD
RIPPLE ROAD, BARKING, ESSEX, ENGLAND

ISBN : 0 85334 579 1

WITH 33 ILLUSTRATIONS AND 88 TABLES

Printed in Great Britain by Galliard (Printers) Ltd., Great Yarmouth

PREFACE

Some designers have tactfully steered their clients away from shell roofs because of their lack of experience in the design of such structures. Because of the complexity of reliable elastic design methods, the designer who has not specialised in this field over at least a few years feels, and usually is, unable to design satisfactory shell structures. A designer has to design a job presented to him with his existing knowledge and a very limited amount of extra study, as the design is usually required in a limited time. He cannot therefore embark upon a two-year programme of study to become proficient at the design of shells; there is also the necessary experience which goes with this particular problem of designing shell roofs. The authors have been involved in the design and construction of shell roofs and in research into these structures and allied problems over many years, and have now produced the tables, which appear at the end of this book, for the use of practical designers, based on this considerable experience. There are existing similar tables but these have very severe limitations for practical designers and some can cause designers to use structurally undesirable shells in practice, as explained in the introduction. Coefficients in past tables cannot always be interpolated satisfactorily and this means that very few shells indeed of the sizes one is asked to design can be designed. Clients usually wish to make full use of their land and not oblige the previous design tables. In addition, none of the past tables deals with north-light shells, which are many times more difficult to design than ordinary shells, nor with ordinary shells with internal feather edges and upstanding edge beams to external shells. All of these types are very commonly used in practice.

The designer can now, with the aid of this book, reliably and rapidly design a large range of practical shell roofs. He is able to make drawings of the outlines of the shells, prepare a Bill of Quantities and from this estimate the cost of the construction. In this connection because of the development of proprietary shuttering the cost of curved shuttering is very little different to the cost of straight shuttering—and there is one instance where the cost of this curved shuttering by the sub-contractor, plus the percentage for the main contractor, was less than the cost of the straight shuttering by the main contractor on the same contract—a contract involving shells and flat *in-situ* roofs and floors.

If the shell contract is subsequently obtained the designer can quickly and easily use these design tables and detail the reinforcement as shown in the example of this book. It is likely that designers using these tables will be able to design many contracts with shell roofs. Should their business then be lucrative enough in this activity they may perhaps designate a member of staff to learn considerably more about shells for dealing with the few cases they may wish to propose designing outside the range of these tables.

Cylindrical shells of the type given in the tables but outside this range can, of course, rapidly be analysed with the digital computer program used for these tables.

The tables have values of forces and moments sufficiently close together to allow linear interpolation so that shells with dimensions which are not integers can be designed, and this is very often necessary in practice. This also means that engineers using metric units and future engineers using S.I. units will find that even if the dimensions in these units are 'rounded off' to be integers and they convert to non-integer British Imperial units then there is no difficulty in using the tables for shells with these non-integer dimensions. The tables can therefore be used for all types of units whether British, American, metric or S.I.

The book gives a theory for the analysis of hyperbolic paraboloidal shell roofs. Because of their frequent use, the hyperbolic paraboloidal shells of the inverted umbrella type are treated in detail. Complete methods of solution of the governing differential equation are given for the case of hypar shells, both with and without perimeter and edge beams. Digital computer programs are also given for the analysis of such shells.

CONTENTS

Plate 1 Large scale test of reinforced concrete hypar, 16 ft (4·88 m) square in plan. This was used in the justification of the theory proposed by the authors in the University of Bradford.

Plate 2 Hyperbolic paraboloidal shell roofs recently constructed over market in Yorkshire.

Plate 3 A symmetrical type of cylindrical shell (barrel vault roof). This particular one is part of a system of six shell roofs, generally 2·48 *in* (63 *mm*) *thick, side by side, each of* 60·9 *ft* (18·570 *m*) *span and* 30·3 *ft* (9·240 *m*) *width. They were designed and constructed by Professor C.B. Wilby in* 1957, *when he was the Development Engineer of Stuart's Granolithic Company Limited. They were possibly unique in having a feature which allowed future loads (for example, due to services) to be hung from them at any location at any future time. This was a specification required by the mechanical engineers and was achieved by providing otherwise unnecessary valley beams with steel tubes of internal diameter of* 0·86 *in* (22 *mm*) *just below the junction between the shell and the valley beam and at about* 3 *ft* (0·914 *m*) *centres longitudinally to accommodate future bolts. Also cadmium plated steel bolts of* 0·74 *in* (19 *mm*) *and* 0·86 *in* (22 *mm*) *diameter were placed through holes in the shell and through steel anchorage plates of* 5·98 *in* (152 *mm*) *square on the top of the shell. Each bolt projected out of the soffit of the shell so that anything could be screwed to it at some future date. The nut and plate were covered with a* 1·96 *in* (50 *mm*) *layer of vermiculite insulation on the top of the shell. The shells were for the Newton Works, Hyde, of the I.C.I., and the design was done with the invaluable help of their Mr W. Hayes.*

Plate 4 North-light shell roofs over a large factory area, designed by Professor C.B. Wilby in 1952 (in his capacity then as the Deputy Chief Engineer of Ferrocon Engineering Company Limited). The factory is the printing works of Messrs. Sharp of Bradford and the architects were Messrs. Chippendale and Edmondson, also of Bradford, Yorkshire.

PART 1

DESIGN TABLES FOR SYMMETRICAL AND NORTH-LIGHT CYLINDRICAL SHELL ROOFS

CHAPTER 1

INTRODUCTION AND NOTES ON TABLES FOR NORTH-LIGHT SHELLS

1.1 INTRODUCTION

Cylindrical north-light shell roofs (Fig. 1.1) are widely used in industrial buildings and are noted for their functional simplicity. The principal dimensions of the shells, namely those of span and width, are decided by architects who take into account aesthetic and planning considerations. The area and disposition of the roof glazing is similarly decided and should take into account the daylight factor necessary for the particular use for which the building is required. In addition there are other dimensions and quantities which affect the analysis of cylindrical shells. There can be innumerable combinations of these and it is practically impossible and unnecessary to provide design tables covering all possible combinations in a book like the present one. The tables given in this book, therefore, are limited to shells which are of practical interest and it is considered that these tables cover the majority of north-light shell roofs encountered in practice. Various dimensions and quantities affecting the analysis of these shells are discussed below, where it is shown how particular values of these have been selected in the preparation of the tables.

1.2 FACTORS AFFECTING ANALYSIS

1.2.1 Number of bays

It is impossible to standardize the number of bays in a building and design tables based on only one or two particular numbers of bays would be of very limited use. However, after analysing roofs with different numbers of bays ranging from one to fifteen, it has been found by the authors that if the bays are of identical dimensions and are identically loaded, the moments, forces and deflections in all the intermediate bays are very nearly the same. From the point of view of practical design, therefore, it can safely be assumed that all the intermediate bays behave identically, and this assumption

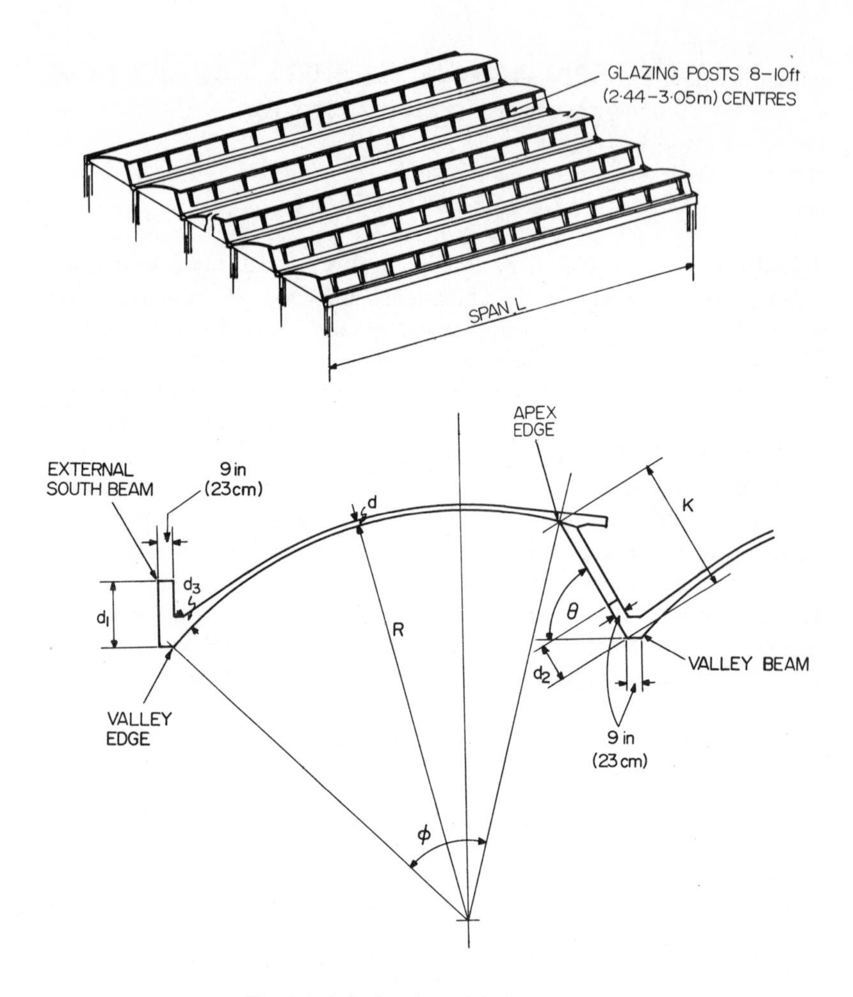

Fig. 1.1 *Cylindrical north-light shell roof.*

has been the normal practice for many designs in the United Kingdom. Thus, to design a multi-bay north-light shell roof, it will only be necessary to know the forces and moments in one intermediate bay and the two external bays. The tables are prepared on the basis of three-bay roofs and the forces and moments are given for all the three bays. In the case of a single-bay roof the shell forces will be similar to those in the external South shell of the tables, and in the case of a two-bay roof the shell forces in one bay will be similar to the external South shell whereas the forces in the other bay will be similar to the external North shell of the tables.

1.2.2 Span

North-light shells are seldom less than 20 ft (6·1 m) in span and this is taken as a lower limit. Shells having spans over 90 ft (27·4 m) would normally require prestressed edge beams and are therefore outside the scope of this book. The tables given, therefore, cover north-light shells with spans ranging from 20 ft to 90 ft (6·1–27·4 m). Spans are increased in steps of 2 ft (0·61 m) and these steps are considered sufficiently close together to allow reasonably accurate interpolation for design purposes.

1.2.3 Width

The width of a shell is usually made equal to one-half the span, and a shell with these proportions is regarded as economical. To allow a greater flexibility in the use of the tables, analysis is given for three different widths for each span less than 60 ft (18·3 m)—one width equal to one-half the span and the other two widths equal to 2 ft (0·61 m) more and 2 ft (0·61 m) less than the first one.

For spans in excess of 60 ft (18·3 m), a span : width ratio of 2 results in large glazing areas. Consequently, this leads not only to lack of uniformity of daylight in the building but also involves additional heating or air-conditioning due to a larger volume of space enclosed under the shells. For this reason, for all spans in excess of 60 ft (18·3 m), tables are given for several widths for each span.

1.2.4 Inclination of glazing

The inclination of the glazing depends upon the latitude of the place of construction. However, inclinations of 60° and 90° to the horizontal are often used and the tables are prepared for these two inclinations only. Most latitudes could, of course, use either of these to facilitate use of the tables.

1.2.5 Area of glazing

A daylight factor of 8% is regarded as an absolute minimum for effective illumination[1]. Apart from the area of glazing, the daylight factor depends also upon the height of the building. For most industrial buildings having glazing inclined at 60° to the horizontal, the daylight factor of approximately 10% is obtained if the inclined glazing area is about 45% of the floor area. The daylight factor drops to approximately 8% if the inclined glazing area is 35% of the floor area. In the tables the area of inclined glazing varies between 35% and 45% of floor area, giving a daylight factor varying from 10% to 8%.

1.2.6 Thickness of shell

Most shells in practice that are designed by British engineers are $2\frac{1}{2}$ in (63·5 mm) thick[2] and this thickness has been adopted in the analysis of all shells given in the tables. Analyses were also carried out for shells of 3 in thickness and these results are expressed such that a shell of 3 in (76·2 mm) thickness is designed by using the tables for shells of $2\frac{1}{2}$ in (63·5 mm) thickness in conjunction with the relevant adjustment factor from Appendix I. The authors would always recommend the use of the tables referring to a shell thickness of $2\frac{1}{2}$ in (63·5 mm) rather than 3 in (76·2 mm), because from their experience, within the range of shells given in the tables, they would consider this to be satisfactory and more economic. Increasing the shell thickness from $2\frac{1}{2}$ in (63·5 mm) is generally an inefficient way of strengthening a shell because it is more efficient to reduce its radius of curvature and/or increase its overall depth. However, shells of 3 in (76·2 mm) thickness seem to have been traditionally used in the U.S.A., and this is why the authors have catered for 3 in (76·2 mm) thick shells.

1.2.7 Radius of shell

The radius of the shell used in the tables is so chosen that the longitudinal compressive stress in the shell is less than one-half of the critical longitudinal stress calculated from the following empirical formula given by Lundgren[3].

$$\sigma_c = \frac{f_c}{1+5\,\dfrac{f_c}{E}\cdot\dfrac{R}{T}}$$

where σ_c = critical stress in the longitudinal direction
f_c = cube strength of concrete
E = Young's modulus for concrete
R = radius of the shell
T = thickness of the shell

In the computation of the critical stress, f_c has been taken as 3000 lb/in^2 ($2{\cdot}07\times10^4$ kN/m^2) and E has been taken as 3×10^6 lb/in^2 ($2{\cdot}07\times10^7$ kN/m^2).

1.2.8 Dimensions of edge and valley beams

The dimensions of edge and valley beams used in the tables are as recommended by Evans and Wilby[2].

1.2.9 Apex and valley thickening

The portions of shells adjoining apex and valley beams are, in practice, thickened so that the maximum thickness of shell at these edges lies between 5 in (127 mm) and 9 in (229 mm) depending upon the span. The weights of these thickenings have been allowed for in the analyses.

1.2.10 Superimposed load

Shells are usually designed to carry snow load and a load of 15 lb/ft^2 (720 N/m^2) of horizontal area is the requirement of British Standard Code of Practice CP3—Chapter 5. However, for simplicity of analysis a superimposed load of 15 lb/ft^2 (720 N/m^2) of surface area is usually adopted and has been used in the tables. Wind forces do not need to be considered for practical purposes as they are predominantly uplifting forces which never overcome the self-weight of a $2\frac{1}{2}$ in (63·5 mm) thick concrete shell.

To make the tables more extensive all the shells given in the tables were also analysed for a snow load of 30 lb/ft^2 (1·44 kN/m^2) of surface area and a table of adjustment factors is given in Appendix II. For example, services and so forth may be light in weight and suspended fairly uniformly beneath the shell and the 30 lb/ft^2 (1·44 kN/m^2) might be adequate for this case of loading.

The weight of roofing felt and insulation was taken as 5·5 lb/ft^2 (263 N/m^2) of surface area and the weight of glazing was taken as 8 lb/ft^2 (383 N/m^2) of the glazed area.

1.3 THEORY

Jenkins' theory[4], otherwise known as the D.J.K. theory, has been used in the analysis of the shells given in the tables.

1.4 ASSUMPTIONS

In addition to the assumptions made by Jenkins in his theory the following assumptions are made:

(i) Elastic shortening of glazing posts is negligible.
(ii) Glazing posts are spaced sufficiently close to each other to allow the post force to be represented by a continuous line load.
(iii) Glazing posts are flexible enough to carry axial forces only and are not subjected to bending moments and transverse forces.
(iv) The weights of valley and apex shell thickenings act as line loads at the respective edges of the shells and a shell is considered to be of uniform thickness throughout.

1.5 BOUNDARY CONDITIONS

The following boundary conditions are assumed:

At Valley Edge (A) of Shell
(i) The rotation of shell is zero.
(ii) The longitudinal strains in shell and edge beam at their junction are equal.
(iii) Vertical deflections of shell edge and edge beam are equal.
(iv) Horizontal deflections of shell edge and edge beam are equal.

At Apex Edge of Shell
(v) Transverse moment is zero.
(vi) Shear force in plane of shell is zero.
(vii) Sum of all shell forces acting at an apex edge in the direction of glazing posts is equal and opposite to the forces in the glazing posts.
(viii) Sum of all forces acting in the direction perpendicular to glazing posts is zero.

(ix) The deflection of an apex edge in the direction of the glazing posts is equal to the deflection of the valley edge of the adjoining shell in the direction of the posts.

CHAPTER 2

NOTES ON TABLES FOR ORDINARY SHELLS

2.1 INTRODUCTION

Two different arrangements of ordinary cylindrical shell roofs are usually used in practice (*see* Figs. 3.4 and 3.7), namely:

(i) Feather-edge shells, that is those having no beams at the internal valleys.
(ii) Shells provided with internal valley beams.

The external edge beams of the feather-edge shells are usually upstanding, whereas the external edge beams of shells with valley beams are partly upstanding and partly downstanding. The depth of internal valley beams, which are always downstanding, is made equal to the depth of the downstanding portion of the external edge beams.

As in the case of north-light shells discussed in Chapter 1, the values which could be taken for the various parameters affecting the analysis of shells are infinitesimal, but it is neither practicable nor necessary to prepare design tables covering all combinations of the parameters. Thus the tables given in this book are limited to those shells which are of practical interest, and it is considered that these tables cover the majority of shell roofs used in practice. Various factors affecting the analysis of these shells are discussed below.

2.2. FACTORS AFFECTING ANALYSIS

2.2.1 Number of bays

It is not possible to standardize the number of bays in a shell roof and any design tables based on one or two bays would be of limited use. However, the analysis of a large number of shells with different numbers of bays indicates that in an ordinary shell roof consisting of identical shells and carrying identical loads, there is very little difference between the forces and moments in all the internal shells of a multi-bay roof. Also, the forces and moments in the internal half of the external shell are very nearly the same as in any completely internal half shell, and the forces and moments in the external half of the external shell are very nearly the same as in a single bay shell. Therefore, for convenience in the design, the analysis of shells given in the tables is based on the following assumptions:

(i) The forces and moments in the external half shell of a roof are the same as in a single bay shell.
(ii) The forces and moments in the internal half of the external shells are the same.

Hence, for the complete design of a multi-bay shell roof, it is sufficient to design only one external-half shell and one internal-half shell, and in a single bay shell the forces in the two halves of the shell will be identical to each other.

2.2.2 Span

Ordinary shell roofs seldom have spans of less than 20 ft (6·1 m), and this is taken as a lower limit in the tables. Shells having spans of over 100 ft (30·5 m) usually require prestressed edge beams and are outside the scope of this book. The tables given, therefore, cover ordinary cylindrical shells with spans ranging between 20 ft (6·1 m) and 100 ft (30·5 m). Spans are increased in steps of 2 ft (0·61 m) and these steps are sufficiently close together to allow fairly accurate interpolation for design purposes.

2.2.3 Width

It is considered that a width of the shell equal to one-half of the span results in a reasonably economical, structurally satisfactory and popular type of shell structure. The tables are therefore prepared on this basis. However, to allow greater flexibility in dimensioning, feather-edge shell tables are given for three different widths for each span—one width equal to half the span and the other two widths being 2 ft (0·61 m) more and 2 ft (0·61 m) less, respectively, than the first one. For shells with valley beams, tables are given for several widths for each span.

2.2.4 Radius of shell

See section 1.2.7.

2.2.5 Thickness of shell

See section 1.2.6.

2.2.6 Dimensions of edge and valley beams

See section 1.2.8.

2.2.7 Shell edge thickening

See section 1.2.9.

2.2.8 Superimposed loads

See section 1.2.10.

2.3 THEORY

Jenkins' theory[4] has been used for the analysis of shells given in the tables.

2.4 ASSUMPTIONS

In addition to the assumptions made in Jenkins' theory it has been assumed in the analysis of shells given in the tables that the weight of the valley thickening acts as a line load at the edge and that the shell is considered to be of uniform thickness throughout.

2.5 BOUNDARY CONDITIONS

The following boundary conditions have been assumed:

At the Crown of Shell

(i) The rotation of the shell is zero.

(ii) The in-plane shear force in the shell is zero.
(iii) The transverse shear force in the shell is zero.
(iv) The horizontal deflection of the shell is zero.

At the Valley Edge of the Shell Adjoining an Edge Beam

(i) The horizontal deflections of the shell and edge beam are equal.
(ii) The vertical deflections of the shell and edge beam are equal.
(iii) The rotation of the shell edge is zero.
(iv) The longitudinal strains in the shell and edge beam at their junction are equal.

At the Valley Edge of a Feather-edge Valley

(i) The vertical shear force at the shell edge is zero.
(ii) The in-plane shear force at the shell edge is zero.
(iii) The horizontal deflection of the shell edge is zero.
(iv) The rotation of the shell edge is zero.

CHAPTER 3

EXPLANATION OF TABLES

I North-light Shell Roof Tables

3.1 DESCRIPTION OF TABLES

The dimensions, loading and the forces and moments induced in cylindrical north-light shell roofs are recorded in Table Nos. N-1 to N-81. In the computations, the weight of glazing including the weight of reinforced concrete glazing posts is assumed to be 8 lb/ft^2 (383 N/m^2) of the total glazing area.

Each table gives the results of the analysis of two different north-light systems, namely, a shell system with the glazing inclined at 60° to the horizontal and a system with the glazing inclined at 90° to the horizontal. For each of the two systems considered in each table, full information is given about the geometry of and the forces and moments in the two external shells and one internal shell. Similarly, the dimensions of and the forces and moments in the two external and one internal beam are given for each of the two systems of shells in each table. It should be emphasized that for a given span and width, the radius of curvature of the shells with a 60° glazing is different to that with a 90° glazing.

Even though the computer programs written for the analysis of shells included the computation of all displacements, forces and moments in the shells and the edge beams, only those forces which are needed for the design of shells are given in the tables. It has been checked that the forces which are not given are insignificant as far as the actual design is concerned.

As pointed out in Chapter 1, all the internal shells and internal beams may be provided with identical reinforcement.

3.2 NOTATION

The notation used for defining the geometry of the two systems of shells and the forces and moments in the shells and edge beams are shown in Figs. 3.1a, b and 3.2a, b, respectively.

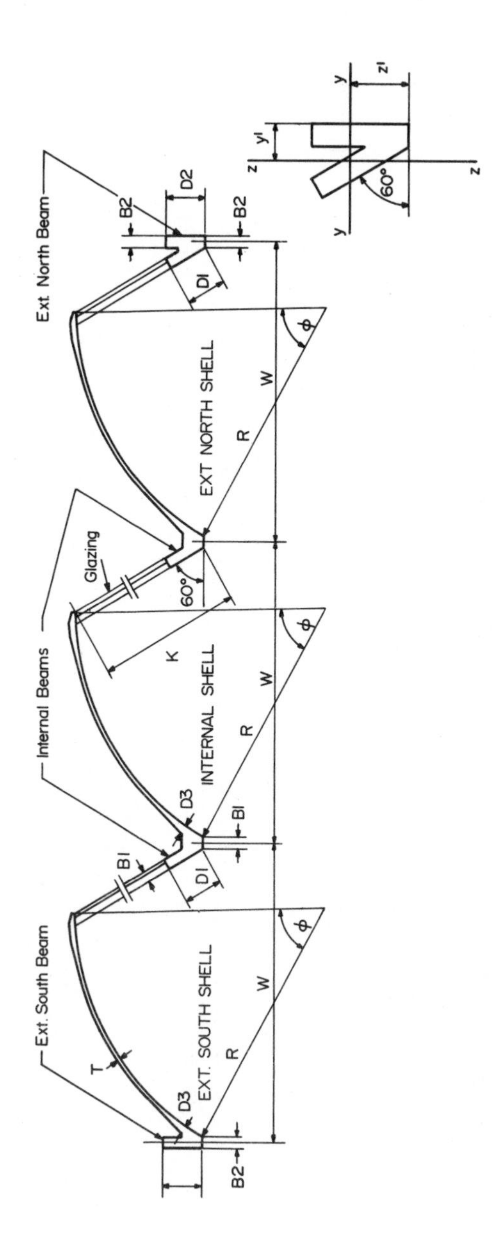

Fig. 3.1a 60° *north-light shell roof.*

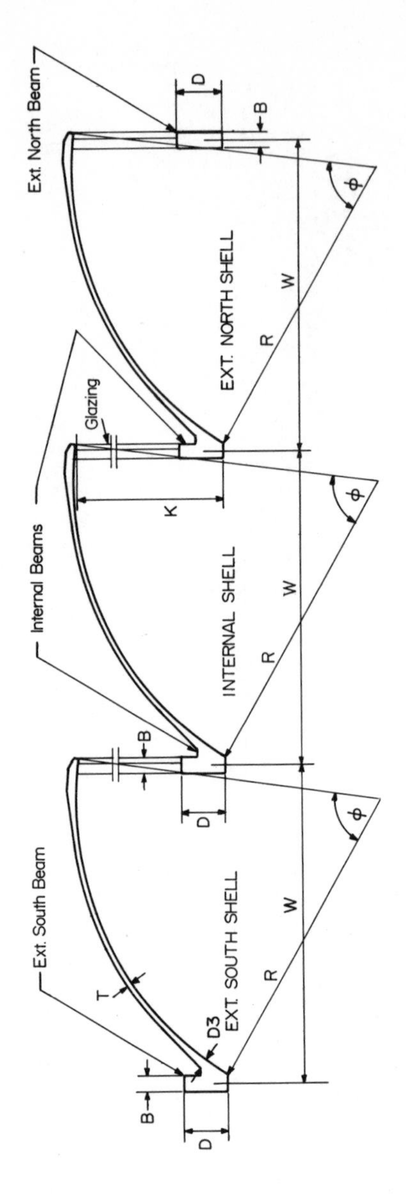

Fig. 3.1b 90° *north-light shell roof.*

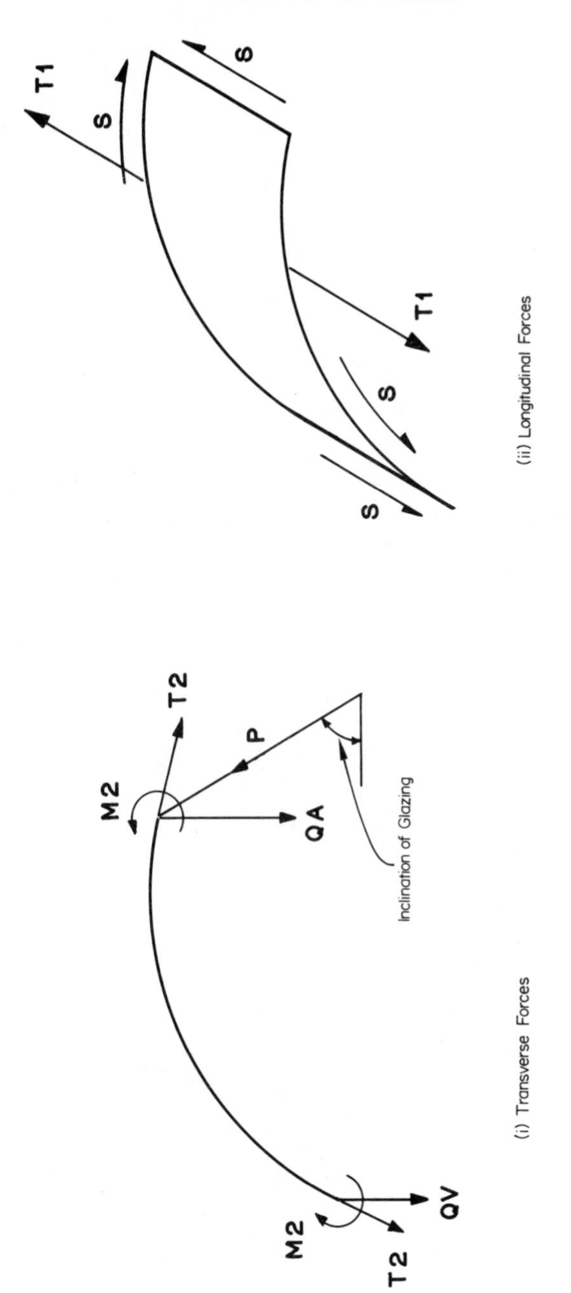

Fig. 3.2a *Shell forces and moments.*

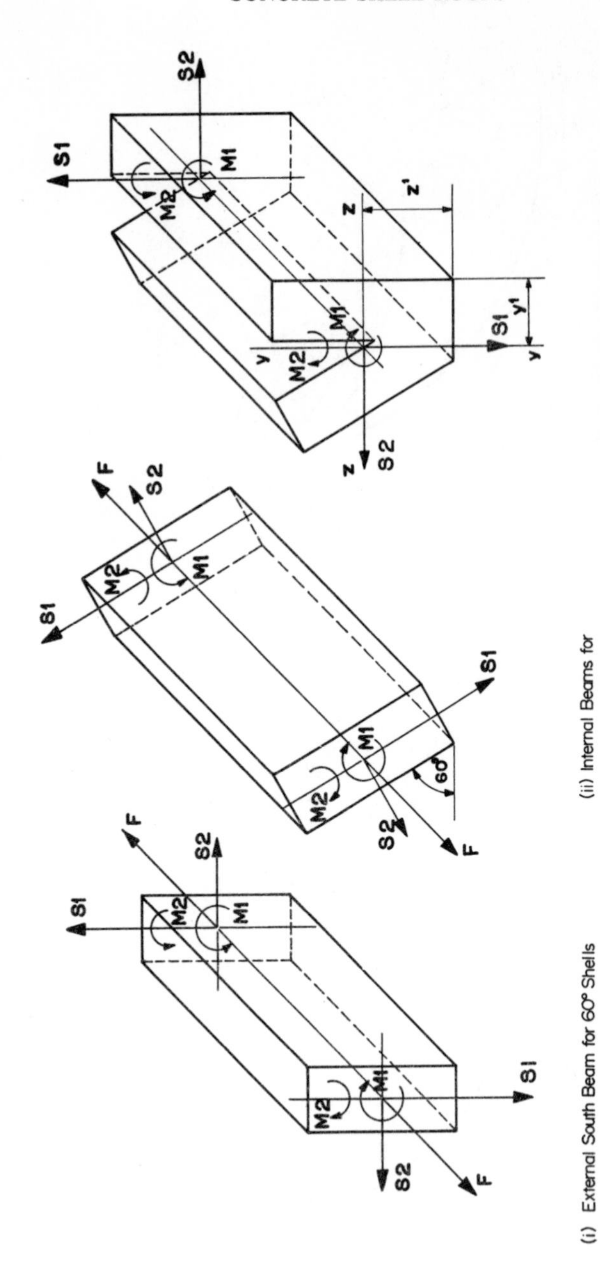

Fig. 3.2b *Edge beam forces and moments.*

3.3 SIGN CONVENTION

The following sign convention has been adopted in the tables:

(*a*) *Shell Forces*

T1, T2—positive for tension.

M2—positive for moment causing tension in the bottom filnes of the shell.

S—positive for shear forces having directions as shown in Fig. 3.2a.

(*b*) *Edge Beam Forces*

F—positive for tension.

M1—positive for moment causing tension in the bottom filnes of the beam.

M2—positive for moment causing tension in filnes nearest to the valley edge of shell and parallel to the depth of beam.
In the case of an external north edge beam where the moment causing tension in the outside face of the beam is taken as positive.

S1, S2—positive for shear forces having directions shown in Fig. 3.2b.

P—positive for compression in the glazing posts.

3.4 UNITS

The units for the dimensions of the shells and edge beams and for the forces in edge beams are printed out in the tables. For forces and moments in the shells the units are as follows:

T1, T2, S—lb/ft.
M2—lb.ft/ft.

3.5 DISTRIBUTION OF FORCES ACROSS THE SHELL

The shell forces *across* the shell (*i.e.* along the shell arc) are given at nine equi-spaced points numbered consecutively from 0 to 8 across the shell. The point 0 corresponds to the valley edge of the shell and the point 8 corresponds to the apex edge. The location of these points on the shell arc is shown in Fig. 3.3.

3.6 DISTRIBUTION OF FORCES ALONG THE SHELL SPAN

3.6.1 Forces in the shell

The shell forces T1, T2 and M2 given in the tables are for the *midspan* section and the shear force S is for the end section of the shell.

The forces at any section along the length of the shell, at a distance x from the midspan can be calculated as follows:

$$(T1)_x = (T1)_{midspan} \cdot \cos \frac{\pi x}{L}$$

$$(T2)_x = (T2)_{midspan} \cdot \cos \frac{\pi x}{L}$$

$$(M2)_x = (M2)_{midspan} \cdot \cos \frac{\pi x}{L}$$

$$(S)_x = (S)_{end} \cdot \sin \frac{\pi x}{L}$$

where L = span of the shell.

The forces T1, T2 and M2 at midspan and the force S at end are those given in the tables. Thus, the force T1 at, say, point 5 across the shell at a distance $L/4$ from midspan would be

$$(T1)_{x = L/4,\ \text{point } 5} = (T1)_{x = 0,\ \text{point } 5} \cdot \cos (\pi/4)$$

3.6.2 Forces in edge and valley beams

The forces F, M1 and M2 given in the tables are for the midspan section of the beam and the forces S1 and S2 are for the end section. The axial force P in the glazing posts is for the midspan section of the shell.

The edge beam forces and the force in the glazing posts at any section along the length of the beam at a distance x from midspan can be calculated as follows:

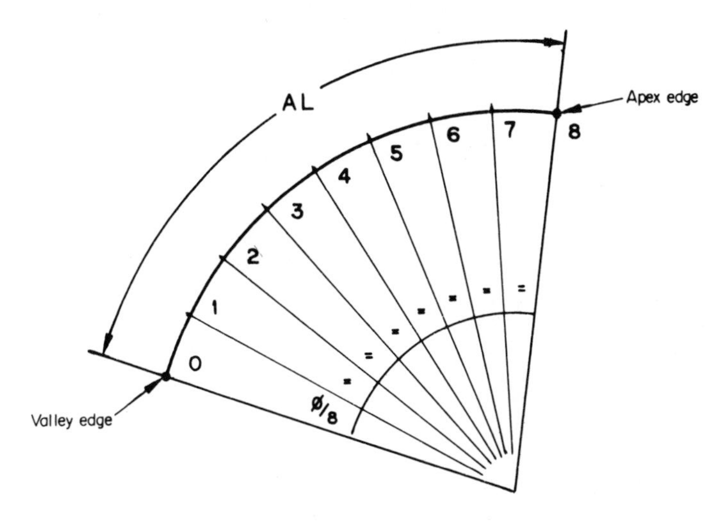

Fig. 3.3 Points across north-light shells at which forces are tabulated.

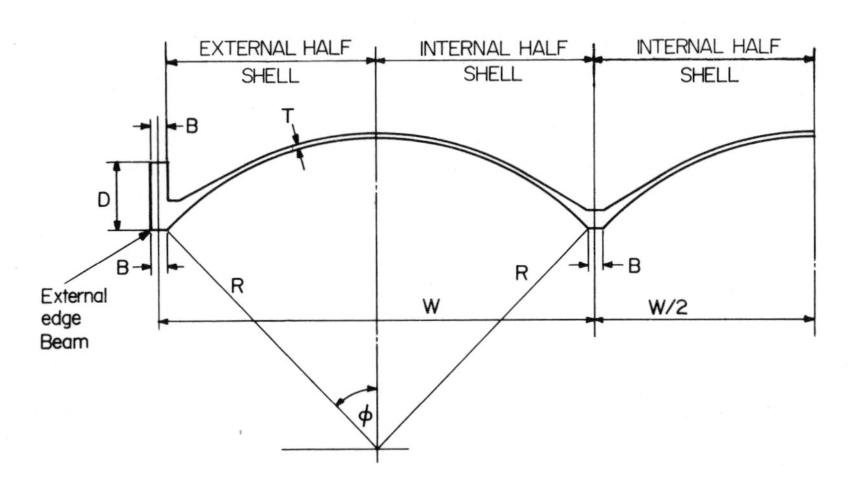

Fig. 3.4 Feather edge cylindrical shell.

$$(F)_x = (F)_{midspan} \cdot \cos \frac{\pi x}{L}$$

$$(P)_x = (P)_{midspan} \cdot \cos \frac{\pi x}{L}$$

$$(M1)_x = (M1)_{midspan} \cdot \cos \frac{\pi x}{L}$$

$$(M2)_x = (M2)_{midspan} \cdot \cos \frac{\pi x}{L}$$

$$(S1)_x = (S1)_{end} \cdot \sin \frac{\pi x}{L}$$

$$(S2)_x = (S2)_{end} \cdot \sin \frac{\pi x}{L}$$

The midspan and end forces on the right-hand sides of the above equations are those given in the tables.

II Ordinary Shell Roof Tables

3.7. DESCRIPTION OF TABLES

The ordinary shell roof tables are divided into two groups: (i) Shell roofs with feather-edge valleys, *i.e.,* with no beams provided at the internal valleys of the shell. In such a roof the external edge beams are taken as upstanding. (ii) Shell roofs with internal valley beams. All the beams in such roofs are taken as downstanding.

3.7.1 Shell roofs with feather-edge valleys

This type of roof is shown in Fig. 3.4. The dimensions, loading and the results of the analysis of these shells are given in Tables Nos. F-1 to F-96.

Each table gives the analysis of an external edge beam, an external-half shell and an internal-half shell. As mentioned in Chapter 2, all internal shells and the internal-halves of the external shells may be provided with identical reinforcement.

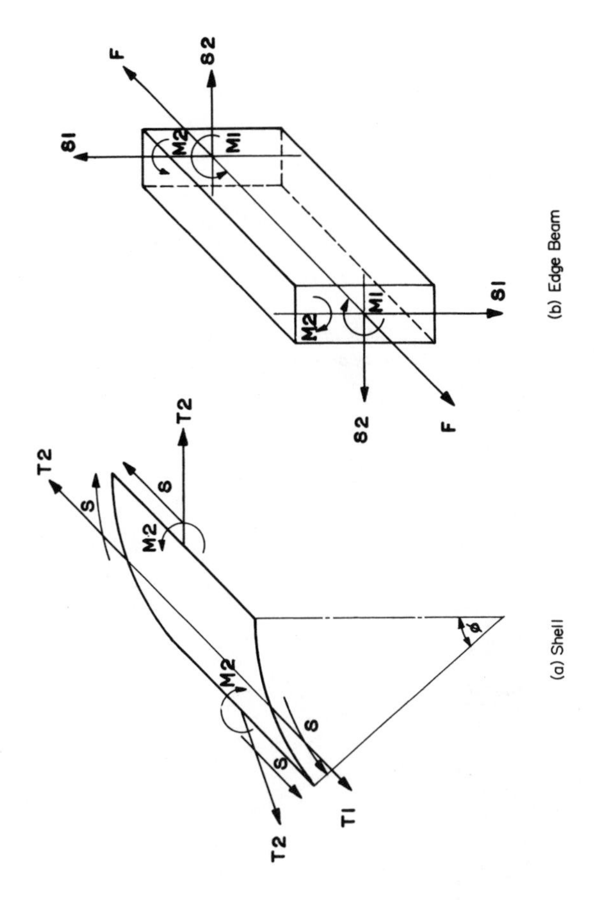

Fig. 3.5 Forces in shells and external edge beams.

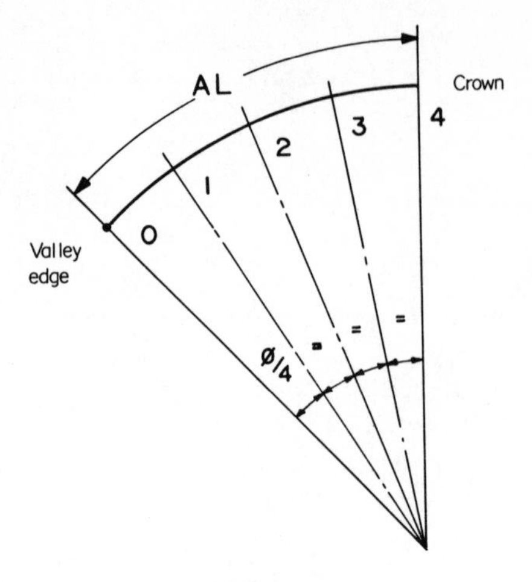

Fig. 3.6 Position of points across an ordinary shell.

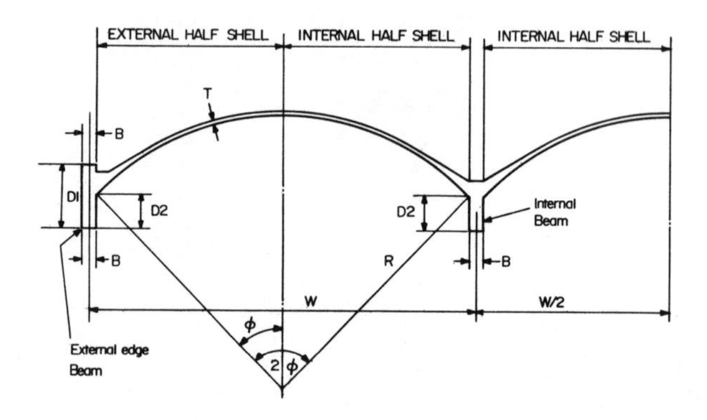

Fig. 3.7 Cylindrical shell with valley beams.

3.7.2 Shell roofs with internal valley beams

This type of shell roof is shown in Fig. 3.5. The dimensions, loading and the results of the analysis of these shells are given in Tables V-1 to V-168.

Each table gives the analysis of an external edge beam, an internal edge beam, an external-half shell and an internal-half shell. It should be noted that the forces and moments given for the internal valley beam are due to the total load coming from shells on *both sides* of the beam.

3.8 NOTATION, SIGN CONVENTION AND UNITS

The sign convention and units used in the ordinary shell roof tables are the same as given for north-light shells in sections 3.2, 3.3 and 3.4. The notation used for forces and moments is shown in Fig. 3.6.

3.9 DISTRIBUTION OF FORCES ACROSS THE SHELL

The shell forces *across* the shell (*i.e.* along the shell arc) are given at five equi-spaced points numbered consecutively from 0 to 4 across one-half of the shell. The point 0 corresponds to the valley edge and the point 4 corresponds to the crown of the shell. The location of these points on the shell arc is shown in Fig. 3.7.

3.10 DISTRIBUTION OF FORCES ALONG THE SHELL SPAN

The distribution of the forces in the shells and the edge and valley beams of the ordinary shell roofs is the same as for the north-light shell roofs described in section 3.6.

CHAPTER 4

EXAMPLE OF DESIGN OF A NORTH-LIGHT SHELL ROOF

It is required to design a cylindrical north-light shell roof consisting of five bays, each 30 ft (9·1 m) wide and with a 60 ft (18·3 m) span. The glazing is inclined at 60° to the horizontal and the load on the shell, inclusive of its own weight, is 50 lb (2400 N/m^2) per sq. ft. The shell is $2\frac{1}{2}$ in (63·5 mm) thick.

The analysis of the north-light shell of the given dimensions is given in Table No. N-58. Other relevant parameters given in this table are (*see* Fig. 3.1 for nomenclature):

R = 27·5 ft (8·38 m)
K = 14·5 ft (4·42 m)
φ = 56·38 ft (17·18 m)
B1 = 0·75 ft (0·228 m)
B2 = 0·75 ft (0·228 m)
D1 = 4·00 ft (1·22 m)
D2 = 3·50 ft (1·07 m)

These quantities completely define the geometry of the shells and edge beams.

As mentioned in Chapter 1, it is only necessary to design the two external shells and one internal shell. The forces and hence the reinforcement in all the three internal bays will be identical.

4.1 DESIGN OF SHELL REINFORCEMENT

The shell reinforcement is designed to resist the following forces:

(i) The longitudinal force T1,
(ii) The shear force, S, and
(iii) The transverse bending moment, M2.

4.2 EXTERNAL SOUTH SHELL

4.2.1 Reinforcement for longitudinal tension, T1

The variation in the longitudinal force T1 across the shell, taken from Table No. N-58 is shown graphically in Fig. 4.1. It is seen from this figure that the shell is subjected to tension over lengths of 5·75 ft (1·75 m) and 5·25 ft (1·60 m) from the valley and apex edges respectively and reinforcement must be provided to resist this tension.

Using mild steel reinforcement, its permissible tensile stress, according to the British Standard Code of Practice C.P. 114 (1957), is 20 000 lb/in^2 ($1{\cdot}38\times10^5$ kN/m^2). The calculation of area and spacing of tensile steel is shown in Table 4.1. .

The maximum longitudinal compression occurs near the mid-point of the shell arc and is approximately equal to 17 000 lb/ft (248 kN/m).

Therefore, the maximum longitudinal compressive stress in the shell $= 17\,000/(12 \times 2{\cdot}5) = 567$ lb/in^2 (3910 kN/m^2).

Using a concrete with a cube strength of 3000 lb/in^2($2{\cdot}07\times10^4$kN/m^2), (USA cylinder strength 2520 psi ($1{\cdot}74\times10^4$ kN/m^2)), the critical longitudinal buckling stress, according to Lundgren, is given by

$$\sigma = \frac{\sigma_c}{1 + \dfrac{5\sigma_c}{E} \cdot \dfrac{R}{T}}$$

where σ = critical stress
σ_c = cube strength of concrete=$1{\cdot}19\times$(USA cylinder strength)
E = Young's modulus for concrete
R, T = radius and thickness, respectively, of the shell

In the present case,

σ_c = 3000 lb/in^2 ($2{\cdot}07\times10^4$ kN/m^2)
E = 3×10^6lb/in^2 ($2{\cdot}07 \times 10^7$ kN/m^2)
R = 27·5 ft = 330 in (8·38 m)
T = 2·5 in (63·5 mm).

TABLE 4.1

Distance from valley edge	For valley tension T1	AT	Bar size and spacing
0 to 2 ft 1 in	41 803 lb/ft	2·09 in^2	3/4 in dia. at 5 in c/c top and bottom
(0 to 0·635 m)	(610·1 kN/m)	(1350 mm^2)	(19·1 mm dia. at 127·0 mm c/c top and bottom)
2 ft 1 in to 4 ft 2 in	21 000 lb/ft	1·05 in^2	3/4 in dia. at 5 in c/c
(0·635 to 1·270 m)	(306 kN/m)	(677 mm^2)	(19·1 mm dia. at 127·0 mm c/c top and bottom)
4 ft 2 in to 5 ft 9 in	8000 lb/ft	0·4 in^2	1/2 in dia. at 5 in c/c
(1·270 to 1·753 m)	(117 kN/m)	(258 mm^2)	(12·7 mm dia. at 127·0 mm c/c top and bottom)
0 to 2 ft 1 in	34 003 lb/ft	1·70 in^2	3/4 in dia. at 6 in c/c top and bottom
(0 to 0·635 m)	(496·2 kN/m)	(1097 mm^2)	(19·1 mm dia. at 152·4 mm c/c top and bottom)
2 ft 1 in to 4 ft 2 in	15 000 lb/ft	0·75 in^2	3/4 in dia. at 6 in c/c
(0·635 to 1·270 m)	(219 kN/m)	(484 mm^2)	(19·1 mm dia. at 152·4 mm c/c top and bottom
4 ft 2 in to 5 ft 9 in	4000 lb/ft	0·2 in^2	3/8 in dia. at 6 in c/c
(1·270 to 1·753 m)	(58 kN/m)	(129 mm^2)	(9·5 mm dia. at 152·4 mm c/c top and bottom)

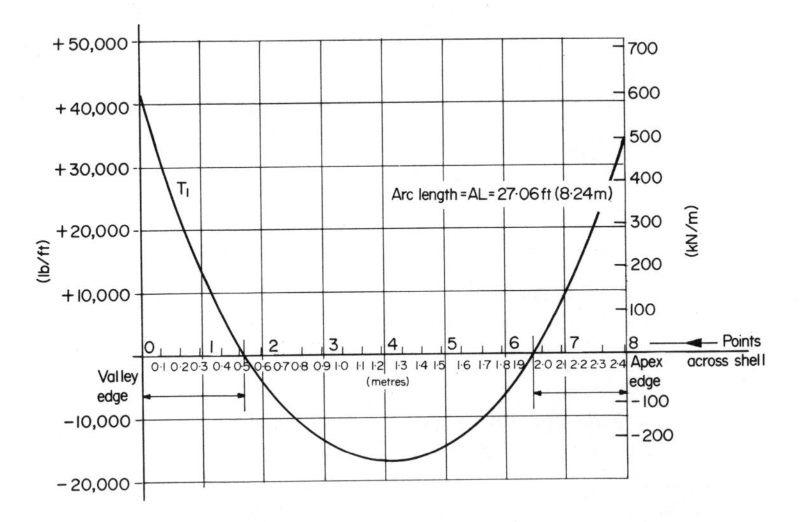

Fig. 4.1 *Variation of* T *across shell.*

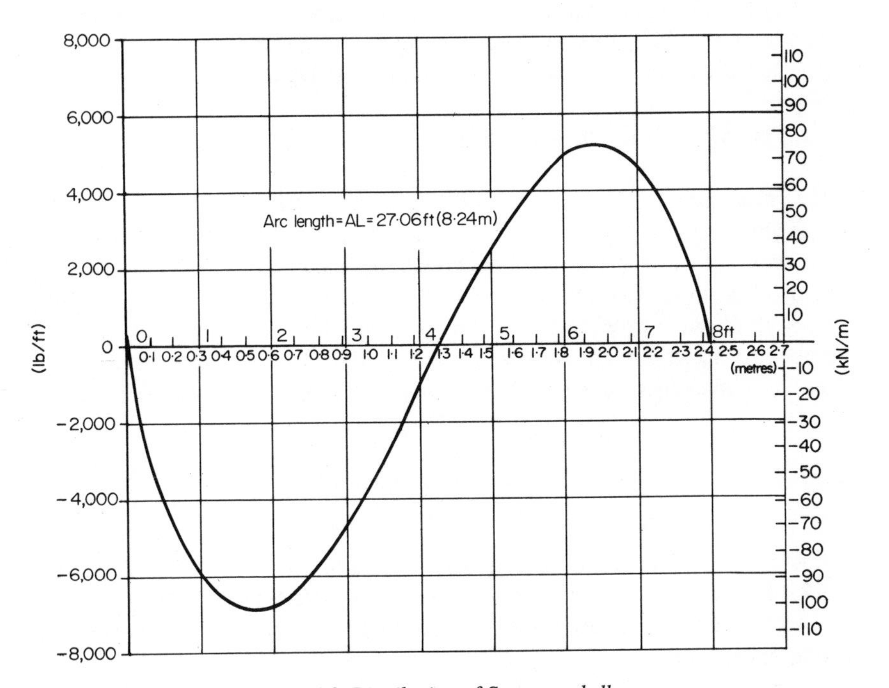

Fig. 4.2 *Distribution of* S *across shell.*

∴ The critical longitudinal buckling stress

$$= \frac{3000}{1 + \frac{5 \times 3000}{3 \times 10^6} \cdot \frac{330}{2 \cdot 5}} = 1809 \text{ lb/in}^2 \, (1 \cdot 247 \times 10^4 \text{ kN/m}^2)$$

Using a load factor of 2, the allowable longitudinal stress = 905 lb/in^2 (6240 kN/m^2).

The actual stress is therefore much less than the allowable stress.

From this theory, therefore, no reinforcement is required in the compression zone as the concrete is quite capable of withstanding the compressive stress induced.

However, to cater for shrinkage and temperature stresses in the shell, provide longitudinally, say, 3/8 in (9·5 mm) diameter bars at 12 in (305 mm) centres over the entire compression zone. These also act as 'spacer' bars between the fabrics and alternate bars are lifted with small concrete blocks to support the top fabric reinforcement.

The tension reinforcements calculated in Table 4.1 are for the midspan section of the shell. This reinforcement may be curtailed along the length of the shell as in the case of a simply supported beam.

4.2.2 Reinforcement for shear force, S

The maximum in-plane shear force in the shell occurs at the end supports and the values of this force are given in Table No. N-58. These values are shown graphically in Fig. 4.2 from which the maximum shear force is found to be 6900 lb/ft (101 kN/m). The shell is usually thickened to 3·5 in (88·9 mm) at its junction with the end supports and this thickening is usually extended to about a tenth of the span from either support. Therefore, taking the shell thickness as 3·5 in (88·9 mm), the maximum shearing stress

$$= \frac{6900}{12 \times 3 \cdot 5} = 165 \text{ lb/in}^2 \, (1140 \text{ kN/m}^2).$$

Assuming a maximum permissible shear stress in concrete of 79 lb/in^2 (545 kN/m^2), the actual shear stress exceeds the permissible one and therefore the authors recommend the provision of reinforcement to resist the total shear force.

Even if the actual shear stress is less than the permissible stress it is normal practice in shell design to provide shear reinforcement at the end section on the assumption that the concrete is not capable of resisting any shear force. The shear reinforcement usually consists of bars inclined at 45° to the longitudinal axis of the shell.

Ignoring the small amount of positive shear force at the valley edge of the shell, it is seen from Fig. 4.2 that the change of sign of shear force occurs at 14·5 ft (4·41 m) from the valley edge. Using mild steel bars with a permissible shear stress of 20 000 lb/in² ($1{\cdot}38 \times 10^5$ kN/m²), the size and spacing of bars at various points across the shell are calculated in Table 4.2.

TABLE 4.2

Distances from valley edge	*Maximum* S	*Area of steel*	*Bar size and spacing*
0–5 ft (0–1·52 m)	– 6800 lb/ft (–99·2 kN/m)	0·338 in² (218 mm²)	3/8 in dia. at 4 in c/c (9·5 mm dia. at 101·6 mm c/c)
5–10 ft (1·52–3·05 m)	– 6900 lb/ft (–100·7 kN/m)	0·345 in² (223 mm²)	3/8 in dia. at 4 in c/c (9·5 mm dia. at 101·6 mm c/c)
10–14·5 ft (3·05–4·41 m)	–4700 lb/ft (–68·6 kN/m)	0·235 in² (152 mm²)	3/8 in dia. at 5 in c/c (9·5 mm dia. at 127·0 mm c/c)
14·5–20 ft (4·41–6·09 m)	4500 lb/ft (65·7 kN/m)	0·225 in² (145 mm²)	3/8 in dia. at 5 in c/c (9·5 mm dia. at 127·0 mm c/c)
20–25 ft (6·09–7·62 m)	5200 lb/ft (75·9 kN/m)	(0·260 in² (168 mm²)	3/8 in dia. at 5 in c/c (9·5 mm dia. at 127·0 mm c/c)
25–27·06 ft (7·62–8·24 m)	4700 lb/ft (68·6 kN/m)	0·235 in² (152 mm²)	3/8 in dia. at 5 in c/c (9·5 mm dia. at 127·0 mm c/c)

The change of sign of shear force at 14·5 ft (4·41 m) from the valley edge indicates that the direction of shear reinforcement should be reversed at this point. Maximum permissible shear force for a 2·5 in (63·5 mm) thick shell $= 79 \times 12 \times 2.5 = 2370$ lb/ft (34·6 kN/m).

Taking a linear distribution of shear force along the span, with zero shear force at midspan, the shear reinforcement from the point of maximum shear at the end on the valley side of the shell must extend over a distance from the end equal to

$$30 - \left(\frac{2370}{6900} \times 30 \right) = 19{\cdot}7 \text{ ft } (6{\cdot}00 \text{ m}).$$

Similarly on the apex side of the shell the reinforcement from the point of maximum shear at the end must extend over a distance from the end equal to

$$30 - \left(\frac{2370}{5200} \times 30 \right) = 16{\cdot}3 \text{ ft } (4{\cdot}97\text{m}).$$

4.2.3 Reinforcement for M2

The reinforcement for the transverse bending moment, M2, usually consists of top and bottom layers of welded wire fabric. As the fabrics used as shell reinforcement are usually light, it is economical to use the same size of fabric both at the top and bottom. For certain large shells, however, it may be more economical to use different sizes of fabrics at the top and bottom. The maximum transverse bending moment for this shell, from Table No. N-58, is equal to −317 lb ft/ft (1410 N.m/m). Using an effective depth of shell, of 1·7 in (43·2 mm), the area of fabric to resist this moment is:

$$\frac{317 \times 12}{30\,000 \times 0{\cdot}872 \times 1{\cdot}7} = 0{\cdot}0855 \text{ in}^2/\text{ft } (181{\cdot}0 \text{ mm}^2/\text{m})$$

Therefore use fabric reference 111, B.S.S. 1221 A., consisting of 10 gauge cross wires at 3 in (76·1 mm) centres and 7 gauge longitudinal wires at 12 in (305 mm) centres and weighing 3·37 lb per yd^2 (1·83 kg/m^2). As the positive bending moment is not too different to the negative bending moment, use the same fabric at both the top and bottom.

Theoretically, the transverse in-plane force T2 and the transverse moment M2 should be used to determine the area of transverse reinforcement. But since the transverse in-plane force is generally very small, the reinforcement obtained on the basis of resisting moments only is usually sufficient.

For example, in this particular case, the maximum transverse force equals

$$T2 = -2210 \text{ lb } (-9830\text{N})$$

$$\therefore \text{ Maximum compressive stress} = \frac{2210}{12 \times 2{\cdot}5} \simeq 74 \text{ lb/in}^2 (510{\cdot}2\text{kN/m}^2)$$

4.3 INTERNAL AND EXTERNAL NORTH SHELLS

The reinforcement for the internal and external north shells may be designed in the same fashion as for the external south shell.

4.4 DESIGN OF BEAM REINFORCEMENT

The reinforcement for the external and internal beams is designed to resist the following forces:

(i) Longitudinal direct force, F, in case of external south and internal beams only;
(ii) The bending moments, M1 and M2; and
(iii) The shearing forces S1 and S2.

4.5 EXTERNAL SOUTH BEAM

Referring to Fig. 4.3; the cross-sectional area of beam

$$= 3{\cdot}5 \times 0{\cdot}75 \text{ ft}^2$$
$$= 378 \text{ in}^2 \ (2{\cdot}44 \times 10^5 \text{ mm}^2)$$

$$I_{11} = \frac{9 \times 42^3}{12} = 55600 \text{ in}^4 \ (2{\cdot}31 \times 10^{10} \text{ mm}^4)$$

$$I_{22} = \frac{42 \times 9^3}{12} = 2548 \text{ in}^4 \ (1{\cdot}06 \times 10^9 \text{ mm}^4)$$

From Table No. N-58,

F = −2498 lb (− 11110 N)
M1 = 278 539 lb/ft (4064·97 kN/m)
M2 = 8516 lb/ft (124·3 kN/m)
S1 = −18 204 lb (− 80 971 N)
S2 = 572 lb (2544 N)

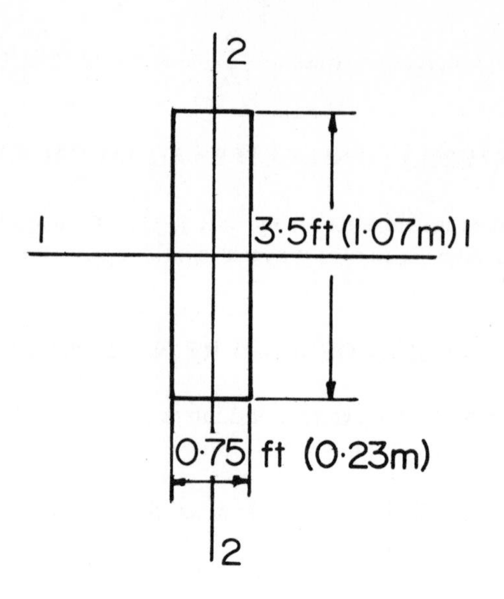

Figure 4.3

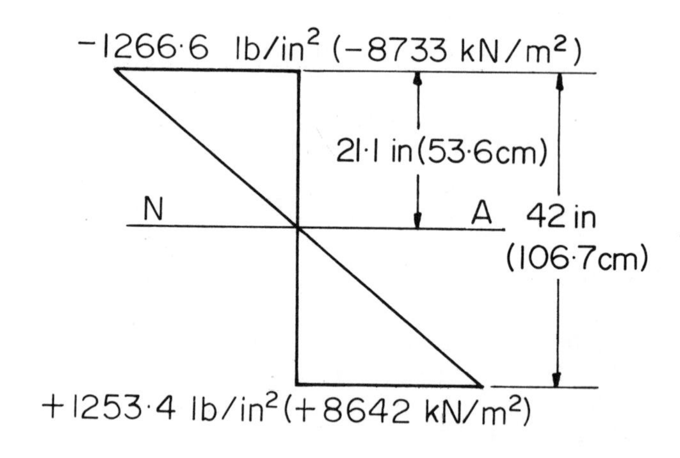

Figure 4.4

$$\therefore \text{Direct stress} = -\frac{2498}{378} = -6\cdot6 \text{ lb/in}^2(-45\cdot5 \text{ kN/m}^2) \text{ comp.}$$

Bending moment about axis 1—1 = 278 539 lb/ft (4064·97 kN/m)

∴ Bending stresses in top and bottom fibres

$$= \pm\frac{278\,538 \times 12 \times 21}{55\,600} \text{lb/in}^2 = \pm\ 1260 \text{ lb/in}^2\ (\pm 8687 \text{ kN/m}^2)$$

∴ The total stresses due to direct force and bending moment due to M1 are:

at top fibre $= -1260 - 6\cdot6 = -1266\cdot6 \text{ lb/in}^2\ (-8733 \text{ kN/m}^2)$
at bottom fibre $= +1260 - 6\cdot6 = +1253\cdot4 \text{ lb/in}^2\ (+8642 \text{ kN/m}^2)$

The variation of these stresses with the depth of the beam is shown in Fig. 4.4.

Depth of neutral axis from top

$$= \frac{1266\cdot6}{2520} \times 42 = 21\cdot1 \text{ in (536 mm).}$$

Therefore, tensile reinforcement is required for the bottom 20·9 in (531 mm) of the depth of beam.

Dividing this depth of tensile zone into say seven strips of 3 in (76·2 mm) depth as shown in Fig. 4.5, the total tensile force and hence the reinforcement required in each strip is calculated in Table 4.3. The permissible stress in steel is taken as 30 000 lb/in² ($2\cdot07 \times 10^5$ kN/m²)

The bending moment about axis 3—2 (Fig. 4.3) = 8516 lb/ft (124·3 kN/m). This moment causes tension on the face of the beam away from the shell. Maximum fibre stresses due to M2 $= \pm \dfrac{8516 \times 4\cdot5 \times 12}{2548}$

$= \pm\ 180 \text{ lb/in}^2\ (1240 \text{ kN/m}^2)$

∴ Total tensile force on the outside half of the beam $= 180 \times 42 \times 4\cdot5 \times \frac{1}{2} = 17\,000$ lb (75 600 N)

$$\text{Area of steel} = \frac{17\,000}{30\,000} = 0\cdot567 \text{ in}^2$$

∴ Provide say 4 Nos. $\frac{1}{2}$ in dia. bars on the outside face of the beam. These

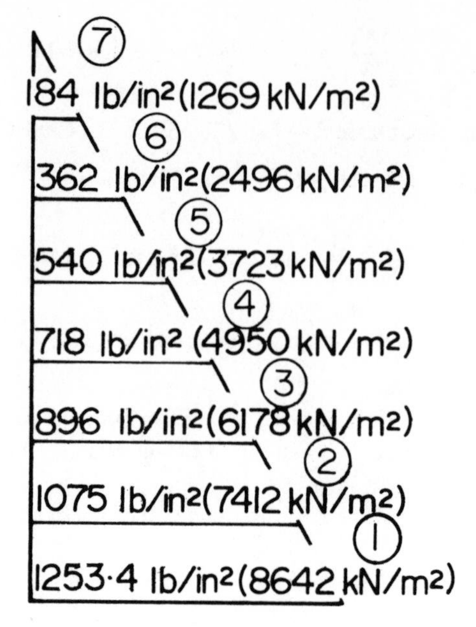

Figure 4.5

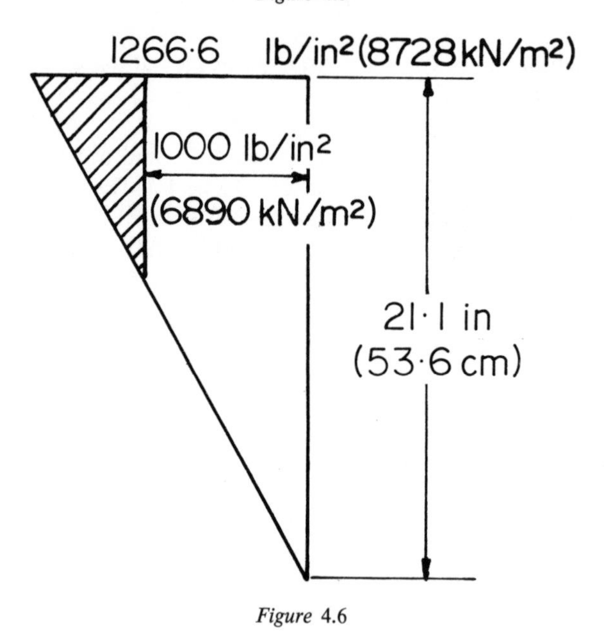

Figure 4.6

bars are in addition to those calculated for the M1 moment. Assuming a maximum permissible compressive stress of 1000 lb/in^2 (6890 kN/m^2) in the concrete, the actual stress due to M1 and F is 266·6 lb/in^2 (1838 kN/m^2) greater than the maximum permissible. Therefore, compressive reinforcement is required at the top sections of the beam.

TABLE 4.3

Strip number	*Total tensile force*		*Area of reinforcement*	*Number and size of bars*
1	$\frac{1253 \cdot 4+1075}{2} \times 3 \times 9$	=31 400 lb (140 000 N)	1·048 in^2 (676·2 mm^2)	2 bars of 7/8 in (22·3 mm) dia.
2	$\frac{1075+896}{2} \times 3 \times 9$	=26 600 lb (118 000 N)	0·887 in^2 (572·3 mm^2)	2 bars of 3/4 in (19·1 mm) dia
3	$\frac{896+718}{2} \times 3 \times 9$	=21 800 lb (97 000 N)	0·729 in^2 (469·1 mm^2)	2 bars of 3/4 in (19·1 mm) dia.
4	$\frac{718+540}{2} \times 3 \times 9$	=17 000 lb (75 600 N)	0·566 in^2 (365·2 mm^2)	2 bars of 5/8 in (15·9 mm) dia.
5	$\frac{540+362}{2} \times 3 \times 9$	=12 200 lb (54 300 N)	0·407 in^2 (262·6 mm^2)	2 bars of 5/8 in (15·9 mm) dia.
6	$\frac{362+184}{2} \times 3 \times 9$	=7360 lb (32 700 N)	0·246 in^2 (158·7 mm^2)	2 bars of 1/2 in (12·7 mm) dia.
7	$\frac{184}{2} \times 2 \cdot 9 \times 9$	= 2400 lb (10 700 N)	0·080 in^2 (51·6 mm^2)	2 bars of 1/2 in (12·7 mm) dia.

Referring to Fig. 4.6 the compressive force for which reinforcement is required

$$= \tfrac{1}{2}\times 266 \cdot 6\times\left(21 \cdot 1-\frac{1000}{1266 \cdot 6}\times 21 \cdot 1\right)\times 9 \text{ lb}$$

$$= 6600 \text{ lb } (29\,400 \text{ N})$$

Allowing a compressive steel stress of 20 000 lb/in^2, area of steel

$$= \frac{6600}{20\,000} = 0{\cdot}33 \text{ in}^2 \ (213 \text{ mm}^2)$$

Therefore provide 2 No. $\frac{1}{2}$ in (12·7 mm) dia. bars near the top surface of the beam.

The size and numbers of bars calculated above are for the midspan section of the beam. These bars may be curtailed at other sections along the length of the beam as for a simply supported beam.

Maximum vertical shear force at end of the beam

$$= S1 = 18\,204 \text{ lb } (80\,971 \text{ N}).$$

This is less than the permissible shear stress of 79 lb/in^2 (545 kN/m^2). However, provide nominal stirrups of 3/8 in (9·5 mm) dia. at 12 in (305 mm) c/c throughout the length of the beam.

4.6 EXTERNAL NORTH AND INTERNAL BEAMS

The reinforcement in the internal and external north beams can be designed in a similar manner to the external south beam. For convenience the sectional properties of the external north beam are given in Table No. N-58.

4.7 GLAZING POSTS

Glazing posts are provided sufficiently close to each other so that the post force may be represented by a continuous line load. In practice, these posts are spaced at not more than 10 ft (3·05 m) centres.

In this example, providing the glazing posts at say 10 ft centres, the total compressive force in each post$=10\times 329 = 3290$ lb (14 600 N). Making the posts 9 in by 9 in (229×229 mm^2) square, only the nominal minimum reinforcement is required.

Treating the posts as a column, provide 4 Nos. $\frac{1}{2}$ in (12·7 mm) dia. bars longitudinally, with $\frac{1}{4}$ in (6·35 mm) dia. stirrups at 6 in (152·4 mm) centres.

REFERENCES

1. Keyte, M.J. and Gloag, H.L. (1959), 'The Lighting of Factories', Factory Building Studies No. 2, Building Research Station, London : H.M.S.O.

2. Evans, R.H. and Wilby, C.B. (1963), 'Concrete—Plain, Reinforced, Prestressed and Shell', London : Arnold.
3. Lundgren, H. (1954), 'Cylindrical Shells', Inst. Danish Civil Engineers, Copenhagen : Danish Technical Press.
4. Jenkins, R.S. (1947), 'Theory and Design of Cylindrical Shell Roofs', London : Ove Arup and Partners.

PART 2

STRUCTURAL ANALYSIS OF HYPERBOLIC PARABOLOIDAL SHELL ROOFS

CHAPTER 5

INTRODUCTION TO HYPAR SHELLS

5.1 HISTORICAL NOTES

Thin shell construction—as understood today—did not become a practical concept until 1924 when Karl Zeiss applied Love's theories to the design of a small semi-elliptical shell roof for the Zeiss works at Jena. Since then there have been rapid developments both in the analysis and design and construction of thin shells. Most attention was, however, given to singly-curved shells.

Doubly curved anti-elastic surfaces such as hyperbolic paraboloids —'hypar' for short—even though known to mathematicians for centuries as a geometric shape were not considered a structural possibility until 1933 when the French engineer Aimond put forward the idea of employing these shapes in the construction of wide-span buildings like aircraft hangars (Fig. 5.1). In 1934 another French engineer, Lafaille, demonstrated the great structural potential of hyperbolic paraboloidal shapes by designing a thin shell structure consisting of four such ruled surface shells supported on a central folded plate frame (Fig. 5.2).

The years prior to World War II instanced a few cases of the use of hyperbolic paraboloidal shell roofs, notably in Yugoslavia. Since World War II innumerable structures employing these shapes have been constructed all over the world. A marked revival of interest in hyperbolic paraboloidal shells during the past two decades is probably due to Felix Candela who employed ingenious combinations of these shapes to produce some of the most spectacular shell structures in the world.

5.2 POTENTIALS AND ECONOMICS OF HYPAR SHELLS

A structure in which aesthetic and functional requirements are satisfied without economic sacrifice is usually regarded as ideal. To be of wider use a particular form of construction should be able to provide:

(*a*) aesthetically pleasing shapes whose flexibility allows sufficient varia-

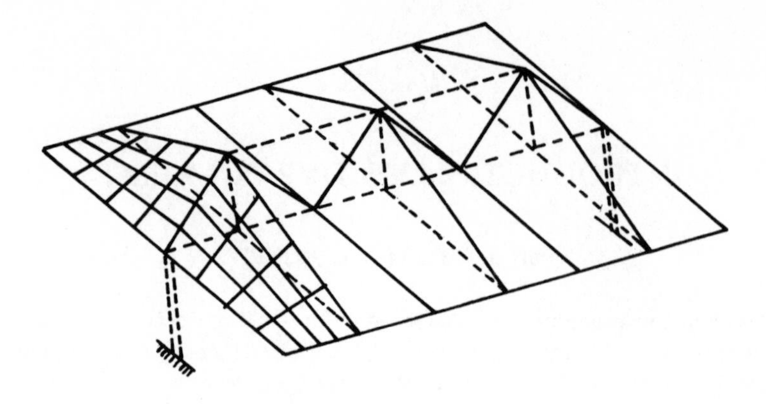

Fig. 5.1 *Wide span buildings using ruled surface shells (Aimond's proposal*—1931).

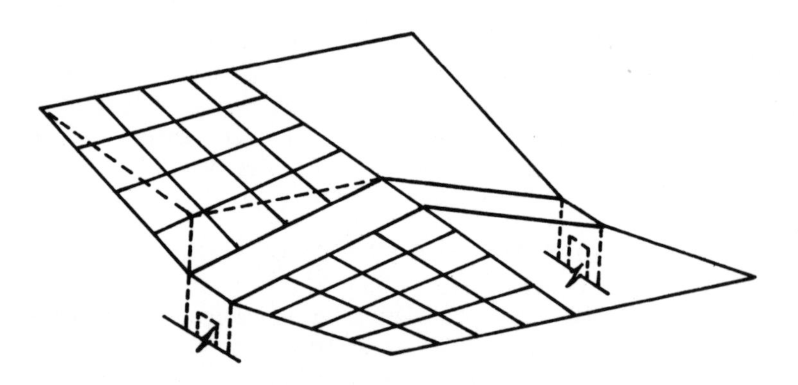

Fig. 5.2 *Wide span buildings using ruled surface shells (Lafaille's proposal*—1933).

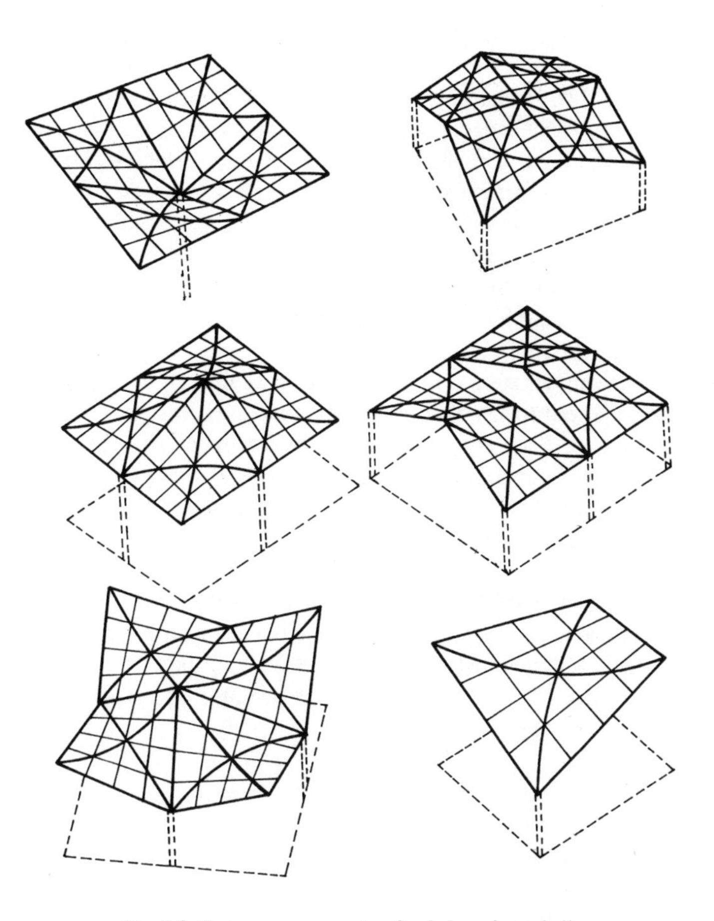

Fig. 5.3 Some arrangements of ruled surface shells.

tions to achieve effective proportions; and

(*b*) economical structural means of covering large areas, probably with a minimum number of supports.

The hypar shell construction offers clean lines, large uninterrupted spans allowing easier architectural planning beneath the roof, avoidance of massive beams, and the elegance of a visually exciting structure of architectural interest. In cases where architectural exhibitionism is of greater significance than economics, the hypar shell structure is of great value. A cursory glance at some of the structures is sufficient to prove that the hypar shell is a very powerful medium for translating architectural conceptions into reality.

Units of hypar surfaces can be arranged in various ways to achieve interesting proportions. Some of the commonly employed arrangements are shown in Fig. 5.3 and it can be seen that the provision for natural lighting is not a difficult problem.

Very little information is at present available on comparative cost analysis of different forms of construction including the hypar shells and it is difficult to assert with any confidence that the hypar shell construction is one of the most economical solutions to the problem of providing an aesthetically attractive roof. The arguments generally put forward in favour of these shells are:

(i) low cost of materials as the quantity of materials required is small;
(ii) straight line geometry of the surface which requires less complicated shuttering than most other types of shells; and
(iii) lighter roof structure requiring fewer columns and resulting in saving in the cost of the substructure.

The stresses induced in a hypar shell under working loads are usually low and the thickness of shell is governed not so much by purely structural considerations as by other factors such as thermal insulation, resistance to weathering, and so forth. Very thin shells, like those designed by Candela, which have a minimum thickness of 16 mm (5/8 in), can only be regarded as exceptions rather than rules. These very thin shells have yet to prove their long-term serviceability and they have not been used in cold, damp climates. Usually hypar shells are made $2\frac{1}{2}$ in (63·5 mm) to 3 in (76·2 mm) thick which is about the same thickness as most cylindrical shells. Thus, compared to cylindrical shells the saving in concrete is negligible. However, the hypar shell requires far less reinforcement and the size of edge beams is usually much smaller compared to those required for cylindrical shells.

The authors consider that the rectilinear geometry of hypar shells as a means of cutting down shuttering costs is over-emphasised. The authors are not aware of any cost figures to prove that the hypar shutterings are cheaper than the various patented shutterings employed for cylindrical shells.

However, it is to be hoped that developments in the design and construction of formwork will result in considerable economy. In developing countries, where the cost of the materials is higher relative to the cost of labour, this type of shell may well be cheaper than any other permanent structural form.

CHAPTER 6

BENDING THEORY OF HYPAR SHELLS

6.1 DEFINITION OF SURFACE

The surface generated by moving a principal parabola ABC (Fig. 6.1) parallel to itself along an inverse principal parabola BOD is a hyperbolic paraboloidal surface. The surface has, therefore, two systems of parabolic generators. The periphery of any section parallel to the plane ACEF is a hyperbola; hence the name hyperbolic paraboloid. As the surface is generated by a translatory motion of one parabola on another inverse parabola, it is usually referred to as a surface of translation.

For the system of coordinate axes as shown in Fig.6.1, the surface can be mathematically defined by the equation

$$\frac{y^2}{b^2} - \frac{x^2}{a^2} = 2\,cz \tag{6.1}$$

where a, b and c are parameters as shown in the figure. This is the equation of a hyperbolic paraboloidal surface of translation.

Referring to Fig. 6.2(a), let the X and Y axes be rotated about the origin through clockwise angles of $(\pi/2 - \omega/2)$ and $(\omega/2)$ respectively and let X_1 and Y_1 be the re-orientated axes. Let the angle ω be such that the loci of the point $z=0$ coincide with the X_1 and Y_1 axes. Then measuring x, y, x_1 and y_1, along X, Y, X_1 and Y_1 respectively,

$$\begin{aligned} z &= 0 \text{ when } x_1 = 0 \\ \text{and } z &= 0 \text{ when } y_1 = 0 \end{aligned} \tag{6.2}$$

Fig. 6.2(b) shows the projection of the surface on the $X - Y$ (and $X_1 - Y_1$) plane. For any point Q lying on the X_1-axis (and hence on the surface) at a distance x_1 from the origin,

$$\begin{aligned} x &= x_1 \sin\frac{\omega}{2} \\ \text{and } y &= x_1 \cos\frac{\omega}{2} \end{aligned} \tag{6.3}$$

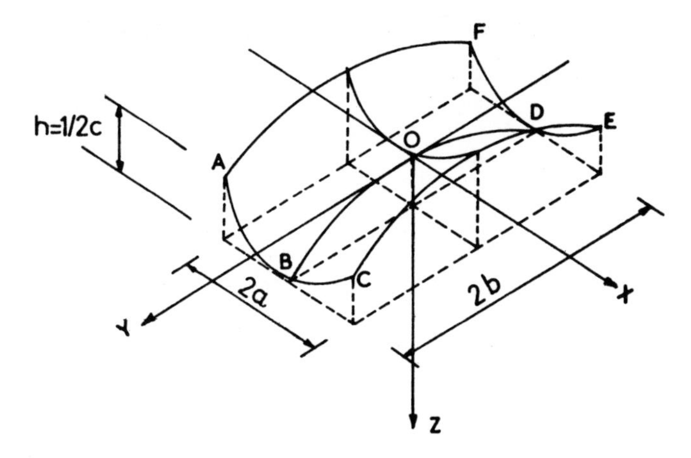

Fig. 6.1 *Hyperbolic paraboloid surface of translation.*

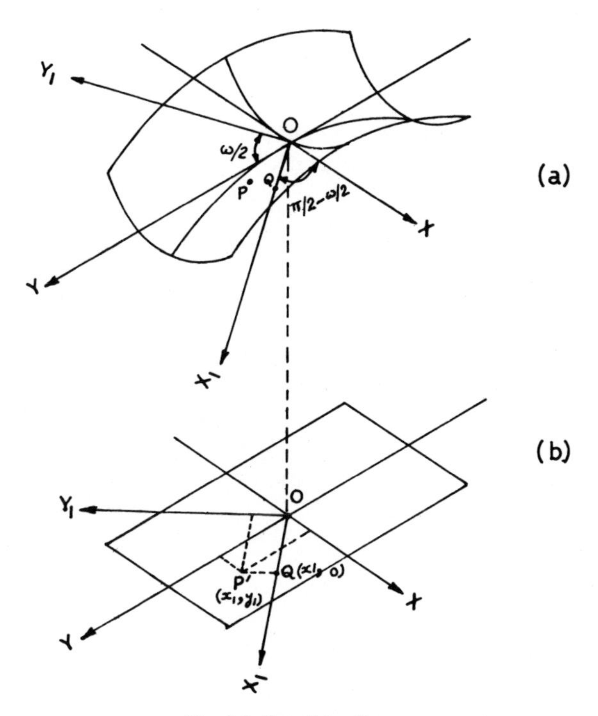

Fig. 6.2 *Rotation of axes.*

Since $z = 0$ when $y_1 = 0$, we have, from equations (6.1) and (6.3)

$$\frac{(x_1 \cos\frac{\omega}{2})^2}{b^2} - \frac{(x_1 \sin\frac{\omega}{2})^2}{a^2} = 0$$

$$\text{or,}\ \tan^2 \frac{\omega}{2} = \frac{a^2}{b^2} \tag{6.4}$$

$$\text{whence}\ \sin \omega = \frac{2\,ab}{a^2 + b^2} \tag{6.5}$$

$$\sin^2 \frac{\omega}{2} = \frac{a^2}{a^2 + b^2} \tag{6.6}$$

Now, for any point $P\ (x_1, y_1)$ lying on the surface

$$x = x_1 \sin \frac{\omega}{2} - y_1 \sin \frac{\omega}{2} \tag{6.7}$$

$$y = x_1 \cos \frac{\omega}{2} + y_1 \cos \frac{\omega}{2} \tag{6.8}$$

From equations (6.1), (6.7) and (6.8)

$$\frac{(x_1 + y_1)^2 \cos^2 \frac{\omega}{2}}{b^2} - \frac{(x_1 - y_1)^2 \sin^2 \frac{\omega}{2}}{a^2} = 2\,cz \tag{6.9}$$

Substituting b^2 from equation (6.4) in equation (6.9) and simplifying, we have

$$4\,x_1\,y_1 \sin^2 \frac{\omega}{2} = 2\,a^2\,cz$$

$$\text{or,}\ z = \frac{2\,x_1\,y_1}{a^2c} \sin^2 \frac{\omega}{2} \tag{6.10}$$

From equations (6.5), (6.6) and (6.10)

$$z = \frac{x_1\,y_1}{abc} \sin \omega$$

$$\text{or,}\ z = k_1\ x_1\ y_1\ \sin \omega \tag{6.11}$$

$$\text{where}\ k_1 = \frac{1}{abc} = \text{constant} \tag{6.12}$$

Equation (6.11) is the simplest second-degree equation defining a hyperbolic paraboloidal surface with reference to the X_1 and Y_1 axes.

In general, taking X_1 and Y_1 as the normal coordinate axes, the suffix can be dropped from the variables in equation (6.11), in which case the equation becomes

$$z = k_1 xy \sin \omega \tag{6.13}$$

It should be noted that the ω is the angle between the coordinate axes. The surface defined by equation (6.13) is usually referred to as a hyperbolic paraboloidal ruled surface, as this can be generated by two systems of straight line generators.

To generate a ruled surface consider axes OX and OY (Fig. 6.3) inclined to each other at an angle ω and lying in the plane XOY. Let B and C be two points on the X and Y axes respectively and A be any other point not lying in the plane XOY. A set of straight lines can now be drawn which intersect OY and AB and which are also parallel to plane XOZ. If a similar set of straight lines are drawn parallel to plane YOZ and intersecting OX and AC, then this set of lines would also intersect the first set of straight lines. A continuous surface joining the points of intersection of the two sets of straight lines is a hyperbolic paraboloidal ruled surface. Any section parallel to the XOZ or YOZ planes is a straight line whereas sections parallel to the diagonals are parabolae. Any section parallel to the XOY plane is a hyperbola.

The surface shown in Fig. 6.3 is given by the following equation

$$z = k\,x\,y \sin \omega \tag{6.14}$$

If the lengths OB and OC are a and b, respectively, and the distance of A from the plane XOY is c, then the value of k can be determined by considering the coordinate of A,

$$c = k\,a\,b \sin \omega$$

$$\therefore\ k = \frac{c}{a\,b \sin \omega} \tag{6.15}$$

The parameter k, even though a constant for a given surface, is not a dimensionless quantity. It is referred to as 'unitary warping' of the surface.

Angle ω between the coordinate axes can have any value, including a right angle in which case $\sin \omega = 1$ and equations (6.14) and (6.15) become

$$z = k\,x\,y \tag{6.16}$$

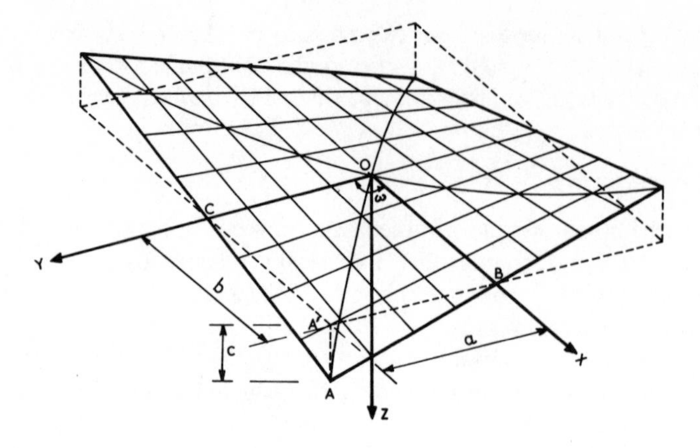

Fig. 6.3 *Development of hyperbolic paraboloidal ruled surface.*

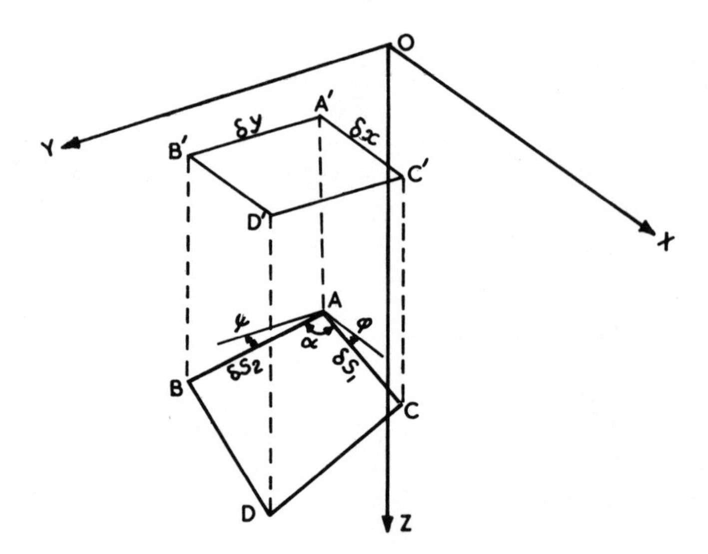

Fig. 6.4 *Angle between intersecting generators.*

where, $k = \frac{c}{ab}$ (6.17)

In subsequent discussion the distance c between the point A and the plane XOY will be referred to as the 'rise' of the shell.

6.2 DEFINITION OF SHALLOW SHELL

The analysis presented here is applicable to thin shallow hyperbolic paraboloidal shells defined by equation (6.16).

The shell is regarded as 'thin' if the thickness is small compared to the lateral dimensions.

It is difficult to define precisely a 'shallow' shell. In general, if the rise of a shell is small compared with its lateral dimensions, it may be regarded as shallow. In such a case the surface area of shell is very nearly equal to its plan area and the total uniformly distributed load on the shell surface can be assumed to be equal to the total uniformly distributed load of equal intensity on the plan area.

According to Candela, this criterion is satisfied if the numerical value of the unitary warping of shell surface is less than or equal to 0·04. It was pointed out in Section 6.1 that the unitary warping, k, is not a dimensionless constant and hence the authors consider that the numerical value of k as suggested by Candela is not truly definitive of the shallowness of a shell. Consider, for example, a hypar shell with plan dimensions of 3 m by 3 m and a rise of 0·6 m. The value of k for this shell is 0·067 m^{-1}. For a shell 10 m by 10 m with a rise of 2 m, the value of k is 0·02 m^{-1}. In both cases the maximum slope of the generators is 1/5 and there can be no possible reason to suppose that the larger shell is any more shallow than the smaller one.

A more rational method of defining a shallow shell is to relate its rise to lateral dimensions.

Consider a shell having plan dimensions of a and b and rise c. Let A be the area of its projection in the X–Y plane and A' be the actual surface area.

Then $A = a\,b$ (6.18)

$$A' = \int_0^a \int_0^b \sqrt{1 + k^2x^2} \;\; \sqrt{1 + k^2y^{2\mathrm{d}xdy}} \qquad (6.19)$$

On integrating the right-hand side of equation (6.19) and simplifying, we get

$$A' = \frac{a^2 b^2}{4\,c^2}\left\{\left(\sinh^{-1}\frac{c}{a} + \frac{c}{a}\sqrt{1+\frac{c^2}{a^2}}\right)\left(\sinh^{-1}\frac{c}{b} + \frac{c}{b}\sqrt{1+\frac{c^2}{b^2}}\right)\right\} \tag{6.20}$$

But $\frac{c}{a}$ and $\frac{c}{b}$ are the ratios of the rise of the shell to the lateral dimensions.

For $\frac{c}{a}$ and $\frac{c}{b}$ less than, say, 0·4, we can write

$$\begin{aligned} \sinh^{-1}\frac{c}{a} &\simeq \frac{c}{a} \\ \sinh^{-1}\frac{c}{b} &\simeq \frac{c}{b} \\ \sqrt{1+\frac{c^2}{a^2}} &\simeq 1 + \tfrac{1}{2}\frac{c^2}{a^2} \\ \sqrt{1+\frac{c^2}{b^2}} &\simeq 1 + \tfrac{1}{2}\frac{c^2}{b^2} \end{aligned} \tag{6.21}$$

Therefore, from equations (6.20) and (6.21)

$$A' = ab + \frac{ab}{4}\left(\frac{c^2}{a^2} + \frac{c^2}{b^2}\right) \tag{6.22}$$

$$\frac{A' - A}{A'} = \frac{\frac{1}{4}\left(\frac{c^2}{a^2} + \frac{c^2}{b^2}\right)}{1 + \frac{1}{4}\left(\frac{c^2}{a^2} + \frac{c^2}{b^2}\right)}$$

It is evident from equation (6.22) that the ratio of the difference in the surface and plan areas to the surface area is dependent only on the maximum slopes of the generators and is independent of the size of the shell. For any given values of these slopes this ratio is a dimensionless constant and hence indicates precisely the shallowness of the shell.

Table 6.1 gives the difference between the surface and plan area expressed as a percentage of surface area for different ratios of a/b and c/a calculated from equations (6.18) and (6.20).

TABLE 6.1

VALUE OF $\left(\frac{A'-A}{A} \times 100\right)$

c/a	a/b				
	1·0	0·8	0·7	0·6	0·5
0·10	0·33	0·27	0·24	0·23	0·21
0·15	0·74	0·61	0·55	0·51	0·47
0·20	1·31	1·08	0·98	0·90	0·82
0·25	2·03	1·67	1·52	1·39	1·28
0·30	2·90	2·39	2·17	1·99	1·83
0·33	3·48	2·87	2·61	2·39	2·20
0·4	5·02	4·15	3·78	3·46	3·18
0·5	7·59	6·31	5·75	5·27	4·84

It is seen from Table 6.1 that even for a ratio c/a of 0·5 the percentage difference between surface and plan areas is less than 8%. However, in addition to this another factor one has to consider in defining a shallow shell is the angle between intersecting generators at any point. This is discussed in Section 6.3, and taking into account the effect of this angle the authors consider that a shell may be considered shallow when the upper limit of the ratio $\left(\frac{A'-A}{A} \times 100\right)$ is less than or equal to 3.

As the ratio decreases the shell tends to behave more like a plate and the 'shell action' becomes smaller. A reasonable lower limit of $\left(\frac{A'-A}{A} \times 100\right)$ of 1·3 is considered by the authors as a minimum if proper advantage of shell action is to be obtained. Thus, a shallow shell theory may be considered applicable to shells where

$$1{\cdot}3 \leqslant \left(\frac{A'-A}{A} \times 100\right) \leqslant 3{\cdot}00$$

6.3 ANGLE BETWEEN INTERSECTING GENERATORS

Re-writing equation (2.16)

$z = kxy$

Let $p = \dfrac{\partial z}{\partial x} = ky$ (6.23)

$q = \dfrac{\partial z}{\partial y} = kx$ (6.24)

$r = \dfrac{\partial^2 z}{\partial x^2} = 0$ (6.25)

$r = \dfrac{\partial^2 z}{\partial y^2} = 0$ (6.26)

$t = \dfrac{\partial^2 z}{\partial x \partial y} = \dfrac{\partial^2 z}{\partial y \partial x} = k$ (6.27)

Figure 6.4 shows a small element $ABCD$ of the shell and its projection $A'B'C'D'$ on the plane $X-Y$ of the orthogonal Cartesian triad (X, Y, Z). The angle between $A'C'$ and $A'B'$, by virtue of equation (6.16), is a right angle. Let α be the angle at A between the generators AB and AC, and φ and Ψ be the angles in the $X-Z$ and $Y-Z$ plane which AC and AB make with OX and OY respectively.

Let $AB = \delta S_2$
$AC = \delta S_1$
$A'B' = \delta y$
$A'C' = \delta x$

Now, $(B'C')^2 = \delta x^2 + \delta y^2$ (6.28)

$BC^2 = (\delta S_1)^2 + (\delta S_2)^2 - 2(\delta S_1)(\delta S_2) \cos \alpha$ (6.29)

But $(\delta S_1)^2 = (\delta x)^2 + (\dfrac{\partial z}{\partial x} \cdot \delta x)^2$

or, $(\delta S_1)^2 = (\delta x)^2 \ (1 + p^2)$ (6.30)

and $(\delta S_2)^2 = (\delta y)^2 + (\dfrac{\partial z}{\partial y} \cdot \delta y)^2 = (\delta y)^2 \ (1+q^2)$

From geometry $BC = B'C'$ (6.31)

Therefore, from equations (6.28) and (6.31)

$$2 \cos \alpha \; \delta x \delta y \sqrt{(1+p^2)(1+q^2)} = p^2 (\delta x)^2 + q^2 (\delta y)^2$$

$$\text{or, } 2 \cos \alpha \sqrt{(1+p^2)(1+q^2)} = p^2 \frac{\delta x}{\delta y} + q^2 \frac{\delta y}{\delta x} \tag{6.32}$$

but in the limit

$$p^2 \frac{\delta x}{\delta y} = pq$$

$$\text{and } q^2 \frac{\delta y}{\delta x} = pq \tag{6.33}$$

Therefore, from equations (6.32) and (6.33)

$$\alpha = \cos^{-1} \left\{ \frac{pq}{\sqrt{(1+p^2)(1+q^2)}} \right\} \tag{6.34}$$

Equation (6.34) gives the angle between the intersecting generators at all points on the shell surface.

If the shell is shallow, angles φ and Ψ will be small and the following approximate relationships will hold:

$$\begin{aligned} \cos \Psi &\approx \cos \varphi \approx 1 \\ \sin \Psi &\approx \tan \Psi \approx p \\ \sin \varphi &\approx \tan \varphi = q \end{aligned} \tag{6.35}$$

Also, for shallow shells, the maximum values of p and q will be small in comparison with unity so that the square and higher powers of these parameters can be ignored. Thus, it can be assumed that

$$p^2 \simeq q^2 \simeq pq = 0 \tag{6.36}$$

From equations (6.34) and (6.36)

$$\begin{aligned} &\cos \alpha = 0 \\ &\text{or } \alpha = \pi/2 \end{aligned} \tag{6.37}$$

The actual values of α for different rise/plan dimension ratios for a shell with square plan are given in Table 6.2 from which it can be seen that for a rise/length ratio of 0·25 the deviation of α from right angle is less than 4%. Hence in the analysis it can be assumed that for shallow shells the angle between intersecting generators is sensibly a right angle.

TABLE 6.2

VALUES OF α FOR DIFFERENT RISE/LENGTH RATIOS

Rise/length	1/3	1/4	1/5	1/6
α	84° 45′	86° 35′	87° 13′	88° 33′

6.4 RELATIONSHIP BETWEEN ACTUAL AND PROJECTED FORCES

Figure 6.5(*a*) shows in-plane forces acting on a small element of a hypar shell and Figure 6.5(*b*) shows the projections of these forces on the projection of the element in the $X-Y$ plane. Figure 6.6 shows the moments and transverse forces acting on the same element and its projection.

Referring to Figure 6.5, let n_x, n_y, n_{xy} be the projections in the $X-Y$ plane of real stresses N_x, N_y and N_{xy} acting on the shell element. Resolving the real forces in the $X-Y$ plane, the following relationships between the real and projected forces are obtained:

$$
\begin{aligned}
n_x &= N_x \sqrt{\frac{1+q^2}{1+p^2}} \\
n_y &= N_y \sqrt{\frac{1+p^2}{1+q^2}} \\
n_{xy} &= n_{yx} = N_{xy} = N_{yx}
\end{aligned}
\qquad (6.38)
$$

Using the approximate relationships for shallow shells, as given by equation (6.36), equation (6.38) can be written as

$$
\begin{aligned}
n_x &\simeq N_x \\
n_y &\simeq N_y \\
n_{xy} &= N_{xy} = n_{yx} = N_{yx}
\end{aligned}
\qquad (6.39)
$$

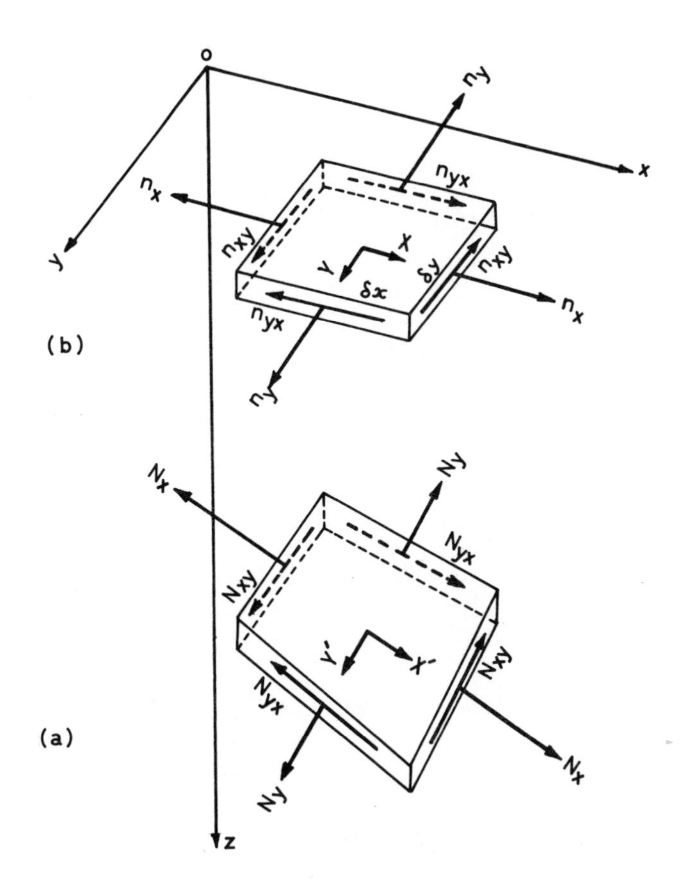

Fig. 6.5 Tangential forces acting on shell element.

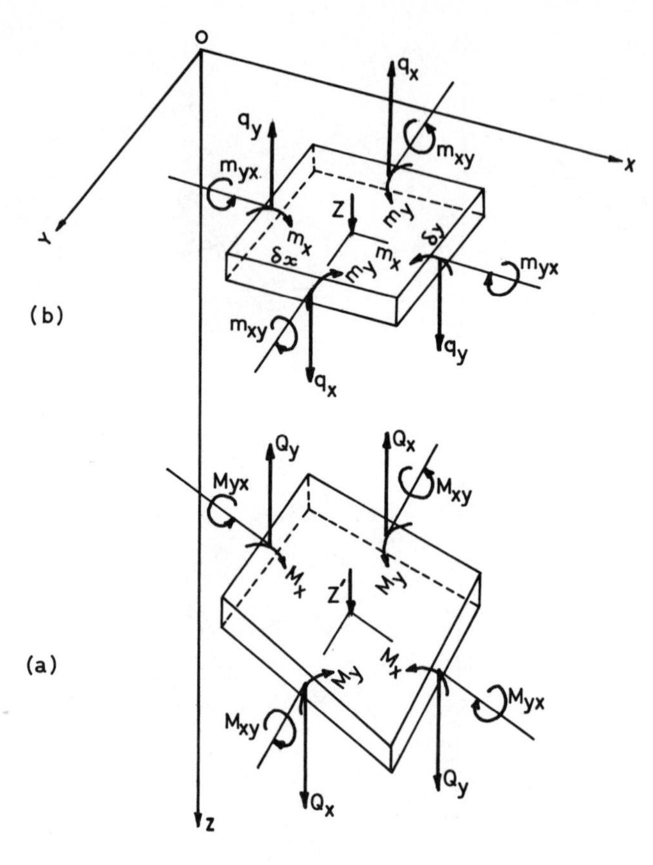

Fig. 6.6 Moments and transverse forces acting on shell element.

Similarly, referring to Figure 6.6, for shallow shells

$$
\begin{aligned}
m_x &\simeq M_x \\
m_y &\simeq M_y \\
m_{xy} &\simeq M_{xy} \\
q_x &\simeq Q_x \\
q_y &\simeq Q_y
\end{aligned}
\tag{6.40}
$$

Having obtained these relationships the projected forces only may be considered in further discussion.

6.5 EQUATIONS OF EQUILIBRIUM

For the shell element shown in Figs. 6.5 and 6.6, the equations of equilibrium are:

$$\frac{\partial n_x}{\partial x} + \frac{\partial n_{xy}}{\partial y} + X = 0 \tag{6.41}$$

$$\frac{\partial n_y}{\partial y} + \frac{\partial n_{yx}}{\partial x} + Y = 0 \tag{6.42}$$

$$\frac{\partial m_x}{\partial x} + \frac{\partial m_{xy}}{\partial y} - q_x = 0 \tag{6.43}$$

$$\frac{\partial m_y}{\partial y} + \frac{\partial m_{yx}}{\partial x} - q_y = 0 \tag{6.44}$$

$$\frac{\partial q_x}{\partial x} + \frac{\partial q_y}{\partial y} + 2t\, n_{xy} + Z - pX - qY = 0 \tag{6.45}$$

where X, Y, Z are components of applied loads along the respective axes, and p, q, and t are partial derivatives defined by equations (6.23), (6.24) and (6.27) respectively.

The five equations of equilibrium (6.41) to (6.45) contain eight dependent variables and, therefore, cannot be solved without three further relationships. These additional relationships can be obtained by considering shell strains and displacements.

6.6 SHELL STRAIN–DISPLACEMENT RELATIONSHIPS

Figure 6.7(*b*) shows a small element *ABCD* of the shell which is displaced to *A' B' C' D'*. The projections of the original and displaced elements on the *x*-*y* plane are shown in Figure 6.7(*a*). Let u' and v' be the displacements of *A* along *AD* and *AB* respectively and *u, v, w* be the projections of the displacements of *A* along the *x*, *y* and *z* axes respectively.

The strains along *AD* and *AB* and the shear strain are:

$$\varepsilon_{AD} = \frac{\partial u'}{\partial s_1} + p \cdot \frac{\partial w}{\partial s_1} \tag{6.46}$$

$$\varepsilon_{AB} = \frac{\partial v'}{\partial s_2} + q \cdot \frac{\partial w}{\partial s_2} \tag{6.47}$$

$$\varepsilon_{\text{shear}} = \gamma_1 + \gamma_2 \tag{6.48}$$

where ε denotes the strain in the direction of subscript and δs_1 and δs_2 are the lengths *AD* and *AB*, respectively.

Incorporating the assumptions of Sections 6.4 and 6.5 for shallow shells, equations (6.46) to (6.48) would become

$$\varepsilon_x = \frac{\partial u}{\partial x} + p \frac{\partial w}{\partial x} \tag{6.49}$$

$$\varepsilon_y = \frac{\partial v}{\partial y} + q \frac{\partial w}{\partial y} \tag{6.50}$$

$$\varepsilon_{xy} = \gamma_1 + \gamma_2 \tag{6.51}$$

Now,
$$\gamma_1 = \frac{\partial v}{\partial x} + q \frac{\partial w}{\partial x} \tag{6.52}$$

$$\gamma_2 = \frac{\partial v}{\partial y} + p \frac{\partial w}{\partial y} \tag{6.53}$$

From equations (6.51) to (6.53)

$$\varepsilon_{xy} = \frac{\partial v}{\partial x} + \frac{\partial u}{\partial y} + p \frac{\partial w}{\partial y} + q \frac{\partial w}{\partial x} \tag{6.54}$$

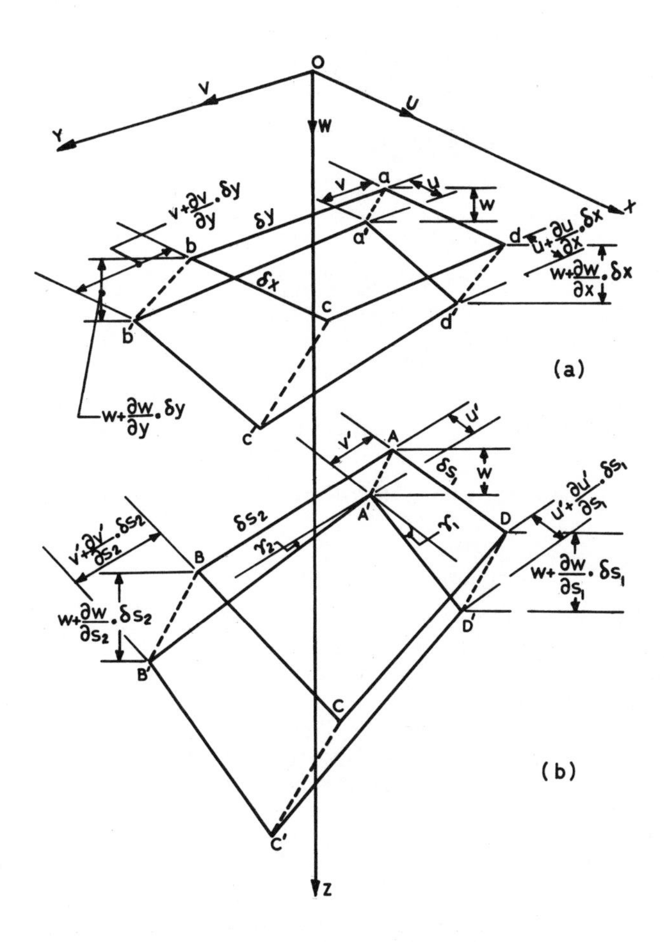

Fig. 6.7 Displacements of shell element.

Subtracting the partial derivative of ε_{xy} with respect to x and y from the second partial derivates of ε_x and ε_y with respect to y and x, respectively, we obtain from equations (6.49), (6.50) and (6.54)

$$\frac{\partial^2 \varepsilon_x}{\partial y^2} + \frac{\partial^2 \varepsilon_y}{\partial y^2} - \frac{\partial^2 \varepsilon_{xy}}{\partial x \partial y} = \frac{\partial p}{\partial y} \cdot \frac{\partial^2 w}{\partial x \partial y} + \frac{\partial q}{\partial x} \cdot \frac{\partial^2 w}{\partial x \partial y} \tag{6.55}$$

but from equations (6.23) and (6.24)

$$\frac{\partial p}{\partial y} = \frac{\partial q}{\partial x} = k$$

Substituting this in equation (6.55)

$$\frac{\partial^2 \varepsilon_x}{\partial y^2} + \frac{\partial^2 \varepsilon_y}{\partial x^2} - \frac{\partial^2 \varepsilon_{xy}}{\partial x \partial y} = 2k \frac{\partial^2 w}{\partial x \partial y} \tag{6.56}$$

Equation (6.56) links the three strains to the deflection w.

6.7 STRESS-STRAIN RELATIONSHIPS

Ignoring Poisson's ratio

$$\left.\begin{aligned} n_x &= Eh \cdot \varepsilon_x \\ n_y &= Eh \cdot \varepsilon_y \\ n_{xy} &= \tfrac{1}{2} Eh\ \varepsilon_{xy} \end{aligned}\right\} \tag{6.57}$$

where h = thickness of shell (assumed constant)
E = Young's modulus for shell material.

6.8 MOMENT–CURVATURE RELATIONSHIPS

Ignoring Poisson's ratio, the relationships between the moments and curvatures of the middle surface of shell are:

$$m_x = -EI \frac{\partial^2 w}{\partial x^2}$$

$$m_y = -EI \frac{\partial^2 w}{\partial y^2} \tag{6.58}$$

$$m_{xy} = -EI\frac{\partial^2 w}{\partial x \partial y}$$

Where, I = moment of inertia of the shell about its middle surface $= \frac{h^3}{12}$ for unit width of shell.

6.9 DERIVATION OF THE COMPATIBILITY EQUATION

Equations of equilibrium (6.41) to (6.45) together with equations (6.49), (6.50), (6.57) and (6.58) form a set of fourteen equations in fourteen dependent variables, namely, n_x, n_y, n_{xy}, m_x, m_y, m_{xy}, q_x, q_y, ε_x, ε_y, ε_{xy}, u, v and w.

For a shallow shell with vertical z-axis and subjected to vertical loads the components X and Y of the load would be zero. Thus, in all further discussion it will be assumed that

$$X = Y = 0 \tag{6.59}$$

From equations (6.56) and (6.57),

$$\frac{1}{Eh}\left(\frac{\partial^2 n_x}{\partial y^2} + \frac{\partial^2 n_y}{\partial x^2} - 2\frac{\partial^2 n_{xy}}{\partial x \partial y}\right) = 2k\frac{\partial^2 w}{\partial x \partial y} \tag{6.60}$$

Let F be a stress function such that

$$n_x = \frac{\partial^2 F}{\partial y^2}$$

$$n_y = \frac{\partial^2 F}{\partial x^2} \tag{6.61}$$

$$n_{xy} = -\frac{\partial^2 F}{\partial x \partial y}$$

Then, from equations (6.60) and (6.61)

$$\frac{1}{Eh}\frac{\partial^4 F}{\partial y^4} + \frac{\partial^4 F}{\partial x^4} + 2\frac{\partial^4 F}{\partial x^2 \partial y^2} = 2k\frac{\partial^2 w}{\partial x \partial y}$$

$$\text{or, } \nabla^4 F = 2\,Ehk\,\frac{\partial^2 w}{\partial x \partial y} \tag{6.62}$$

$$\text{where } \nabla^2 \equiv \frac{\partial^2}{\partial x^2} + \frac{\partial^2}{\partial y^2} \tag{6.63}$$

Putting $X = Y = 0$ in equations (6.41) to (6.45), we obtain, from equations (6.41) to (6.45), (6.49), (6.50), (6.56) and (6.58), after elimination and simplification,

$$EI\left(\frac{\delta^2 w}{\partial x^4} + \frac{\partial^4 w}{\partial y^4} + 2\,\frac{\partial^4 w}{\partial x^2 \partial y^2}\right) + 2k\,\frac{\partial^2 F}{\partial x \partial y} = Z$$

$$\text{or, } \nabla^4 w + \frac{2k}{EI}\,\frac{\partial^2 F}{\partial x \partial y} = \frac{Z}{EI} \tag{6.64}$$

The two linear simultaneous fourth-order partial differential equations (6.62) and (6.64) define the complete elastic behaviour of the shell. These can be solved as such or may be combined into a single eighth-order equation in one dependent variable.

Let φ be a function of stresses such that

$$F = 2\,Ehk\,\frac{\partial^2 \varphi}{\partial x \partial y}$$

$$\text{and, } w = \nabla^4 \varphi \tag{6.65}$$

By so choosing φ, equation (6.62) is identically satisfied and equation (6.64) becomes

$$\nabla^8 \varphi + \frac{4\,Ehk^2}{EI}\,\frac{\partial^4 \varphi}{\partial x^2 \partial y^2} = \frac{Z}{EI}$$

$$\text{or, } \nabla^8 \varphi + 4\beta^4\,\frac{\partial^4 \varphi}{\partial x^2 \partial y^2} = \frac{Z}{EI} \tag{6.66}$$

$$\text{where } \beta^4 = \frac{k^2 h}{I} \tag{6.67}$$

A single eighth-order equation in terms of the stress-function φ is thus obtained.

Alternatively, a single equation can also be obtained in terms of w. Let F_1 be a function such that

$$F_1 = \frac{\partial^2 F}{\partial x \partial y} = - n_{xy} \tag{6.68}$$

Differentiating equation (6.62) with respect to x and y and using equation (6.68),

$$\nabla^4 F_1 - 2\, Ehk \; \frac{\partial^4 w}{\partial x^2 \partial y^2} = 0 \tag{6.69}$$

From equations (6.64) and (6.68)

$$\nabla^4 w + \frac{2k}{EI} F_1 = \frac{Z}{EI} \tag{6.70}$$

Eliminating F_1 from equations (6.69) and (6.70),

$$\nabla^8 w + \frac{4k^2 h}{I} \frac{\partial^4 w}{\partial x^2 \partial y^2} = \frac{1}{EI} \nabla^4 \; (Z)$$

$$\text{or, } \nabla^8 w + \frac{4\beta^4 \, \partial^4 w}{\partial x^2 \partial y^2} = \frac{1}{EI} \nabla^4 Z \tag{6.71}$$

Either of the two equations of compatibility (6.66) and (6.71) can be used for the elastic analysis of shallow hyperbolic paraboloidal shells whose surface is defined by equation (6.17).

CHAPTER 7

ANALYSIS OF INVERTED UMBRELLA TYPE OF SHELLS

7.1 ASSUMPTIONS

The governing differential equation (6.71) developed in Chapter 6 is applicable to hyperbolic paraboloidal ruled surface shells with any given set of boundary conditions. As mentioned in Chapter 5, there can be innumerable combinations of shell elements to give different overall shapes to the structures and resulting in different sets of boundary conditions. It is therefore not possible to consider in a book like the present one all the possible types of boundaries and attention has therefore been focused on the boundary conditions pertaining to one of the most common types of hyperbolic paraboloidal shell—the inverted umbrella shell shown in Fig. 5.4(*a*).

The following assumptions are made:

(1) The shell is rectangular or square in plan and is symmetrical about the two principal axes passing through the central column.

(2) The shell is subjected to uniformly distributed load over the entire face.

(3) In the case where shells are provided with edge and spine beams, the flexural rigidity of such beams is assumed to be negligible. In practice the cross-sectional dimensions of the perimeter and edge beams are usually small and it is reasonable to assume that these are not capable of resisting any bending moments.

7.2 BOUNDARY CONDITIONS

Figure 7.1(*a*) shows the plan of an inverted umbrella shell supported by a central column. As the shell is symmetrical about the principal axes passing through the column and is subjected to symmetrical uniformly distributed load, it is sufficient to consider only one quadrant of the complete structure. Fig. 7.1(*b*) shows the plan of one of the shell quadrants and the co-ordinate

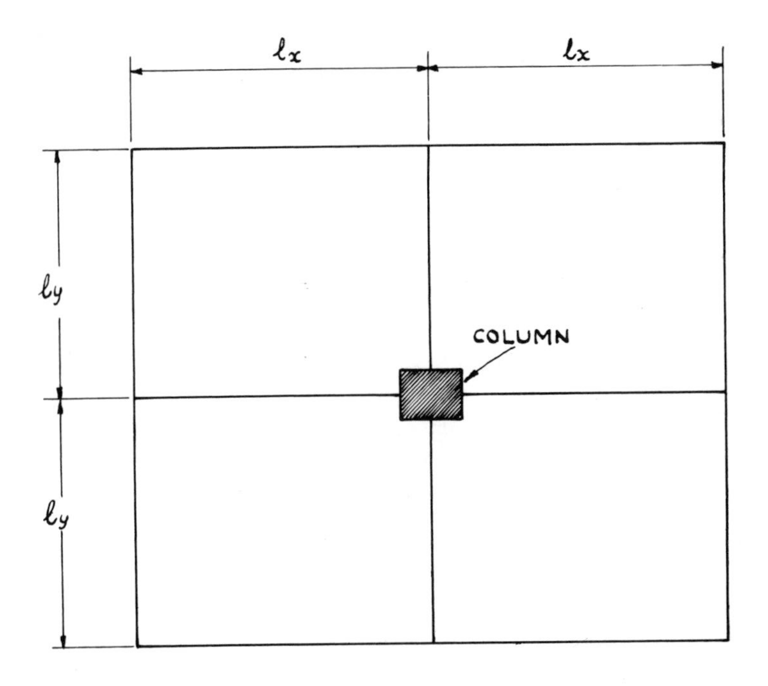

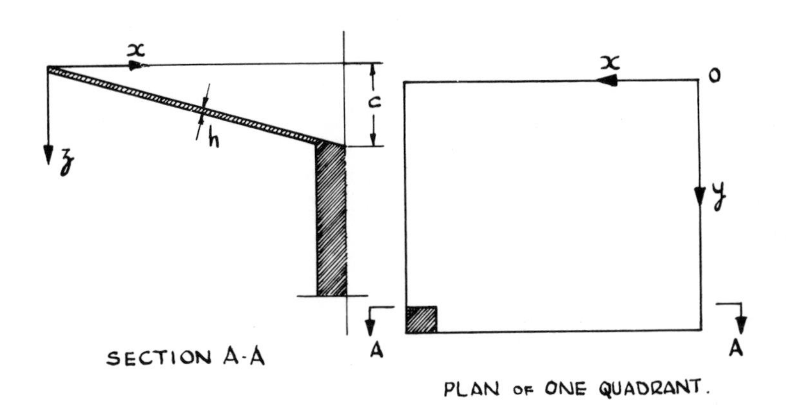

Fig. 7.1 Inverted umbrella shell showing co-ordinate axes: (a) plan of inverted umbrella shell roof; (b) co-ordinate axes.

axes chosen. Let l_x and l_y be the plan dimensions of the quadrant along the x and y axes respectively, and c and h be the rise and thickness, respectively, of the shell.

The boundary conditions to be satisfied depend on whether any edge and spine beams are provided in the structure. Two cases are considered:

(*a*) Shells with no beams.
(*b*) Shells provided with beams along the perimeter and valleys.

The boundary conditions for these two cases are given below.

7.2.1 Shells without any beams

The boundary conditions in this case are:

At $x = l_x$ and $y = l_y$,

$$w = 0 \tag{7.1}$$

At $x = 0$,

$$n_x = 0 \tag{7.2}$$
$$n_{xy} = 0 \tag{7.3}$$
$$m_x = 0 \tag{7.4}$$
$$V_x = 0 \tag{7.5}$$

where V_x = equivalent transverse shear in x-direction

$$= \frac{\partial m_x}{\partial x} + 2\frac{\partial m_{xy}}{\partial y} \tag{7.6}$$

At $y = 0$,

$$n_y = 0 \tag{7.7}$$
$$n_{xy} = 0 \tag{7.8}$$
$$m_y = 0 \tag{7.9}$$
$$V_y = 0 \tag{7.10}$$

$$\text{where } V_y = \frac{m_y}{\partial y} + 2\frac{\partial m_{xy}}{\partial x} \tag{7.11}$$

At $x = l_x$,

$$\frac{\partial \omega}{\partial x} = 0 \tag{7.12}$$
$$q_x = 0 \tag{7.13}$$
$$n_{xy} = 0 \tag{7.14}$$
$$u = 0 \tag{7.15}$$

At $y = l_y$,

$$\frac{\partial \omega}{\partial x} = 0 \tag{7.16}$$

$$q_y = 0 \tag{7.17}$$

$$n_{xy} = 0 \tag{7.18}$$

$$v = 0 \tag{7.19}$$

7.2.2 Shells with edge and spine beams

It was assumed in section 7.1 that the edge and spine beams are not capable of resisting bending moments. This implies that all the boundary conditions given for shells without beams, with the exception of those relating to the in-plane shear forces, are also applicable to shells with edge and spine. The beams are assumed to be capable of resisting direct forces and hence the in-plane shear forces at shell edges will not be zero, and the strains in the shell edges will be equal to the axial strains in the adjoining beams. Hence, for shells with edge and spine beams, equations (7.3), (7.8), (7.14), and (7.18) are replaced by the following equations:

At $x = 0$,

$$v = \frac{\int_0^y (n_{xy})\,dy}{E\,(ar_1)} \tag{7.20}$$

At $y = 0$,

$$u = \frac{\int_0^x (n_{xy})\,dx}{E\,(ar_2)} \tag{7.21}$$

At $x = l_x$,

$$v = \frac{\int_0^y (n_{xy})\,dy}{E\,(ar_3)} \tag{7.22}$$

At $y = l_y$,

$$u = \frac{\int_0^x (n_{xy})\, dx}{E\,(ar_4)} \tag{7.23}$$

Where (ar_1), (ar_2), (ar_3), (ar_4) are the cross-sectional areas of the beams at $x = 0$, $y = 0$, $x = l_x$, $y = l_y$ respectively and E is Young's modulus.

7.3 ANALYSIS OF SHELLS WITHOUT EDGE BEAMS

The analysis of the umbrella shell shown in Fig. 7.1 is carried out by considering several cases in which the shell quadrant is extended beyond its boundaries and assuming different sets of boundary conditions for these cases, and then combining these various cases to satisfy the actual boundary conditions for the quadrant given by equations (7.1) to (7.19) in order to obtain the final results.

7.3.1 Case (1)

In this case the shell is extended by a distance l_y in the y-direction and assumed to be supported along the two edges parallel to the x-axis and free along the other two edges, as shown in Fig. 7.2(1). The origin and orientation of the axes is the same as for the original quadrant shown in Fig. 7.1. This extended shell with the assumed support conditions is considered to be carrying a uniformly distributed load over the whole of its surface. The boundary conditions specified in this case are:

At $y = \pm l_y$,

$$w = 0 \tag{7.24}$$
$$m_y = 0 \tag{7.25}$$
$$n_{xy} = 0 \tag{7.26}$$
$$v = 0 \tag{7.27}$$

At $x = 0$ and $x = l_x$

$$m_x = 0 \tag{7.28}$$
$$V_x = 0 \tag{7.29}$$
$$n_{xy} = 0 \tag{7.30}$$
$$n_x = 0 \tag{7.31}$$

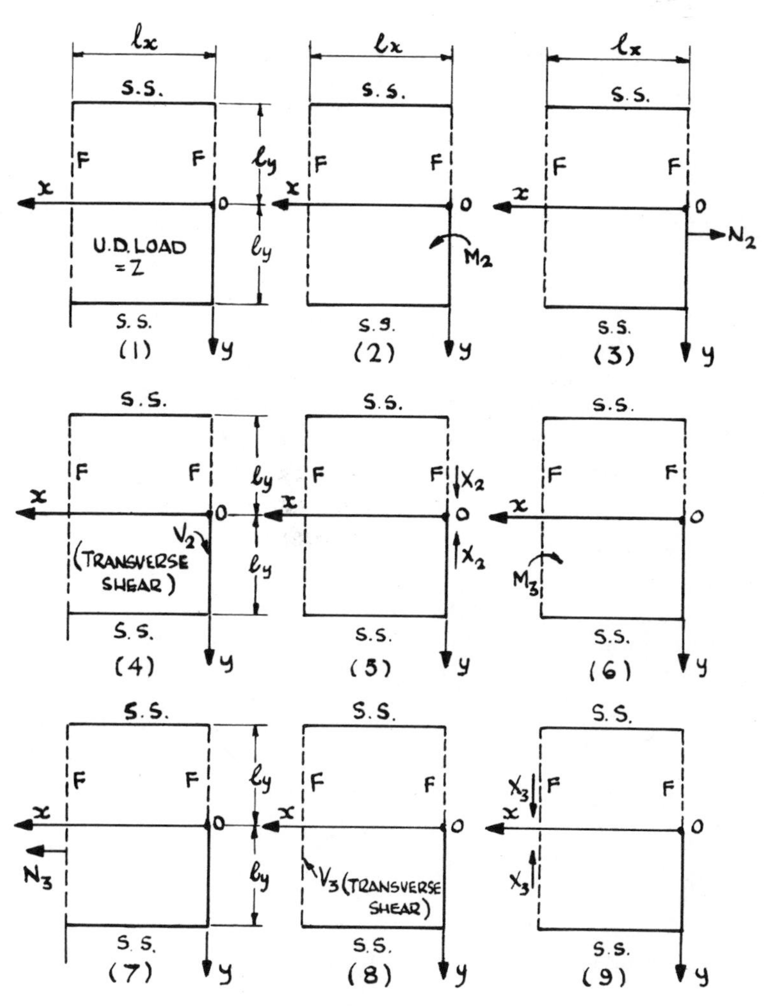

S S. = SIMPLY SUPPORTED EDGE
F = FREE EDGE
ALL EDGE FORCES & MOMENTS APPLIED ALONG ENTIRE EDGE.

Fig. 7.2 Edge forces applied in cases (1) to (9).

Complementary function

The solution of the homogeneous equation (6.71) for this case is taken as

$$w_1 = \sum_j B_j e^{\lambda_j x} \cos(my) \tag{7.32}$$

$$\text{Where } m = \frac{j\pi}{2l_y} \tag{7.33}$$

$j = 1, 3, 5, \ldots..$
B_j = Arbitrary constants
λ_j = roots of the characteristic equation.

The subscript 1 with w refers to Case (1) that is being considered. Substituting w_1 from equation (7.32) into the governing equation (6.71), the following eighth-degree characteristic equation is obtained,

$$(\lambda_j^2 - m^2)^4 - 4b^4 m^2 \lambda_j^2 = 0 \tag{7.34}$$

This equation has eight roots. Derivation of the roots is given in Appendix I, where it is shown that four of the roots are real and the other four complex. For the present purpose, therefore, the roots of the equation derived in Appendix I in terms of the various constants appearing in equation (7.34) can be written as

$$\begin{aligned} (\lambda_j)_{1,2} &= \pm J_1 \\ (\lambda_j)_{3,4} &= \pm J_2 \\ (\lambda_j)_{5,6,7,8} &= \pm (J_3 \pm iK_3) \\ \text{Where } i &= \sqrt{-1} \end{aligned} \tag{7.35}$$

Substituting these roots in equation (7.32)

$$\begin{aligned} w_1 = \sum_j \Big[& B_1 e^{+J_1 x} + B_2 e^{-J_1 x} + B_3 e^{+J_2 x} + B_4 e^{-J_2 x} \\ & + B_5 e^{(J_3 + iK_3)x} + B_6 e^{(J_3 - iK_3)x} + B_7 e^{-(J_3 + iK_3)x} \\ & + B_8 e^{-(J_3 - iK_3)x} \Big]_j \cos my \end{aligned} \tag{7.36}$$

Where B_5, B_6, B_7 and B_8 are arbitrary complex constants.

$$\left.\begin{aligned} \text{But } e^{(J_3 \pm iK)\,x} &= e^{J_3x}(\cos K_3 \pm i\, sin\ K_3x) \\ \text{and } e^{-(J_3 \pm iK_3)x} &= e^{-J_3x}(\cos K_3x \pm i \sin K_3x) \end{aligned}\right\} \tag{7.37}$$

Substituting equations (7.37) in equation (7.36) and simplifying,

$$\begin{aligned} w_1 = \sum_J \Big[& A^1{}_1\, e^{J_1x} + A^1{}_2\, e^{-J_1x} + A^1{}_3\, e^{J_2x} + A^1{}_4\, e^{-J_2x} \\ & + A^1{}_5\, e^{J_3x} \cos K_3x + A^1{}_6\, e^{-J_3x} \cos K_3x \\ & + A^1{}_7\, e^{J_3\,x} \sin K_3x + A^1{}_8\, e^{-J_3x} \sin K_3x \Big]_j \cos my \end{aligned} \tag{7.38}$$

Where $A_1, A_2, \ldots, A_8$ are all real and the superscript 1 with these constants refers to Case (1). Having obtained w_1 from equation (7.38) other forces, moments and displacements in the shell can be obtained from the appropriate relations given in Chapter 6.

Any force, moment or displacement, f, can be expressed in the following general form for the jth term of the series:

$$\begin{aligned} f_j^1 = R_j \Big[& A^1{}_1 a_1 e^{J_1x} + A^1{}_2 a_2 e^{-J_1x} + A^1{}_3 a_3 e^{J_2x} \\ & + A^1{}_4 a_4 e^{-J_2x} + A^1{}_5 a_5 e^{J_3x} + A^1{}_6 a_6 e^{-J_3x} \\ & + A^1{}_7 a_7\, e^{J_3x} + A^1{}_8\, a_8\, e^{-J_3x} \Big]_j \end{aligned} \tag{7.39}$$

Where R = a multiplier which is a function of y

a_1, a_2, a_3, a_4x = constants in terms of J_1, J_2, J_3 and J_4

a_5, a_6, a_7, a_8 = functions of x.

The expressions for R, a_1, $a_2 \ldots a_8$, for various forces and moments in the shell are given in Table 7.1.

Particular integral

The uniformly distributed load on the shell surface can be expressed in the form

$$Z = \frac{8z_0}{\pi^2 mn} \sum_m \sum_n \cos my \sin nx \tag{7.40}$$

Where z_0 = applied load per unit area

$$m = \frac{j\pi}{2l_y}$$

$$n = \frac{j\pi}{l_x}$$

Take the particular integral in the form

$$w_0 = \Sigma\Sigma B_{mn} \sin nx \cos my \tag{7.41}$$

Where B is a constant and the subscript 0 with w refers to the particular integral.

Substituting Z and w_0 from equations (7.40) and (7.41) in equation (6.71), the following expression for B is obtained.

$$B = \frac{8z_0}{\pi^2} \quad \frac{(m^2 + n^2)^2}{EI[\,(m^2 + n^2)^4 + 4b^4 m^2 n^2]} \tag{7.42}$$

Using appropriate relationships given in Chapter 6, the various forces, moments and displacements due to the particular integral are:

$$w_0 = B \sin nx \cos my \tag{7.43}$$

$$\left[\frac{\partial w}{\partial x}\right]_0 = nB \cos nx \cos my \tag{7.44}$$

$$\left[\frac{\partial w}{\partial y}\right]_0 = -mB \sin nx \sin my \tag{7.45}$$

$$(m_x)_0 = EI\, n^2\, B \sin nx \cos my \tag{7.46}$$

$$(m_y)_0 = EI\, m^2\, B \sin nx \cos my \tag{7.47}$$

$$(m_{xy})_0 = EI\, mn\, B \cos nx \sin my \tag{7.48}$$

$$(V_x)_0 = EI\,(n^2 + 2m^2)\, B \cos nx \cos my \tag{7.49}$$

$$(V_y)_0 = -\,EI\, m\,(2n^2 + m^2)\, B \sin nx \sin my \tag{7.50}$$

$$(n_{xy})_0 = H_1 \sin nx \cos my \tag{7.51}$$

$$(n_x)_0 = H_1 \frac{m}{n} \cos nx \sin my \tag{7.52}$$

$$(n_y)_0 \quad = H_1 \frac{n}{m} \cos nx \sin my \tag{7.53}$$

$$(u)_0 \quad = \int \frac{(n_x)_0}{Eh} dx - kyw_0 \tag{7.54}$$

$$(v)_0 \quad = \int \frac{(n_y)_0}{Eh} dy - kyw_0 \tag{7.55}$$

$$\text{Where } H_1 = \left\{ \frac{EI}{2k} \ (m^2+n^2)^2 B - \frac{8}{2 k\pi^2} z_0 \right\} \tag{7.56}$$

Determination of the constants of integration

The $A_1^1, A_2^1 \ldots A_8^1$ can be determined by satisfying the boundary equations (7.24) to (7.31). The forces and displacements on the left-hand sides of these equations represent the sum of the complementary function and the particular integral. Thus, eight linear algebraic equations are obtained for the determination of the constants A_1^1 to A_8^1. These equations can be expressed as

$$K \{A^1\} + \{P^1\} = 0 \tag{7.57}$$

Where K is the matrix of the coefficients of the constants A_1^1 to A_1^8 of the complementary function, $\{A^1\}$ is the vector of the constants and $\{P^1\}$ is the particular integral vector.

Once these constants of integration are found from equation (7.57) all the forces, moments and displacements in the shell can be obtained by adding the complementary function to the particular integral. The solution thus obtained is for the shell shown in Fig. 7.2(1) and not for the original shell quadrant. Various other cases have to be considered and their results superimposed so as to satisfy the boundary condition equations (7.1) to (7.19) before the final results for the quadrant are obtained. Also, as the coefficients matrix K occurs in some of the other cases to be considered, equation (7.57) may not be solved at this stage.

7.3.2 Cases (2) to (9)

The shell in all these cases is taken to be the same as in Case (1) but carrying no superimposed load but subjected to forces and moments applied at each edge in turn as shown in Fig. 7.2 (2 to 9).

The edge forces considered for different Cases are:

Case (2) — a moment $m_x = m_2$ applied at the edge $x = 0$
Case (3) — a force $n_x = n_2$ applied at $x = 0$
Case (4) — a force $V_x = V_2$ applied at $x = 0$
Case (5) — a force $n_{xy} = X_2$ applied at $x = 0$
Case (6) — a moment $m_x = m_3$ applied at $x = l_x$
Case (7) — a force $n_x = n_3$ applied at $x = l_x$
Case (8) — a force $V_x = V_3$ applied at $x = l_x$
Case (9) — a force $n_{xy} = X_3$ applied at $x = l_x$

Each of the applied edge forces is taken in the form of a series in y corresponding to the function of y of the complementary function for the variable considered. Thus the force M_2 at $x = 0$ is taken as

$$M_2 = \sum_j (F_j)_2 \cos my \tag{7.58}$$

where $(F_j)_2$ represents an arbitrary constant. The subscript 2 refers to Case (2).

Also,

$$N_2 = \sum_j (F_j)_3 \sin my \tag{7.59}$$

All the other edge forces are similarly expressed in terms of arbitrary constants F_j.

In each of the Cases (2) to (9) the form of the complementary function remains the same as in Case (1) except that the arbitrary constants A_1 to A_8 will be different in each case. The boundary conditions to be satisfied in each case also remain the same as in Case (1) with the exception of one boundary condition in each case relating to the applied edge force. Thus, considering Case (2) the boundary condition given by equation (7.28) is replaced by

$$m_x = M_2 \text{ at } x = 0 \tag{7.60}$$

Each case is now solved in turn for the constants A_1 to A_8 by satisfying the boundary conditions. However, as the applied edge forces are in terms of the arbitrary constants F_m, the constants A_1 to A_8 can only be determined in terms of F_m. Designating the constants A_1 to A_8 for the various Cases by superscripts corresponding to the Case considered, for example A_1^2, A_2^2... A_8^2 to represent the constants for Case (2), the boundary condition equations in each case take the form

$$[K]\ \{A^p\} = \{P^p\} \tag{7.61}$$

Where $[K]$ is the coefficient matrix as defined for Case (1), $\{A^p\}$ is the vector of the constants $A_1, A_2, \ldots A_8$ for the pth case and $\{P^p\}$ is the load vector of the pth case. All but one element of the vector P^p will be zero. In each Case K will be an 8×8 matrix, and A and P will be 8×1 vectors. Thus for Case (2), equation (7.61) can be written as

$$[K] \begin{Bmatrix} A^2_1 \\ A^2_2 \\ A^2_3 \\ A^2_4 \\ A^2_5 \\ A^2_6 \\ A^2_7 \\ A^2_8 \end{Bmatrix} = \begin{Bmatrix} (F_j)_2 \\ 0 \\ 0 \\ 0 \\ 0 \\ 0 \\ 0 \\ 0 \end{Bmatrix} \tag{7.62}$$

Solving equation (7.62) A^2_1 to A^2_8 can be determined in terms of $(F_j)_2$. Cases (3) to (9) can similarly be solved for the constants A in terms of F_j. As all these sets of equations for the Cases (1) to (9) are similar, these can be written in a combined form as

$$[K]\ [A] = [P] \tag{7.63}$$

Where

$$[A] = \begin{bmatrix} A^1_1 & A^2_1 & \cdots & A^9_1 \\ A^1_2 & A^2_2 & & A^9_2 \\ A^1_3 & A^2_3 & & A^9_3 \\ \cdot & \cdot & & \cdot \\ \cdot & \cdot & & \cdot \\ \cdot & \cdot & & \cdot \\ \cdot & \cdot & & \\ A^1_8 & A^2_8 & & A^9_8 \end{bmatrix} \tag{7.64}$$

and

$$[P] = \begin{bmatrix} -P_1 & (F_j)_2 & 0 & 0 & 0 & 0 & 0 & 0 & 0 \\ -P_2 & 0 & (F_j)_3 & 0 & 0 & 0 & 0 & 0 & 0 \\ -P_3 & 0 & 0 & (F_j)_4 & 0 & 0 & 0 & 0 & 0 \\ -P_4 & 0 & 0 & 0 & (F_j)_5 & 0 & 0 & 0 & 0 \\ -P_5 & 0 & 0 & 0 & 0 & (F_j)_6 & 0 & 0 & 0 \\ -P_6 & 0 & 0 & 0 & 0 & 0 & (F_j)_7 & 0 & 0 \\ -P_7 & 0 & 0 & 0 & 0 & 0 & 0 & (F_j)_8 & 0 \\ -P_8 & 0 & 0 & 0 & 0 & 0 & 0 & 0 & (F_j)_9 \end{bmatrix} \quad (7.65)$$

7.3.3 Cases (10) to (17)

In all these Cases the original shell quadrant shown in Fig. 7.1 is extended a distance of l_x along the x-axis only, so that the plan dimensions of the new shell would be $2l_x$ and l_y as shown in Fig. 7.3. This extended shell is considered supported along $x = \pm l_x$ and free along the edges $y = 0$ and $y = l_y$. The origin and orientation of the co-ordinate axes are the same as for the original quadrant. The following boundary conditions are specified along the edges:

At $x = \pm l_x$,

$$w = 0 \quad (7.66)$$

$$m_y = 0 \quad (7.67)$$

$$n_{xy} = 0 \quad (7.68)$$

$$u = 0 \quad (7.69)$$

At $y = 0$

$$m_y = 0 \quad (7.70)$$

$$V_y = 0 \quad (7.71)$$

$$n_{xy} = 0 \quad (7.72)$$

$$n_y = 0 \quad (7.73)$$

At $y = l_y$,

$$m_y = 0 \quad (7.74)$$

$$V_y = 0 \quad (7.75)$$

$$n_{xy} = 0 \quad (7.76)$$

$$n_y = 0 \quad (7.77)$$

The shell in each Case is assumed to be acted upon by a force or moment applied along one of the edges only.

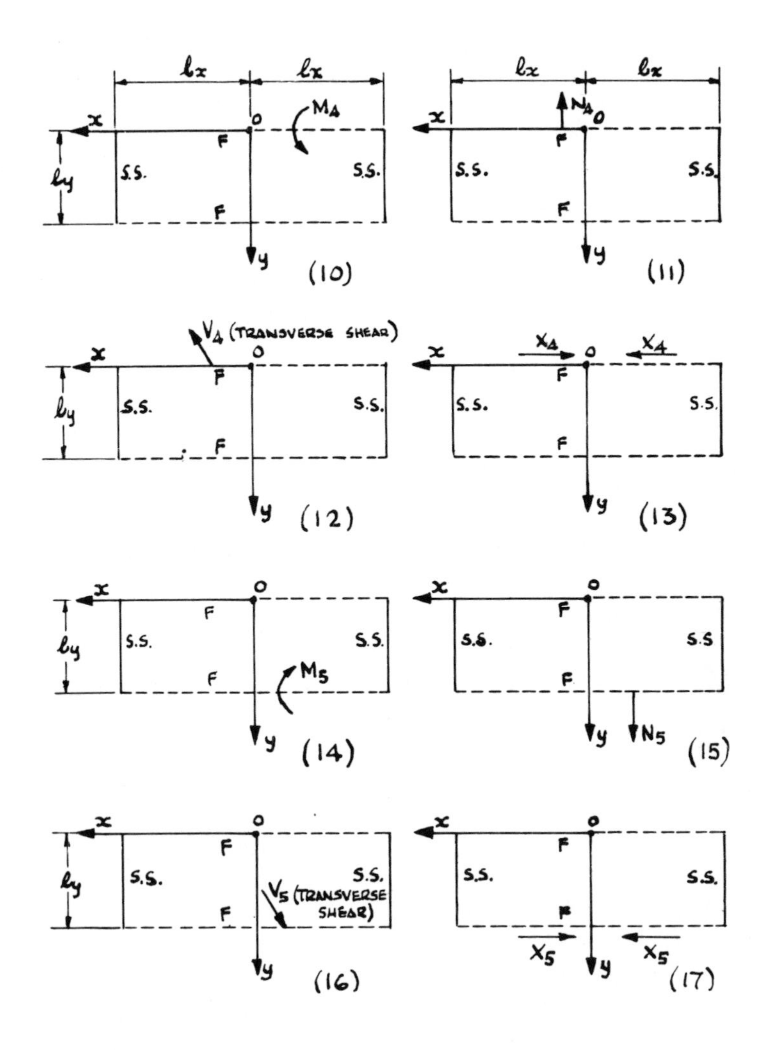

S. S. = SIMPLY SUPPORTED EDGE

F = FREE EDGE

ALL EDGE FORCES & MOMENTS APPLIED ALONG ENTIRE EDGE.

THE ORIGIN & AXES IN ALL CASES ARE SAME.

Fig. 7.3 *Edge forces applied in cases* (10) *to* (17).

Complementary function

The complementary function of the homogeneous equation (6.71) in each of the Cases considered is taken as

$$w = \sum_j A_j \, e^{\mu_j y} \cos\left(\frac{j\pi}{2l_x}\right)x \tag{7.78}$$

Putting $n = \dfrac{j\pi}{2l_x}$ and substituting equation (7.78) in equation (6.71)

$$(\mu^2_j - n^2)^4 - 4b^4\mu^2_j n^2 = 0 \tag{7.79}$$

As in Case (1), the solution of the above polynomial gives four real and four complex roots. Let these roots be

$$\left.\begin{aligned} (\mu_j)_{1,2} &= \pm J_4 \\ (\mu_j)_{3,4} &= \pm J_5 \\ (\mu_j)_{5,6,7,8} &= \pm (J_6 \pm iK_6) \end{aligned}\right\} \tag{7.80}$$

Adopting the procedure described in Section 7.3.1, any force, moment or displacement in the shell considering j^{th} term of the series can be represented as

$$\begin{aligned} f_j = R_j \Big[& A_1 b_1 e^{J_4 y} + A_2 b_2 e^{-J_4 y} + A_3 b_3 e^{J_5 y} + A_4 b_4 e^{-J_5 y} \\ & + A_5 b_5 e^{J_6 y} + A_6 b_6 e^{-J_6 y} + A_7 b_7 e^{J_6 y} + A_8 b_8 e^{-J_6 y} \Big] \end{aligned} \tag{7.81}$$

Where R_j is a multiplier which is a function of x, b_1 to b_8 are either constants or functions of y and A_1 to A_8 are arbitrary real constants. The expressions for R_j, b_1, b_2,... b_8 for various forces and moments in the shell are given in Table 7.2.

Now forces M_4, N_4, V_4, X_4, M_5, N_5, V_5 and X_5 are applied at the edges as shown in Fig. 7.3. A similar procedure to the one adopted for Cases (2) to (9) is employed and the shell in each Case is analysed so that the boundary conditions given by equations (7.66) to (7.77) are satisfied. Account is taken of the effect of the applied edge force on one of the boundary conditions, as discussed in Section 7.3.2.

7.3.4 Determination of constants for Cases (10) to (17)

Expressing the applied edge forces in the form of trigonometrical series, as in Cases (2) to (9), the resulting boundary equations for Cases (10) to (17) can be expressed as

$$[K_1]\ [A_1] = [Q] \tag{7.82}$$

where K_1 = coefficient matrix

$$[A_1] = \begin{bmatrix} A^{10}_{1} & A^{11}_{1} & \cdots\cdots & A^{17}_{1} \\ A^{10}_{2} & A^{11}_{2} & \cdots\cdots & A^{17}_{2} \\ A^{10}_{3} & A^{11}_{3} & \cdots\cdots & A^{17}_{3} \\ & & & \\ A^{10}_{8} & A^{11}_{8} & \cdots\cdots & A^{17}_{8} \end{bmatrix} \tag{7.83}$$

$$[Q] = \begin{bmatrix} (F_j)_{10} & & & & & & & \\ & (F_j)_{11} & & & & 0 & & \\ & & (F_j)_{12} & & & & & \\ & & & (F_j)_{13} & & & & \\ & & & & (F_j)_{14} & & & \\ & & & & & (F_j)_{15} & & \\ & 0 & & & & & (F_j)_{16} & \\ & & & & & & & (F_j)_{17} \end{bmatrix} \tag{7.84}$$

The superscripts with A's and F's represent Cases (10) to (17). As the F's in matrix $[Q]$ are unknown, the constants A_1 to A_8 can only be determined in terms of F's.

7.3.5 Determination of the forces and moments applied to shell edges in Cases (2) to (17)

Any force, moment or displacement, f, in the inverted umbrella shell shown in Fig. 7.1 can be obtained by superimposing on one another the corresponding forces, moments and displacements obtained in Cases (1) to (17). There is a particular integral, however, in Case (1) and this is also taken into account. Thus, considering the jth term of the series,

$$f_j = \sum_{i=1}^{17} f_j^i + (P.I.)_j \tag{7.85}$$

where f_j^i = the value of the jth term in Case (i).
$(P.I.)$ = particular integral in Case (1).

Considering, for example, the bending moment m_x in the inverted umbrella shell quadrant along $x = 0$, the jth term of the series, from Tables 7.1 and 7.2, becomes

$$\begin{aligned}(m_x)_{x=0} = -\Big[\Big\{ & EI \cos my \sum_{i=1}^{9} \Big(J^2{}_1 A^i{}_1 + J^2{}_1 A^i{}_2 + J^2{}_2 A^i{}_3 \\ & + J^2{}_2 A^i{}_4 + C^2{}_1 \cos 2\omega \cdot A^i{}_5 + C^2{}_1 \cos 2\omega \cdot A^i{}_6 \\ & + C^2{}_1 \sin 2\omega \cdot A^j{}_1 - C^2{}_1 \cdot \sin 2\omega\, A^j{}_8 \Big)\Big\} \\ & + \Big\{ EIn^2 \sum_{i=10}^{17} \Big(e^{J_4 y} A^i{}_1 + e^{-J_4 y} A^i{}_2 + e^{J_5 y} A^i{}_3 \\ & + e^{-J_5 y} A^i{}_4 + e^{J_6 y} \cos K_6 y \cdot A^i{}_5 + e^{-J_6 y} \cos K_6 y \cdot A^i{}_6 \\ & + e^{J_6 y} \cdot \sin K_6 y \cdot A^i{}_7 + e^{-J_6 y} \sin K_6 y \cdot A^i{}_8 \Big)\Big\}\Big] \\ & + [\text{Part. Int.}]\end{aligned} \tag{7.86}$$

It was pointed out earlier that the constants of integration, A, in Cases (2) to (17) can only be obtained in terms of the unknown applied forces and moments. Thus constants A^2 contain the applied moment M_2, A^3 contain the applied in-plane force N_2, and so on. Other forces, moments and displacements in the inverted umbrella shell at any point can be similarly expressed.

One of the boundary conditions for the inverted umbrella shell, for example, requires that the bending moment m_x is zero at x=0 (equation 7.4). Hence the right-hand side of equation (7.86) must be zero for all values of y. Similarly, by satisfying equations (7.2) to (7.19) a total of sixteen equations in the sixteen unknown applied forces and moments are obtained. However, it is noticed in equation (7.86) that the right-hand side contains trigonometric and exponential functions of y and because of this it would be impossible to satisfy the boundary condition for *all* values of y. This difficulty may be overcome by making the average value of $(m_x)_x = 0$ along the edge to be zero. This method has been used by Gerard in the analysis of hyperbolic paraboloid shells. Thus, instead of satisfying $(m_x)_{x=0} = 0$ for all values of y, one can satisfy

$$\int_0^{l_y} (m_x)_{x=0}\, dy = 0 \tag{7.87}$$

Other boundary conditions given by equations (7.2) to (7.19) can be similarly modified and by adopting the procedure given above, sixteen linear algebraic simultaneous equations are obtained and the solution of these equations gives the values of the forces and moments applied in Cases (2) to (17). Once these are determined the results of the seventeen cases can be superimposed to obtain the required moments, forces and displacements in the inverted umbrella shell.

7.4 ANALYSIS OF SHELLS WITH EDGE AND SPINE BEAMS

The method described in Section 7.3 for the analysis of shells without edge beams is also applicable to the analysis of shells with beams of small cross-sectional dimensions to which the assumptions mentioned in Section 7.1 are applicable. The actual analysis will be different from that of shells without edge beams only in so far as the boundary conditions are different.

Using the method developed in Section 7.3, inverted umbrella shells with edge beams can be analysed as follows:

(i) Analyse the Cases (1) to (17) as in Section 7.3.
(ii) Form the equations for the determination of the unknown applied forces and moments by taking the boundary conditions applicable to this type of shells as given in Section 7.2.
(iii) Solve these equations and substitute the solutions in expressions for forces, moments and deflections given in Tables 7.1 and 7.2.
(iv) Superimpose the results thus obtained on one another and add the sum to the particular integral in order to obtain the required forces, moments and deflections in the umbrella shells with edge beams.

7.5 COMPUTER PROGRAMS

Computer programs written in Atlas Autocode for the two types of shells given in Sections 7.3 and 7.4 are given in Appendices II and III respectively. The relevant notes on the order in which the data is to be read are also given in these Appendices.

TABLE 7.1

COMPLEMENTARY FUNCTION FOR CASES (1) TO (9)

$$f = R \cdot \{A_1 a_1 e^{J_1 x} + A_2 a_2 e^{-J_1 x} + A_3 a_3 e^{J_2 x} + A_4 a_4 e^{-J_2 x} + A_5 a_5 e^{J_3 x} + A_6 a_6 e^{-J_3 x} + A_7 a_7 e^{J_3 x} + A_8 a_8 e^{-J_3 x}\}$$

f	R	a_1	a_2	a_3	a_4
w	$\cos(my)$	1	1	1	1
$\partial w/\partial x$	$\cos(my)$	J_1	$-J_1$	J_2	$-J_2$
m_x	$-EI\cos(my)$	J_1^2	J_1^2	J_2^2	J_2^2
V_x	$-EI\cos(my)$	$(J_1^3 - 2m^2J_1)$	$-(J_1^3 - 2m^2J_1)$	$(J_2^2 - 2m^2J_2)$	$-(J_2^2 - 2m^2J_2)$
$\partial w/\partial y$	$-m\sin(my)$	1	1	1	1
m_y	$EI\,m^2\cos(my)$	1	1	1	1
V_y	$-EIm\sin(my)$	$m^2 - 2J_1^2$	$m^2 - 2J_1^2$	$m^2 - 2J_2^2$	$m^2 - 2J_2^2$
m_{xy}	$EIm\sin(my)$	J_1	$-J_1$	J_2	$-J_2$
n_{xy}	$\frac{EI}{2k}\cos(my)$	$(J_1^2 - m^2)^2$	$(J_1^2 - m^2)^2$	$(J_2^2 - m^2)^2$	$(J_2^2 - m^2)^2$
n_x	$-\frac{EIm}{2k}\sin(my)$	$\frac{(J_1^2 - m^2)^2}{J_1}$	$-\frac{(J_1^2 - m^2)^2}{J_1}$	$\frac{(J_2^2 - m^2)^2}{J_2}$	$-\frac{(J_2^2 - m^2)^2}{J_2}$
n_y	$\frac{EI}{2mk}\sin(my)$	$J_1(J_1^2 - m^2)^2$	$-J_1(J_1^2 - m^2)^2$	$J_2(J_2^2 - m^2)^2$	$-J_2(J_2^2 - m^2)^2$

TABLE 7.1—*contd.*

$$C_1 = \sqrt{J_3^2 + K_3^2},\ \omega_1 = \tan^{-1}\frac{K_3}{J_3}$$

a_5	a_6	a_7	a_8
$\cos(K_3x)$	$\cos(K_3x)$	$\sin(K_3x)$	$\sin(K_3x)$
$C_1\cos(K_3x+\omega_1)$	$C_1\cos(K_3x-\omega_1)$	$C_1\sin(K_3x+\omega_1)$	$C_1\sin(K_3x-\omega_1)$
$C_1^2\cos(K_3x+2\omega_1)$	$C_1^2\cos(K_3x-2\omega_1)$	$C_1^2\sin(K_3x+2\omega_1)$	$C_1^2\sin(K_3x-2\omega_1)$
$C_1^3\cos(K_3x+3\omega_1)$	$C_1^3\cos(K_3x-3\omega_1)$	$C_1^3\sin(K_3x+3\omega_1)$	$C_1^3\sin(K_3x-3\omega_1)$
$-2m^2C_1\cos(K_3x+\omega_1)$	$-2m^2C_1\cos(K_3x-\omega_1)$	$-2m^2C_1\sin(K_3x+\omega_1)$	$-2m^2C_1\sin(K_3x-\omega_1)$
$\cos(K_3x)$	$\cos(K_3x)$	$\sin(K_3x)$	$\sin(K_3x)$
$\cos(K_3x)$	$\cos(K_3x)$	$\sin(K_3x)$	$\sin(K_3x)$
$m^2\cos(K_3x)$	$m^2\cos(K_3x)$	$m^2\sin(K_3x)$	$m^2\sin(K_3x)$
$-2C_1^2\cos(K_3x+2\omega_1)$	$-2C_1^2\cos(K_3x-2\omega_1)$	$-2C_1^2\sin(K_3x+\omega_1)$	$-2C_1^2\sin(K_3x-2\omega_1)$
$C_1\cos(K_3x+\omega_1)$	$C_1\cos(K_3x-\omega_1)$	$C_1\sin(K_3x+\omega_1)$	$C_1\sin(K_3x-\omega_1)$
$C_1^4\cos(K_3x+4\omega_1)$ $-2m^2C_1^2\cos(K_3x+2\omega_1)$ $+m^4\cos(K_3x)$	$C_1^4\cos(K_3x-4\omega_1)$ $-2m^2C_1^2\cos(K_3x-2\omega_1)$ $+m^4\cos(K_3x)$	$C_1^4\sin(K_3x+4\omega_1)$ $-2m^2C_1^2\sin(K_3x+2\omega_1)$ $+m^4\sin(K_3x)$	$C_1^4\sin(K_3x-4\omega_1)$ $-2m^2C_1^2\sin(K_3x-2\omega_1)$ $+m^4\sin(K_3x)$
$C_1^3\cos(K_3x+3\omega_1)$ $-2m^2C_1\cos(K_3x+\omega_1)$ $+\frac{m^4}{C_1}\cos(K_3x-\omega_1)$	$C_1^3\cos(K_3x-3\omega_1)$ $-2m^2C_1\cos(K_3x-\omega_1)$ $+\frac{m^4}{C_1}\cos(K_3x-\omega_1)$	$C_1^3\sin(K_3x+3\omega_1)$ $-2m^2C_1\sin(K_3x+\omega_1)$ $+\frac{m^4}{C_1}\sin(K_3x-\omega_1)$	$C_1^3\sin(K_3x-3\omega_1)$ $-2m^2C_1\sin(K_3x-\omega_1)$ $+\frac{m^4}{C_1}\sin(K_3x-\omega_1)$
$C_1^5\sin(K_3x+5\omega_1)$ $-2m^2C_1^3\cos(K_3x+3\omega_1)$ $+m^4C_1\cos(K_3x+\omega_1)$	$C_1^5\sin(K_3x-5\omega_1)$ $-2m^2C_1^3\cos(K_3x-3\omega_1)$ $+m^4C_1\cos(K_3x-\omega_1)$	$C_1^5\sin(K_3x+5\omega_1)$ $-2m^2C_1^3\sin(K_3x+3\omega_1)$ $+m^4C_1\sin(K_3x+\omega_1)$	$C_1^5\sin(K_3x-5\omega_1)$ $-2m^2C_1^3\sin(K_3x-3\omega_1)$ $+m^4C_1\sin(K_3x-\omega_1)$

TABLE 7.2

COMPLEMENTARY FUNCTION FOR CASES (10) TO (17)

$$f = R \cdot \{ A_1 a_1 e^{J_4 y} + A_2 a_2 e^{-J_4 y} + A_3 a_3 e^{J_5 y} + A_4 a_4 e^{-J_5 y} + A_5 a_5 e^{J_6 y} + A_6 a_6 e^{-J_6 y} + A_7 a_7 e^{J_6 y} + A_8 a_8 e^{-J_6 y} \}$$

f	R	a₁	a₂	a₃	a₄
w	$\cos(nx)$	1	1	1	1
$\partial w/\partial y$	$\cos(nx)$	J_4	$-J_4$	J_5	$-J_5$
m_y	$-EI\cos(nx)$	J_4^2	J_4^2	J_5^2	J_5^2
V_y	$-EI\cos(nx)$	$(J_4^3-2n^2J_4)$	$-(J_4^3-2n^2J_4)$	$(J_5^3-2n^2J_5)$	$-(J_5^3-2n^2J_5)$
$\partial w/\partial x$	$-n\sin(nx)$	1	1	1	1
m_x	$EIn^2\cos(nx)$	1	1	1	1
V_x	$-EIn\sin(nx)$	$n^2-2J_4^2$	$n^2-2J_4^2$	$n^2-2J_5^2$	$n^2-2J_5^2$
m_{xy}	$EIn\sin(nx)$	J_4	$-J_4$	J_5	$-J_5$
n_{xy}	$\frac{EI}{2k}\cos(nx)$	$(J_4^2-n^2)^2$	$(J_4^2-n^2)^2$	$(J_5^2-n^2)^2$	$(J_5^2-n^2)^2$
n_y	$\frac{-EIn}{2k}\sin(nx)$	$\frac{(J_4^2-n^2)^2}{J_4}$	$-\frac{(J_4^2-n^2)^2}{J_4}$	$\frac{(J_5^2-n^2)^2}{J_5}$	$-\frac{(J_5^2-n^2)^2}{J_5}$
n_x	$\frac{EI}{2nk}\sin(nx)$	$J_4(J_4^2-n^2)^2$	$-J_4(J_4^2-n^2)^2$	$J_5(J_5^2-n^2)^2$	$-J_5(J_5^2-n^2)^2$

TABLE 7.2—*contd.*

$$C_2 = \sqrt{J_6^2 + K_6^2},\ \omega_2 = \tan^{-1} \frac{K_6}{J_6}$$

a_5	a_6	a_7	a_8
$\cos(K_6 y)$	$\cos(K_6 y)$	$\sin(K_6 y)$	$\sin(K_6 y)$
$C_2 \cos(K_6 y + \omega_2)$	$C_2 \cos(K_6 y - \omega_2)$	$C_2 \sin(K_6 y + \omega_2)$	$C_2 \sin(K_6 y - \omega_2)$
$C_2^2 \cos(K_6 y + 2\omega_2)$	$C_2^2 \cos(K_6 y - 2\omega_2)$	$C_2^2 \sin(K_6 y + 2\omega_2)$	$C_2^2 \sin(K_6 y - 2\omega_2)$
$C_2^3 \cos(K_6 y + 3\omega_2)$	$C_2^3 \cos(K_6 y - 3\omega_2)$	$C_2^3 \sin(K_6 y + 3\omega_2)$	$C_2^3 \sin(K_6 y - 3\omega_2)$
$-2n^2 C_2 \cos(K_6 y + \omega_2)$	$-2n^2 C_2 \cos(K_6 y - \omega_2)$	$-2n^2 C_2 \sin(K_6 y + \omega_2)$	$-2m^2 C_2 \sin(K_6 y - \omega_1)$
$\cos(K_6 y)$	$\cos(K_6 y)$	$\sin(K_6 y)$	$\sin(K_6 y)$
$\cos(K_6 y)$	$\cos(K_6 y)$	$\sin(K_6 y)$	$\sin(K_6 y)$
$n^2 \cos(K_6 y)$	$n^2 \cos(K_6 y)$	$n^2 \sin(K_6 y)$	$n^2 \sin(K_6 y)$
$-2C_2^2 \cos(K_6 y + 2\omega_2)$	$-2C_2^2 \cos(K_6 y - 2\omega_2)$	$-2C_2^2 \sin(K_6 y + 2\omega_2)$	$-2C_2^2 \sin(K_6 y - 2\omega_2)$
$C_2 \cos(K_6 y + \omega_2)$	$C_2 \cos(K_6 y - \omega_2)$	$C_2 \sin(K_6 y + \omega_2)$	$C_2 \sin(K_6 y - \omega_2)$
$C_2^4 \cos(K_6 y + 4\omega_2)$	$C_2^4 \cos(K_6 y - 4\omega_2)$	$C_2^4 \sin(K_6 y + 4\omega_2)$	$C_2^4 \sin(K_6 y - 4\omega_2)$
$-2n^2 C_2^2 \cos(K_6 y + 2\omega_2)$	$-2n^2 C_2^2 \cos(K_6 y - 2\omega_2)$	$-2n^2 C_2^2 \sin(K_6 y - 2\omega_2)$	$-2n^2 C_2^2 \sin(K_6 y - 2\omega_2)$
$+n^4 \cos(K_6 y)$	$+n^4 \cos(K_6 y)$	$+n^4 \sin(K_6 y)$	$+n^4 \sin(K_6 y)$
$C_2^3 \cos(K_6 y + 3\omega_2)$	$C_2^3 \cos(K_6 y - 3\omega_2)$	$C_2^3 \sin(K_6 y + 3\omega_2)$	$C_2^3 \sin(K_6 y - 3\omega_2)$
$-2n^2 C_2 \cos(K_6 y + \omega_2)$	$-n^2 \cos(K_6 y - \omega_2)$	$-2n^2 C_2 \sin(K_6 y + \omega_2)$	$-2n^2 C_2 \sin(K_6 y - \omega_2)$
$+\frac{n^4}{C_2} \cos(K_6 y - \omega_2)$	$(2C_2 - \frac{n^2}{C_2})$	$+\frac{n^4}{C_2} \sin(K_6 y - \omega_2)$	$+\frac{n^4}{C_2} \sin(K_6 y - \omega_2)$
$C_2^5 \cos(K_6 y + 5\omega_2)$	$C_2^5 \cos(K_6 y - 5\omega_2)$	$C_2^5 \sin(K_6 y + 5\omega_2)$	$C_2^5 \sin(K_6 y - 5\omega_2)$
$-2n^2 C_2^3 \cos(K_6 y + 3\omega_2)$	$-2n^2 C_2^3 \cos(K_6 y - 3\omega_2)$	$-2n^2 C_2^3 \sin(K_6 y + 3\omega_2)$	$-2n^2 C_2^3 \sin(K_6 y - 3\omega_2)$
$+n^4 C_2 \cos(K_6 y + \omega_2)$	$+n^4 C_2 \cos(K_6 y - \omega_2)$	$+n^4 C_2 \sin(K_6 y + \omega_2)$	$+n^4 C_2 \sin(K_6 y - \omega_2)$

APPENDICES

APPENDIX I

ROOTS OF AN EIGHTH DEGREE POLYNOMIAL EQUATION

Consider an eighth degree polynomial equation in λ, of the form:

$$(\lambda^2 - n^2)^4 - 4b^4n^2\lambda^2 = 0' \qquad \text{(A1.1)}$$

where b and n are real positive constants.

Putting $\lambda^2 = x$ in equation (A1.1) and expanding, the following quartic equation in x is obtained:

$$x^4 - 4n^2x^3 + 6n^4x^2 - 4n^2(n^4+b^4)x+n^8 = 0 \qquad \text{(A1.2)}$$

Let P and Q be two constants.
Add $(Px+Q)^2$ to both sides of equation (A1.2)

$$x^4-4n^2x^3 + (6n^4+P^2)x^2-4n^2(n^4+b^4) + 2PQx+(n^8+Q^2)$$
$$= P^2x^2 + 2PQx + Q^2 \qquad \text{(A1.3)}$$

Let R be another constant such that

Left-hand side of equation (A1.3) $= (x^2 - 2n^2x + R)^2$
or $x^4 - 4n^2x^3 + (6n^4+P^2)x^2-4n^2(n^4+b^4) + 2PQx + (n^8+Q^2)$
$$= x^4-4n^2x^3 + (4n^4+2R)x^2 - (4n^2R)x+R^2 \qquad \text{(A1.4)}$$

Now, comparing the coefficients of similar powers of x on both sides of equation (A1.4)

$$6n^4 + P^2 = 4n^4 + 2R \qquad \text{(A1.5)}$$

$$4n^2(n^4 + b^4) + 2PQ = 4n^2R \qquad \text{(A1.6)}$$

$$n^8 + Q^2 = R^2 \qquad \text{(A1.7)}$$

From equations (A1.5) to (A1.7) the constants P and Q can be expressed in terms of the other constants R, b and n as follows:

$$P^2 = -2n^4 + 2R \tag{A1.8}$$

$$PQ = n^2R - n^2(n^4+b^4) \tag{A1.9}$$

$$Q^2 = R^2 - n^8 \tag{A1.10}$$

The three equations (A1.8) to (A1.10) in the two unknowns P and Q are satisfied if, and only if

$$n^2R - n^2(n^4 + b^4) = (-2n^4 + 2R)(R^2 - n^8) \tag{A1.11}$$

Expanding both sides of equation (A1.11) and rearranging, the following cubic equation in R is obtained:

$$R^3 - (3n^4)R^2 + (3n^8 + 4n^4b^4)R - (n^{12} + 4n^8b^4 + 2n^4b^8) = 0 \tag{A1.12}$$

Putting

$$S = R - n^4 \tag{A1.13}$$

in equation (A1.12), the following reduced cubic equation is obtained:

$$S^3 + (4n^4b^4)S - (2n^4b^8) = 0 \tag{A1.14}$$

Equation (A1.14) has one real root given by

$$S = \left\{\frac{2n^4b^8}{2} + \sqrt{\frac{4n^8b^{16}}{4} + \frac{64}{27}n^{12}b^{12}}\right\}^{1/3} + \left\{\frac{2n^4b^8}{2} - \sqrt{\frac{4n^8b^{16}}{4} + \frac{64}{27}n^{12}}\right\}^{1/3}$$

$$\text{or } S = (n^4b^8)^{1/3}\left[\left\{1 + 1 + \frac{64\,n^4}{27\,b^4}\right\}^{1/3} + \left\{1 - 1 + \frac{64\,n^4}{27\,b^4}\right\}^{1/3}\right] \tag{A1.15}$$

As n and b are real and positive, the first expression within the square bracket on the right-hand side of equation (A1.15) will be real and positive and the second expression will be real and negative. Also, the absolute value of the first expression will be greater than the absolute value of the second expression.

Thus, irrespective of the magnitudes of b and n, the value of S given by equation (A1.15) will be real and positive.

From equations (A1.8), (A1.10), and (A1.13)

$$P^2 = 2^S \tag{A1.16}$$

$$Q^2 = S^2 + 2Sn^4 \tag{A1.17}$$

Since S and n are real and positive, it follows from equations (A1.16) and (A1.17) that

$$\begin{aligned} &P > 0 \\ &Q > S > 0 \end{aligned} \tag{A1.18}$$

Also, from equation (A1.9),

$$\begin{aligned} &PQ > 0 \\ &n^2R > n^2(n^4 + b^4) \\ \text{or } &R > n^4 + b^4 \end{aligned} \tag{A1.19}$$

but from equation (A1.13)

$$\begin{aligned} &R = S + n^4 \\ \therefore\ &S + n^4 > n^4 + b^4 \\ \text{or }\ &S > b^4 \end{aligned} \tag{A1.20}$$

From equation (A1.3) and (A1.4)

$$\begin{aligned} &(Px + Q)^2 = (x^2 - 2n^2x + R)^2 \\ \therefore\quad &(x^2 - 2n^2x + R - Px - Q)(x^2 - 2n^2x + R + Px + Q) = 0 \quad \text{(A1.21)} \\ \text{If}\quad &x^2 - 2n^2x + R - Px - Q = 0 \\ \text{then}\quad &x = \tfrac{1}{2}\{(2n^2+P) \pm \sqrt{(2n^2 + P)^2 - 4(R - Q)}\} \\ &\ \ = \tfrac{1}{2}\{(2n^2 + P) \pm \sqrt{4n^4 + 4n^2P + P^2 - 4R + 4Q}\} \end{aligned}$$

Substituting R and P^2 from equations (A1.13) and (A1.16) respectively,

$$\begin{aligned} &x = \tfrac{1}{2}\{(2n^2 + P) \pm \sqrt{4n^4 + 4n^2P + 2S - 4S - 4n^4 = 4Q}\} \\ \text{or}\quad &x = \tfrac{1}{2}\{(2n^2 + P) \pm \sqrt{4n^2P + 4Q - 2S}\} \end{aligned} \tag{A1.22}$$

$Q > S$, the expression under the square root sign on the right-hand side of equation (A1.22) will be positive so that the two values of x given by this equation will be real.

It can be proved that the two values of x given by equation (A1.22) are also positive. This will be the case if

$$2n^2 + P > \sqrt{4n^2P+4Q-2S}$$

Squaring both sides,

$$4n^4+4n^2P+P^2>4n^2P+4Q-2S$$
$$4n^4+P^2>4Q-2S$$

But, $P^2 = 2S$

$\therefore$ $4n^4+2S>4Q-2S$

or $4n^4+4S>4Q$

$$n^4+S>Q$$

Squaring both sides again and substituting the value of Q^2 from equation (A1.17),

$$n^8+2n^4S + S^2>S^2 + 2Sn^4$$

or $n^8>0$

Therefore, for all positive values of n, as is the case, the two values of x given by the equation (A1.22) will be real and positive.

Now $\lambda^2 = x$

$\therefore$ $\lambda = \sqrt{x}$

Four of the eight roots of equation (A1.1) will be real. Let these four roots be

$$\lambda_1 = J_1$$
$$\lambda_2 = -J_1$$
$$\lambda_3 = J_2$$
$$\lambda_4 = -J_2$$

where,

$$J_1 = \tfrac{1}{2}\{(2n^2+P) + \sqrt{4n^2P+4Q-2S}\}^{1/2} \qquad \text{(A1.23)}$$

$$J_2 = \tfrac{1}{2}\{(2n^2+P) - \sqrt{4n^2P+4Q-2S}\}^{1/2} \tag{A1.24}$$

As P, Q and S are known in terms of the coefficients b and n from equations (A1.15), (A1.16) and (A1.17), the four real roots of equation (A1.4) are fully determined.

Now considering the second factor on the left-hand side of equation (A1.21)

$$x^2 - 2n^2x + R - Px + Q = 0$$

The solution of which is

$$\begin{aligned} x &= \tfrac{1}{2}\{(2n^2-P) \pm \sqrt{(2n^2-P)^2 - 4(R+Q)}\} \\ &= \tfrac{1}{2}\{(2n^2-P) \pm \sqrt{4n^4 - 4n^2P + P^2 - 4R - 4Q}\} \\ &= \tfrac{1}{2}\{(2n^2-P) \pm \sqrt{4n^4 - 4n^2P + 2S - 4S - 4n^4 - 4Q}\} \end{aligned}$$

$$\text{or } x = \tfrac{1}{2}\{(2n^2-P) \pm - \sqrt{4n^2P - 2S - 2Q}\} \tag{A1.25}$$

As n, P, Q are all positive, the expression within the square-root sign on the right-hand side of equation (A1.26) will be negative so that the two values of x given by this equation will be complex.

$$\therefore \quad x = U \pm iV \tag{A1.26}$$

$$\begin{aligned} \text{where } U &= \tfrac{1}{2}(2n^2 - P) \\ V &= \tfrac{1}{2}\sqrt{4n^2P + 2S + Q} \\ i &= \sqrt{-1} \end{aligned} \tag{A1.27}$$

Therefore, the other four roots of equation (A1.1) will be complex. Let these roots be λ_5, λ_6, λ_7 and λ_8.

$$\begin{aligned} \text{Then } \lambda_5 &= (U+iV)^{1/2} \\ \lambda_6 &= (U-iV)^{1/2} \\ \lambda_7 &= -(U-iV)^{1/2} \\ \lambda_8 &= -(U+iV)^{1/2} \end{aligned}$$

Let $\lambda_{5,6,7,8} = \pm(J_3 \pm iK_3)$

$$\text{Then } \pm(J_3 + iK_3) = \pm(U \pm iV)^{1/2} \tag{A1.28}$$

To determine J_3 and K_3, consider the first root

$$\lambda_5 = J_3 + iK_3 = (U + iV)^{1/2}$$

Let $U = T\cos\theta$

$V = T\sin\theta$

Then $T = \sqrt{U^2 + V^2}$

$\tan\theta = V/U$

or $\theta = \tan^{-1} V/U$

$$(U + iV)^{1/2} = T^{1/2}(\cos\theta + i\sin\theta)^{1/2}$$

$$= T^{1/2}\left(\cos\frac{\theta}{2} + i\sin\frac{\theta}{2}\right)$$

(De Moirve's Theorem)

$$\therefore\ J_3 + iK_3 = T^{1/2}\left(\cos\frac{\theta}{2} + i\sin\frac{\theta}{2}\right) \tag{A1.29}$$

Equating the real and the imaginary parts on both sides of equation (A1.29)

$$J_3 = T^{1/2}\cos\frac{\theta}{2}$$

$$= 4\sqrt{U^2 + V^2}\cos\left(\tfrac{1}{2}\tan^{-1}\frac{V}{U}\right)$$

$$\text{or, } J_3 = \left\{4\sqrt{\tfrac{1}{4}(4n^4+4S+Q)}\right\}\cdot\cos\left\{\tfrac{1}{2}\tan^{-1}\left(\frac{\sqrt{4n^2P+2S+Q}}{2n^2-P}\right)\right\} \tag{A1.30}$$

Similarly,

$$K_3 = \left\{4\sqrt{\tfrac{1}{4}(4n^4+4S+Q)}\right\}\cdot\sin\left\{\tfrac{1}{2}\tan^{-1}\left(\frac{\sqrt{4n^2P+2S+Q}}{2n^2-P}\right)\right\} \tag{A1.31}$$

The four complex roots of equation (A1.1) are thus fully determined. The eight roots of equation (A1.1) are

$$\lambda_1 = +J_1$$
$$\lambda_2 = -J_1$$
$$\lambda_3 = +J_2$$
$$\lambda_4 = -J_2$$
$$\lambda_5 = +J_3 + iK_3$$
$$\lambda_6 = +J_3 - iK_3$$
$$\lambda_7 = -J_3 + iK_3$$
$$\lambda_8 = -J_3 - iK_3$$

where J_1, J_2, J_3 and K_3 are all real quantities given by equations (A1.23), (A1.24), (A1.30) and (A1.31) respectively.

APPENDIX II

COMPUTER PROGRAM FOR THE ANALYSIS OF INVERTED UMBRELLA SHELLS WITHOUT EDGE BEAMS

The computer program given in this Appendix is for the analysis of hyperbolic paraboloidal shells of the inverted umbrella type, not provided with edge beams and carrying uniformly distributed load on the whole area. The program is written in the Atlas Autocode and is suitable for the Atlas or any other computer with the Atlas Autocode compiler. The program requires a storage of 75 blocks of 512 words and the computation time on the Atlas computer is 3·5 minutes. The total output is of the order of 300 lines.

The data, and the order in which it is to be punched, is as follows:

E (Young's modulus of shell material) in lb/in^2
lx (length, in x-direction, of one quadrant) in ft
ly (length in y-direction of one quadrant) in ft
c (rise of shell) in ft
h (thickness of shell) in inches
z (u.d. load) in lb/ft
***Z

The program then causes the computer to perform the necessary computations and to print out the values of w, N_x, N_y, N_{xy}, M_x, M_y, M_{xy}, V_x, and V_y at the 81 points x = 0, 1/8 lx, 1/4 lx, 3/8 lx,......lx and y = 0, 1/8 ly, 1/4 ly, 3/8 ly, ...ly,

The output is in a tabulated form.

```
begin
real     Epsi, E, lx, ly, c, hin, h, m, n, k, I, EI, b4, Z, x, y, det, J1, J2, J3,
         J4, J5, cJ6, K3, K6, C1, C2, wom1, wom2, X, Y, w, nx, ny, nxy,
         mx, my, mxy, Vx, Vy, 1, q, c ja, jb, jc, jd
integer  i1, i2, i3, i4, i5, i6, i7, i8, i9
```

```
array   CH(0:8), U, V, CONV(1:8), K1, K2, Q2, A2(1:8, 1:8), RC1,
        RC2, RV1, c RV2, PI(1:11), pl, p2(1:11, 1:8), Q1, A1(1:8, 1:9),
        L(1:16, 1:16), Y1, c S(1:16), A(1:17, 1:8)

routine spec solve polynomial (integer, n, F, arrayname a, u, v, CONV)
routine spec const mult (integer i, arrayname R, realname m)
routine spec var mult (integer i, arrayname R, realname m, y)
routine spec cf coef (integer i, arrayname C, realname m, x)
routine spec simp coef (integername i, j)
routine spec conmul a
routine spec conmul b
routine spec varmul a
routine spec varmul b
routine spec cfa
routine spec cfb
real fn spec fora (integer i, integername j)
real fn spec forb (integer i, integername j, p)
real fn spec forc (integer i, integername j, p)
read (Epsi, lx, ly, c, hin, Z)
E = 144Epsi ; h = hin/12 ; I = I = h†3/12 ; EI = E*I
k = c/(1x*1y) b4 = 12k²/(h²)
m = ½*π/ly;n = ½*π/lx
comment solution of characteristic equations.
cycle i1 = 1, 1, 2
if il = 1 then 1 = m; if il = 2 then il = n
cycle i2 = 1, 2, 7
CH(i2) = 0
repeat
CH(0) = 1; CH(2) = −41²;  CH(4) = 6*(1†4); CH(6) = − 41²*(1†4 + b4)
CH(8) = 1†8
solve polynomial (8, 10, CH, U, V, CONV)
X = 100 ; Y = 100
cycle i2 = 1, 1, 4
X=X+1 if V(i2)=0; Y=Y+1 if V(i2+4)=0
repeat
→ 1 if X = 104 and Y = 100; → 2 if X = 100 and Y = 104
newline; caption solution $ of $ characteristic $ eqns $ faulty.
caption check $ data.
stop
1:      ja = ½*(1U(1)1+1U(2)1); jb = ½*(1U(3)1+1U(4)1); jc = 0; jd = 0
```

```
cycle i2=5, 1, 8
jc=jc+|U(i2)|/4; jd=jd+|V(i2)|/4
repeat
→ 20
2:          ja=½*(1U(5)1+1U(6)1); jb=½*(1U(7)1+1U(8)1); jc=0;
jd=0 cycle i2=1, 1, 4
jc=jc+1U(i2)1/4;  jd=jd+(1V(i2)1)/4
repeat
20:         → 21 if i1 = 2
            J1=ja; J2=jb; J3=jc; K3=jd
            → 22
21:         J4=ja ; J5=jb ; J6=jc ; K6=jd
22:         C1=radius (J3, K3)
repeat
C1=radius (J3, K3); C2=radius (J6, K6); woml=arctan (J3, K3);
wom 2=arctan (J6, K6)
comment formation of matrix K (named K1) for cases (1) to (9)
null (K1)
cycle i2=1,1,2
if i2=1 then x=0 ; if i2=2 then x=1x
cfa
cycle i1=1,1,8
K1(4i2–3, i1)=pl(3, i1); K1(4i2–2, i1)=p1(4,i1)
K1(4i2–1, i1)=p1(9, i1); K1(4i2, i1)=p1(10, i1)
repeat
repeat
comment formation of matrix K′ (named K2) for cases (10) to (17)
null (K2)
cycle i=1, 1, 2
if i2=1 then y=0; if i2=2 then y=1y
cfb
cycle i1=1, 1, 8
K2(4i2–3,i1)=p2(6,i1);K2(4i2–2,i1)=p2(7,i1);K2(4i2–1,i1)=p2(9,i1)
K2(4i2,i1)=p2(11,i1)
repeat
repeat
comment determination of constants of integration in cases (2) to (17)
comment due to unit applied forces
invert(K3,K1,det)
if det 0 then → 4
```

```
newline
caption Determinant $ of $ K1 zero. Check $ data
stop
4:          invert(K4,K2,det)
            if det 0 then → 5

caption Determination $ of $ K2 $ zero. Check $ data.
stop
5: comment formation of matrix Q(namedQ1)
null(Q1)
cycle i1=1,1,2
if i1=2 then → 6
x=0; y=0; → 7
6: x=1x; y=0
7: part integ (1,x,y)
if i1=1 then → 8 ; if i1=2 then → 9
8 : i2=1 ; i3=2 ; i4=3 ; i5=4 ; → 10
9 : i2=5 ; i3=6 ; i4=7 ; i5=8
10:     Q1(i2,1)=PI(3);Q1(i3,1)=PI(4);Q1(i4,1)=PI(9);Q1(i5,1)=
        PI(10)

repeat
Cycle i1=1,1,8
Q1 (i1,i1+1)=4/π
repeat
comment formation of matrix Q' (named Q2)
cycle il = 1,1,8
Q2(i1,i1)=4/π
repeat
matrix mult (A1,K3,Q1)  ;  matrix mult (A2, K4, Q2)
cycle i1=1,1,9
cycle i2=1,1,8
A(i1,i2)=A1(i2,i1) ; if i1+9 > 17 then→ 11
A(i1+9,i2)=A2(i2,i1)

11: repeat
    repeat
comment formation of boundary equations for the actual shell.
comment numerical integration is performed by Simpson's Rule
null (L)
```

```
x=0  ;  cfa  ;  conmul a
cycle i1=0,1,16
simp coef(i1,i9)
y=i1*1y/16
varmul a
cycle i2=1,1,8
cycle i3=1,1,8
L(1,i2)=L(1,i2)+(RC1(3)*RV1(3)*p1(3,i3)*A1(i3,i2+1))*i9*1y/48
L(2,i2)=L(2,i2)+(RC1(4)*RV1(4)*p1(4,i3)*A1(i3,i2+2))*i9*1y/48
L(3,i2)=L(3,i2)+(RC1(9)*RV1(9)*p1(9,i3)*A1(i3,i2+2))*i9*1y/48
L(4,i2)=L(4,i2)+(RC1(10)*RV1(10)*p1(10,i3)*A1(i3,i2+1))*i9*1y/48
repeat
repeat
repeat
conmul b ; varmul b ;
cycle i1=0,1,16
simp coef (i1,i9)
y=i1*1y/16 ; cfb
cycle i2=9,1,16
cycle i3=1,1,8
14=i2-8
L(1,i2)=L(1,i2)+(RC2(3)*RV2(3)*p2(3,i3)*A2(i3,i4))*i9*1y/48
L(2,i2)=L(2,i2)+(RC2(4)*RV2(4)*p2(4,i3)*A2(i3,i4))*i9*1y/48
L(3,i2)=L(3,i2)+(RC2(9)*RV2(9)*p2(9,i3)*A2(i3,i4))*i9*1y/48
L(4,i2)=L(4,i2)+(RC2(10)*RV2(10)*p2(10,i3)*A2(i3,i4))*i9*1y/48
repeat
repeat
repeat
y=0 ; conmul a ; varmul a
cycle i1=0,1,16
simp coef (i1,i9)
x=i1*1x/16;    cfa
cycle i2=1,1,8
cycle i3=1,1,8
L(5,i2)=L(5,i2)+(RC1(6)*RV1(6)2p1(6,i3,i2+1))*i9*1x/48        i4=i2+1
L(6,i2)=L(6,i2)+(RC1(7)*RV1(7)*p1(7,i3)*A1(i3,i4))*i9*1x/48
L(7,i2)=L(7,i2)+(RC1(9)*RV1(9)*p1(9,i3)*A1(i3,i4))*i9*1x/48
L(8,i2)=L(8,i2)+(RC1(11)*RV1(11)*p1(11,i3)*A1(i3,i4))*i9*1x/48
repeat
repeat
```

```
repeat
conmul b ; cfb
cycle il=0,1,16
simp coef (i1,i9) ; x=i1*1x/16 ; varmul b
cycle i2=9,1,16
cycle i3=1,1,8
L(5,i2)=L(5,i2)+(RC2(6)*RV2(6)*p2(6,i3)*A2(i3,i2-8))*i9*1x/48
i4=i2-B
L(6,i2)=L(6,i2)+(RC2(7)*RV2(7)*p2(7,i3)*A2(i3,i4))*i9*1x/48
L(7,i2)=L(7,i2)+(RC2(9)*RV2(9)*p2(9,i3)*A2(i3,i4))*i9*1x/48
L(8,i2)=L(8,i2)+(RC2(11)*RV2(11)*p2(11,i3)*A2(i3,i4))*i9*1x/48
repeat
repeat
repeat
x=1x ; cfa ; conmul a
cycle i1=0,1,16
simp coef (i1,i9) ; y=i1*1y/16 ; varmul a
cycle i2=1,1,8
cycle i+1,1,8
i4=i2+1
L(9,i2)=L(9,i2)+(RC1(4)*RV1(4)*p1(4,i3)*A1(i3,i4))*i9*1y/48
L(10,i2)=L(10,i2)+RC1(4)*RV1(4)*p1(4,i3)*A1(i3,i4))*i9*1y/48
L(11,i2)=L(11,i2)+(RC1(2)*RV1(2)*p1(2,i3)*A1(i3,i4))*i9*1y/48

repeat
repeat
repeat
conmul b ; varmul b
cycle i1=0,1,16
simp coef (i1,i9) ; y=i1*1y/16 ; cfb
cycle i2=9,1,16
cycle i3=1,1,8
L(9,i2)=L(9,i2)+(RC2(4)*p2(4,i3)*A2(i3,i2-8))*i9*1y/48
L(10,i2)=L(10,i2)+(RC2(9)*RV2(9)*p2(9,i3)*A2(i3,i2-8))*i9*1y/48
L(11,i2)=L(11,i2)+(RC2(2)*RV2(2)*p2(2,i3)*A2(i3,i2-8))*i9*ly/48
repeat
repeat
repeat
y=1y ; conmul a ; varmul a
cycle i1=0,1,16
```

```
simp coef (i1,i9) ; x=i1*1x/16 ; cfa
cycle i2=1,1,8
cycle i3=1,1,8
i4=i2+1
L(12,i3)=L(12,i3)+(RC1(7)*RV1(7)*p1(7,i3)*A1(i3,i4))*i9*1y/48
L(13,i3)=L(13,i3)+(RC1(9)*RV1(9)*p1(9,i3)*A1(i3,i4))*i9*1y/48
L(14,i3)=L(14,i3)+(RC1(5)*RV1(5)*p1(5,i3)*A1(i3,i4))*i9*1y/48

repeat
repeat
repeat
begin
array D,F(1:16,1:16),D1,F1(1:16)
null (D) ; null (F)
cycle i1=1,1,16
D1(i1)=0   ;   F1(i1)=0
repeat
cycle i1=0,1,16
y=il*ly/16   ;   conmul   a   ;   varmul   a
cycle i2=0,1,16
simp coef (i2,i5) ; x=i2*1x/16 ; cfa
cycle i3=1,1,8
cycle i4=1,1,8
D(i1,i3)=D(i1,i3)+(RC1(10)*RV1(10)*p1(10,i4)*A1(i3,i4+1))*i5*1x/48
repeat
D1(i1)=D1(i1)+(RC1(10)*RV1(10)*p1(10)*A1(i3,1))*i5*1x/48
repeat
repeat
cfb ; conmul b
cycle i2=0,1,16
simp coef (i2,i5) ; x=i2*1x/16 ; varmul b
cycle i2=0,1,16
cycle i4=1,1,8
D(i1,i3)=D(i1,i3)+(RC2(10)*RV(10)*p2(10,i4)*A2(i4,i3–8))*i5*1x/48
repeat
repeat
repeat
x=1x ; conmul a ; varmul a ; cfa
cycle i2=1,1,8
cycle i3=1,1,8
```

```
D(i1,i2)=D(i1,i2)–k*y*(RC1(1)*RV1(1)*p1(1,i3)*A1(i3,i2+1))
repeat
D1(i1)=D1(i1)–k*y*(RC1(1)*RV1(1)*p1(1,i2)*A1(i2,1))
repeat
conmul b   ;   varmul b   ;   cfb
cycle i2=9,1,16
cycle i3=1,1,8
D(i1,i2)=D(i1,i2)–k*y*(RC2(1)*RV2(1)*p2(1,i3)*A2(i3,i2–8))
repeat
repeat
repeat
cycle 11x611,06
x=i1*i=,/0 ; conmul a ; cfa
cycle i2=0,01,16
simp coef (i2,i5) ; y=i2*1y/16
varmul a
cycle i3=1,1–8
cycle i4=1,1,8
F(i1,i3)=F(i1,i3)+(RC1(11)*RV1(11)2p1(11,i4)*A1(i4,i3+1))*i5*ly/48
repeat
F1(i1)=F1(i1)+(RC1(11)*RV1(11)*p1(11,i3)*A1(i3,1))*i5*1y/48
repeat
repeat
conmul b ; varmul b
cycle i2=0,1,16
simp coef (i2,i5) ; y=i2*1y/16 ; cfb
cycle i3=9,1,16
cycle i4=1,1,8
F(i1,i3)=F(i1,i3)+(RC2(11)*RV2(11)*p2(11,i4)*A2(i4,i3–8))*i5*1y/48
repeat
repeat
repeat
y=ly
cycle i2=1,1,16
if i2⩾9 then → 20
conmul a ; varmul a ; cfa
cycle i2 3=1,1,8
F(i1,i2)=F(i1,i2)–k*x*(RC1(1)*RV1(1)*p1(1,i2)*A1(i2,1))
repeat
F(i1,i2)=F(i1,i2)–k*x*(RC2(1)*RV2)*p2(1)*p2(1,i3)*A2(i3,i2–8))
```

```
repeat
21:repeat
repeat
cycle i1=1,1,16
cycle i2=0,1,16
simp coef (i2,i5)
L(15,i1)=L(15,i1)+i5*D(i2,i1)*1y/48
L(16,i1)=L(16,i1)+i5*F(i2,i1)*1x/48
repeat
repeat
cycle i1=0,1,16
simp coef (i1,i5)
S(15)=S(15)+i5*D1(i1)*1y/48 ; S(16)=S(16)+i5*F1(i1)*1x/48
end


x=0 ; cfa ; conmul a
cycle i1=0,1,16
simp coef(i1,i9) ; y=i1*1y/16 ; varmul a
cycle i1=1,1,14
S(i1)=0
repeat
x=0 ; cfa ; conmul a
cycle i1=0,1,16
simp coef(i1,i9) ; y=i1*1y/16 ; varmul a
cycle i3=1,1,8
1=A1(i3,1) ; q=i9*1y/48
S(5)=S(5)+8(RC1(6)*RV1(6)*p1(6,i2)*1)*q
S(6)=S(6)+(RC1(7)*RV1(7)*p1(9,i2)*1)*q
S(7)=S(7)+(RC1(9)*RV1(9)*p1(9,i2)*1)*q
S(8)=S(8)+(RC1(11)*RV1(11)*p1(11,i2)*1)*q
repeat
repeat
x=1x ; cfa ; conmul a
cycle i2=1,1,8
S(9)=S(9)+(RC1(4)*RV1(4)*p1(4,i2)*A1(i2,1))*i9*1y/48
S(10)=S(10)+(RC1(2)*RV1(2)*p1(2,i2)*A1(i2,1))*i9*1y/48
S(11)=S(11)+(RC1(2)*RV1(2)*p1(2,i2)*A1(i2,1))*i9*1y/48
repeat
```

```
repeat
y=1y
conmul a ; varmul a
cycle i1=0,1,16
simp coef (i1,i9) ; x=i1*1x/16
cfa
cycle i2=1,1,8
S(12)=S(12)+(RC1(7)*RV1(7)*p1(7,i2)*A1(i2,1))*i9*1x/48
S(13)=S(13)+(RC1(9)*RV1(9)*p1(9,i2)*A1(i2,1))*i9*1x/48
S(14)=S(14)+(RC1(5)*RV1(5)*p1(5,i2)*A1(i2,1))*i9*1x/48
repeat
repeat
cycle i1=1,1,16
S(i1)=S(i1)
repeat

eqn solve (L,s,det)
if det 0 then → 50
newline; caption Determinant of $
caption final $ equations $ zero. Check data
stop
50: comment determination on forces and moments in shell
new page ; newlines (3) ; caption RESULTS
newlines (2) ; spaces (6) ; caption Y $$$$$$$ w $$$$$$$$$ Nx $$$$$$$$
caption Ny $$$$$$$ Nxy $$$$$$$ Mx $$$$$$$$ My $$$$$$$$ Mxy
$$$$$$$$
caption Vx $$$$$$$$ Vy
newline
cycle i1=0,1,8
x=i1*1x/8 ; newline ; caption x $=$ ; print (x,2,2); newline

cycle i2=0,1,8
y=i2*1y/8; newline; spaces (4); print (y,2,2); spaces (4)
w=0; mx=0; nx=0;
ny=0; nxy=0; mx=0; my=0; mxy=0; Vx=0; Vy=0
conmul a ; varmul a ; cfa
cycle i3=1,1,8
w=w+fora (1,i3); nx=nx+fora(10,i3); ny=ny+fora (11,i3)
nxy=nxy+fora(9,i3); mx=mx+fora(3,i3); my=my+fora(6,i3)
mxy=mxy+fora(8,i3); Vx=Vx+fora(4,i3); Vy=Vy+fora(7,i3)
```

```
repeat
cycle i3=2,1,9
cycle i4=1,1,8
w=w+forb(1,i3,i4); nx=nx+forb(10,i3,i4); ny=ny+forb(11,i3,i4)
nxy=nxy+forb(9,i3,i4); mx=mx+forb(3,i3,i4); my=my+forb(6,i3,i4)
mxy=mxy+forb(8,i3,i4); Vx=Vx+forb(4,i3,i4); Vy=Vy+forb(7,i3,i4)
repeat
repeat
conmul b ; varmul b ; cfb
cycle i3=10,1,17
cycle i4=1,1,8
w=w+forc(1,i3,i4); nx=nx+forc(11,i3,i4);
ny=ny+forc(10,i3,i4); nxy=nxy+forc(9,i3,i4); mx=mx+forc(6,i3,i4);
mxy=mxy+forb(8,i3,i4); Vx=Vx+forb(4,i3,i4); Vy=forb(7,i3,i4)
repeat
repeat
conmul b ; varmul b ; cfb
cycle i3=10,1,17
cycle i4=1,1,8
w=w+forc(1,i3,i4); nx=nx+forc(11,i3,i4); mx=mx+forc(6,i3,i4)
ny=ny+forc(10,i3,i4); nxy=nxy+forc(9,i3,i4); mx=mx+forc(6,i3,i4)
mxy=mxy+forc(8,i3,i4) ; my=my+forc(3,i3,i4)
Vx=Vx+forc(7,i3,i4); Vy=Vy+forc(4,i3,i4)
repeat
repeat
print (12w,2,4) ; spaces(2) ; print(nx,4,2) ; spaces(2) ; print (ny,4,2)
spaces(2) ; print(nxy,4,2) ; spaces(2) ; print(mx,4,2) ; spaces(2)
print(my,4,2); spaces(2) ; print(mxy,4,2) ; spaces(2)
print(Vx,4,2) ; spaces(2) ; print (Vy,4,2)
repeat
repeat
stop
routine const mult (integer i, arrayname R, realname m)
integer u
integer array a(1:11)
cycle u=1,1,11
a(u)=u
repeat
if i=1 then → 1
cycle u=2,1,4
```

```
a(u)=u+3 ; a(u+3)=u
repeat
a(10)=11 ; a(11)=10
1:        R(a(1))=1 ; R(a(2))=1 ; R(a(3))=-EI
          R(a(4))=-EI ; R(a(5))=-m ; R(a(6))=EI*m²
          R(a(7))=-EI*m ; R(a(8))=EI*m ; R(a(9))=½ EI/k
          R(a(10))=-EI*m/(2k); R(a(11))=EI/(2m*k)
return
end
```

```
routine var mult (integer i, arrayname R, realname m,y)
real d,r
integer u
integer array a(1:11)
cycle u=1,1,11
a(u)=u
repeat
if i=1 then → 1
cycle u=2,1,4
a(u)=u+3 ; a(u+3)=u
repeat
a(10)=11 ; a(11)=10
1:d=cos(my*y) ; r=sin(m*y)
cycle u=1,1,4
R(a(u))=d ; R(a(u+4))=r
repeat
R(a(6))=d ; R(a(9))=d ; R(a(10))=r ; R(a(11))=r
return
end
routine cf coef (integer i, arrayname C, realname m,x)
integer u,v,z
integer array a (1:11)
real p,q,r,s,t,l,e
array b(1:4, 0:5)
if i=1then → 1; if i = 2 then → 2
1:    p=J1 ; q=J2 ; r=J3 ; s=K3*x; t=C1 ; 1=woml
      → 3
2:    p=J4 ; q=J5 ; r=J6 ; s=K6*x ; t=C2 ; 1=wom2
```

```
3:      cycle u=1,1,11
a(u)=u
repeat
if i=1 then → >
cycle u=2,1,4
a(u)=u+3 ; a(u+3)=u
repeat
a(10)=11 ; a(11)=10
4: cycle u=1,1,4
C(a(1),u)=1 ; C(a(5),u)=1 ; C(a(6),u)=1
repeat
cycle u=2,6,8
C(a(u),1)=p ; C(a(u),2)=−p ; C(a(u)=q ; C(a(u),4)= −q
repeat
cycle u=1,1,2
C(a(3),u)p² ; C(a(3),u+2)=q²
repeat
C(a(4),1)=p*(p²−2m²); C(a(4),2)=−C(a(4),1)
C(a(4),3)=q*(q²−2m²); C(a(4),4)=−C(a(4),3)
cycle u=1,1,2
C(a(7),u)=m²−2p²; C(a(7),u+2)=m²−2q²;
C(a(9),u)=(p²−m²)†2; C(a(9),u+2)=(q²−m²)†2
C(a(10),u)=C(a(9),u)/p; C(a(10),u+2)=C(a(9),u+2)/q
C(a(11),u)=p*C(a(9),u); C(a(11),u+2)=q*C(a(9),u)
repeat
cycle u=9,1,10
C(a(u),2)=−C(a(u),2); C(a(u),4)=−C(a(u),4)
repeat
cycle u=1,1,2
cycle v=0,1,5
if u=1 then z=1; if u=2 then z=−1
b(u,v)=(t†v)*cos(s+z*v*1); b(u+2,v=(t†v)*sin(s+z*v*1)
repeat
repeat
cycle u=1,1,3
cycle v=5,1,8
C(a(u),v)=b(v−4,u−1)
repeat
repeat
cycle u=5,1,8
```

```
C(a(4),u)=b(u–4,3)–2m²*b(u–4,1)
repeat
cycle u=5,1,6
cycle v=5,1,8
C(a(u),v)=b(v–4,0)
repeat
repeat
cycle u=5,1,8
C(a(7),u)=m²*b(u–4,0)–2b(u–4,2); C(a(8),u)=b(u–4,1)
C(a(9),u)=b(u–4,4)–2m²*b(u–4,2)+(m†4)*b(u–4,0)
C(a(10),u)=b(u–4,3)–2m²*b(u–4,1)
C(a(11),u)=b(u–4,5)–2m²*b(u–4,3)+(m†4)*b(u–4,1)
repeat
C(a(10),5)=C(a(10),5)+(m†4)*cos(s–1)/t
C(a(10),6)=C(a(10),6)+(m†4)*sin(s–1)/t
C(a(10),7)=C(a(10),7)+(m†4)*sin(s–1)/t
C(a(10),8)=C(a(10),8)+(m†4)*sin(s–1)/t
cycle u=1,1,11
cycle v=1,1,8
if v=1 then e=exp(p*x); if v=2 then e=exp(–p*x)
if v=3 then e=exp(q*x); if v=4 then e=exp(–q*x)
if v=5 or v=7 then e=exp(r*x)
if v=6 or v=8 then e=exp(–r*x)
C(a(u),v)=e*C(a)U),v)
repeat
repeat
return
end
```

```
routine part integ (integer i, realname x,y)
real B,H1,snx,smy,cnx,cmy
integer u
B=8Z*((m²+n²)†2)/(π²*EI*(((m²+n²)†4)+4b4*m²*n²))
snx=sin(n*x); cnx=cos(n*x)
PI(1)=B*snx; PI(2)=n*B*cnx; PI(3)=EI*n²*  B* snx
PI(4)=EI*n*(n² + 2m²)*B*cnx;   PI(5)=–m*b*snx;   PI(7)=–EI*m*(2n²c
    +m²)*B*snx; PI(6)=EI*m²*B*snx; PI(8)EI*m*n*B*cnx
H1=EI*((m²+n²)†2)*B/(2k)–8Z/(2k*π²))
```

```
PI(9)=H1*snx; PI(10)=H1*m*cnx/n; PI(11)=H1*n*cnx/m
if i=1 then → 1
smy=sin(m*y); cmy=cos(m*y)
cycle u=1,1,4
PI(u)=PI(u)–*cmy
repeat
cycle u=5,1,11
PI(u)=PI(u)*smy
repeat
PI(6)=PI(6)*cmy/smy ; PI(9)=PI(9)*cmy/smy
1: return
end
routine simp coef (integername i,j)
j=2 fracpt (i/2)
if i=0 or i=16 then j=1
return
end

routine conmul a
const mult (1,RC1,m)
return
end
routine conmul b
const mult (2,RC2,n)
return
end

routine varmul a
var mult(1,RV1,m,y)
return
end

routine varmul b
var mult (2,RV2,n,x)
return
end
routine cfa
cf coef (1,p1,m,x)
return
end
```

```
routine cfb
cf coef(2,p2,n,y)
return
end
real fn fora(integer i,integername j)
result=RC1(i)*RV1(i)*A1(j,1)*p1(i,j)
end
real fn forb (integer i, integername j,p)
result=RC1(i)*RV1(i)*A(j,p)*S(j−1)*p1(i,p)
end
real fn forc(integer i, integername j,p)
result=RC2(i)*RV2(i)*A(j,p)*S(j−1)*p²(i,p)
end
```

```
Routine solve polynomial 2(integer n,F,array name a,u,v,CONV)
        integer L,i,j,m; array h,b,c,d,e(−4:2n+1)
        real t,K,p,p′,q,q′,ps,ps′,qs,qs′,pt,pt′,qt,qt′,s,s′,rev:
        rec,rec′,r,r′
   array fn g(addr(q),1)
      fault 1,2 → 99; → 98
   99: caption
      EXP $ OR $ DIV $ OVERFLOW $$$$$$ H $$$$$$
      BBB $$$$ O $$$$$/$$$$$$$C D $$$$$$ E $$$$$$ T$ ETC
      cycle i=−4,1,41
      newline; print f1(h(i),8); print f1(b(i),8)
      print f1(c(i),8); print f1(d(I),8); print f1(e(i),8);
      print f1(g(i),8)
      repeat
      return
   98: L=20
      i=addr(h(0)); 101,50,−,αi
      i=addr(b(0)); 101,51,−,αi
      i=addr(c(0)); 101,52,−,αi
      i=addr(d(0)); 101,53,−,αi
      i=addr(e(0)); 101,54,−,αi
      101,56,−,αn; 104,56,−,αn
      cycle j=−4,1,−1
      b(j)=0; c(j)=0; d(j)=0; e(j)=0; h(j)=0
```

```
        repeat
        cycle j=0,1,n
        h(2j)=a(j); h(2j+1)=0
        repeat
        t=1; K=.1*F
   1: → 2 unless h(2n)=0
        u(n)=0; v(n)=0; CONV(n)=K
        n=n−1; 122,56,0,2; → 1
   2: return if n=0
        ps=0; qs=0; pt=0; qt=0; s=0
        ps′=0; qs′=0; pt′=0; qt′=0
        rev=1; K=.1*F
        →3 unless n=1
comment r=−h(1)/h(0)
        1505,50,0,2; 1576,50,0,0; 1556,0,−,r; →4
comment Scale coefficients.
     3: cycle j=0,2,2n
        s=s+log(|h(j)|) unless h(j)=0
        repeat
        s=exp(−s/(n+1))
        cycle j=0,2,2n
comment h(j)=h(j)*s
        101,55,−,αj; 1504,0,−,s
        1542,50,55,0; 1556,50,55,0
        repeat
        →5 unless|h(2)*h(2n)|<|h(2n−2)*h(0)|
comment Reverse order of coefficients.
     15: t=−t; m=2int pt(½n)
        cycle j=0,2,m
        s=h(j); h(j)=h(2n−j); h(2n−j)=s
        s=h(j+1); h(j+1)=h(2n−j+1); h(2n−j+1)=s
        repeat
     5:  →6 if qs=0
        p=ps; q=qs; p′=ps′; q′=qs′; → 7
     6:  →8 unless|h(2n−4)|<K
        q=1; p=−2; q′=0; p′=0; →9
comment rec=1/h(n−2), q=h(n)*rec, p=(h(n−1)−q*h(n−3))*rec
     8:  1524,0,0,1; 1576,50,56,−4; 1556,0,−,rec
        1542,50,56,0; 1556,0,−,q; 1543,50,56,−6
        1500,50,56,−2; 1542,0,−,rec; 1556,0,−,p
```

```
      9:  →10 if n=2
          r=0; r′=0
comment Iterate.
      7: cycle i=1,1,L
comment Bairstow method.
         cycle j=0,2,2n
comment b(j)=h(j)—p*b(j—1)—q*b(j—2),c(j)=b(j)—p*c(j—1)—q*c(J—2)
         101,55,—,αj; 1504,51,55,—4; 1542,0,—,p
         1556,90,0,0; 1505,51,55,—4; 1542,0,—,q
         1501,90,0,0; 1500,50,55,0; 1556,51,55,0
         1504,52,55,—2; 1542,0,—,p; 1556,90,0,0
         1505,52,55,—4; 1542,0,—,q; 1501,90,0,0
         1500,51,55,0; 1556,52,55,0
         repeat
         →11 if|h(2n—2)|<K or|b(2n—2)|<K
         →12 if|h(2n—2)*K|⩽|b(2n—2)|
comment b(n)=h(n)—q*b(n—2)
         1505,0,—,q; 1542,51,56,—4
         1500,50,56,0; 1556,51,56,0
    11: →10 if |b(2n)|⩽ |h(2n)*K|
comment Newton-Raphson method.
    12:   cycle j=0,2,2n
comment d(j)=h(j)+r*d(j—1),e(j)=d(j)+r*e(j—1)
         101,55,—,αj; 1504,0,—,r; 1542,53,55,—2
         1500,50,55,0; 1556,53,55,0; 1504,0,—,r
         1542,54,55,—2; 1500,53,55,0; 1556,54,55,0
         repeat
         → 4 if |d(n)|<|K*h(n)|
comment c(n—1)=—p*c(n—2)—q*c(n—3)
         1504,0, —,p; 1542,52,56,—4; 1556,90,0,0
         1505,0,—,q; 1542,52,56,—6
         1501,90,0,0; 1556,52,56,—2
comment s=c(n—2)²—c(n—1)*c(n—3)
         1542,52,56,—6; 1556,90,0,0; 1504,52,56,—4
         1542,52,56,—4; 1501,90,0,0; 1556,0,—,s
         → 13 if |s|<K
comment rec=1/s
         1524,0,0,1; 1576,0,—,s; 1556,0,—,rec
comment p=p+(b(n—1)*c(n—2)—b(n)*c(n—3))*rec
         1504,51,56,0; 1542,52,56,—6; 1556,90,0,0
```

```
        1504,51,56,–2; 1542,52,56,–4; 1501,90,0,0
        1542,0,–,rec; 1500,0,–,p; 1556,0,–,p
comment q=q–(b(n–1)*c(n–1)–b(n)*c(n–2))*rec
        1504,51,56,–2; 1542,52,56,–2; 1556,90,0,0
        1504,51,56,0; 1542,52,56,–4; 1501,90,0,0
        1542,0,–,rec; 1500,0,–,q; 1556,0,–,q; → 14
comment p=p–2,q=q*(q+1)
    13:  1525,0,0,2; 1500,0,–,p; 1556,0,–,p
         1524,0,0,1; 1500,0,–,q
         1542,0, –,q; 1556,0,–,q
comment  r=r–1,r=r+1–d(n)/e(n–1)unless e(n–1)=0
    14:  →23 if|e(2n–2)| < K
         1505,53,56,0; 1576,54,56,–2
         1500,0,–,r; 1556,0,–,r; → 24
    23:  1525,0,0,1; 1500,0,–,r; 1556,0,–,r
    24:  repeat
         ps=pt; qs=qt; pt=p; qt=q
         ps′=pt′; qs′=qt′; pt′=p′; qt′=q′
         K=10K if rev < 0
         rev=–rev; → 15
comment  Linear.
comment  r=1/r if t < 0
     4:  25 if t > 0
         1524,0,0,1; 1576,0,–,r; 1556,0,–,r
    25:  u(n)=r; v(n)=0
         CONV(n)=K; n=n–1; 122,56,0,2
         cycle j=0,1,2n+1
         h(j)=d(j)
         repeat
         → 2
comment  Quadratic.
    10:  → 17 unless t < 0
comment  q=1/q, p=p*q
         1524,0,0,1; 1576,0,–,q; 1556,0,–,q
         1542,0,–,p; 1556,0,–,p
comment  p=–½p, s=q–p²
    17:  1525,0,0,½; 1542, 0,–,p; 1556,0,–,p
         1543,0,–,p; 1500,0,–,q; 1556,0,–,s
         → 18 if s ≤ 0
         u(n)=p; u(n–1)=p
```

```
      s=sqrt(s); v(n)=s; v(n−1)=−s; → 19
18:   s=sqrt(−s)
      → 20 if p ⩾0
      u(n)=p−s; → 21
20:   u(n)=s+p
21:   u(n−1)=q/u(n)
      v(n)=0; v(n−1)=0
19:   CONV(n)=K; CONV(n−1)=K; n=n−2; 122,56,0,4
      cycle j=0,1, 2n+1
      h(j)=b(j)
      repeat
      →2
      end
```

APPENDIX III

COMPUTER PROGRAM FOR THE ANALYSIS OF INVERTED UMBRELLA SHELLS WITH EDGE BEAMS

The computer program given in this Appendix is for the analysis of hyperbolic paraboloidal shells of the inverted umbrella type, provided with edge and valley beams, subject to uniformly distributed load. The shell is assumed to be symmetrical about the two orthogonal axes passing through the column. The perimeter edge beams are assumed to be of constant cross-section whereas the spine beams are assumed to have a varying width and depth with maximum values at their junction with the column.

The data and the order in which it is to be punched is as follows:

- E (Young's modulus of shell material) in lb/in^2
- l_x (length, in x-direction, of one quadrant) in ft
- l_y (length, in y-direction, of one quadrant) in ft
- c (rise of shell) in ft
- L (thickness of shell) in inches
- z (u.d. load) in lb/ft^2
- b_1,d_1 (width and depth, respectively, of perimeter beam parallel to the y-axis) in ft
- b_2,d_2 (width and depth, respectively, of perimeter beams parallel to the x-axis) in ft
- b_3,d_3 (width and depth, respectively, of the spine beams parallel to the y-axis, at their junction with perimeter beams) in ft
- b_4,d_4 (width and depth, respectively, of the spine beams parallel to the y-axis, at their junction with the column) in ft
- b_5,d_5 (width and depth, respectively, of the spine beams parallel to the x-axis, at their junction with perimeter beams) in ft
- b_6,d_6 (width and depth, respectively, of the spine beams parallel to the x-axis, at their junction with column) in ft

The program causes the computer to do the necessary calculations and print out deflections, forces and moments in the shell at the nodes of the grid obtained by dividing the sides of one quadrant of the shell into eight equal parts.

```
begin
real   Epsi,E,lx,ly,c,hin,h,m,n,k,I,EI,b4,Z,x,y,det,J1,J2,J3,J4,J5,c
       J6,K3,K6,C1,C2,wom1,wom2,X,Y,w,nx,ny,nxy,mx,my,mxy,Vx,
       Vy,l,q,c
       ja,jb,jc,jd

integer i1,i2,i3,i4,i5,i6,i7,i8,i9
array  CH(0:8),U,V,CONV(1:8),K1,K2,Q2,A2(1:8,1:8),RC1,RC2,RV1,c
       RV2,PI(1:11),p1,p2(1:11,1:8),Q1,A1(1:8,1:9),L(1:16,1:16),Y1,c
       S(1:16),A(1:17,1:8)
routine spec solve polynomial (integer n,F, arrayname a,u,v,CONV)
routine spec const mult (integer i, arrayname R, real name m)
routine spec var mult (integer i, arrayname R, real name m,y)
routine spec cf coef (integer i, arrayname C, real name m,x)
routine spec simp coef (integername i,j)
routine spec conmul a
routine spec conmul b
routine spec varmul a
routine spec varmul b
routine spec cfa
routine spec cfb
real fn spec fora(integer i, integername j)
real fn spec forb(integer i, integername j,p)
real fn spec forc(integer i, integername j,p)
read (Epsi,1x,1y,c,hin,Z)
E=144Epsi; h=hin/12; I=h†3/12; EI=E*I; k=c/(lx*ly)
b4=12k²/(h²)
m=½*π/ly; n=½*π/lx
comment solution of characteristic equations.
cycle il=1,1,2
if i1=1 then 1=m; if i1=2 then 1=n
cycle i2=1,2,7
CH(i2)=0
repeat
CH(0)=1; CH(2)=−41²; CH(4)=6*(1†4); CH(6)=−41²*
  (1†4+b4)
CH(R)=1†8
```

```
solve polynomial (8,10,CH,U,V,CONV)
X=100; Y=100
cycle i2=1,1,4
X=X+1 if V(i2)=0; Y=Y+1 if V(i2+4)=0
repeat
→ 1 if X=104 and Y=100; →2 if X=100 and Y=104
newline ; caption solution $ of $ characteristic $ eqns $ faulty.
caption check $ data.
stop
1: ja=½*(|U(1)|+|U(2)|) ; jb=½*(|U(3)|+|U(4)|) ; jc=0 ; jd=0
cycle i2=5,1,8
jc=jc+|U(i2)|/4 ; jd=jd+|V(i2)|/4
repeat
→ 20
2: ja=½*(|U(5)|+|U(6)|) ; jb=½*(|U(7)|+|U(8)|) ; jc=0 ; jd=0
cycle i2=1,1,4
jc=jc+|U(i2)|4 ; jd=jd+|V(i2)|)/4
repeat

20:   →21 if il=2
      J1=ja ; J2=jb ; J3=jc ; K3=jd
      →22
21:   J4=ja ; J5=jb ; J6=jc ; K6=jd
22:   C1=radius (J3,K3)
repeat
C1=radius (J3,K3) ; C2=radius(J6,K6) ; wom1=arctan(J3,K3);
wom2=arctan(J6,K6)
comment formation of matrix K (named K1) for cases (1) to (9)
null (K1)
cycle i2=1,1,2
if i2=1 then x=0 ; if i2=2 then x=1x
cfa
cycle i1=1,1,8
K1(4i2−3,i1)=p1(3,i1) ; K1(4i2−2,i1)=p1(4,i1)
K1(4i2−1,il)=p1(9,il) ; K1(4i2,il)=p1(10,il)
repeat
repeat
comment formation of matrix K′ (named (K2)) for cases (10) to (17)
null (K2)
cycle i2=1,1,2
```

```
if i2=1 then y=0 ; if i2=2 then y=ly
cfb
cycle i1=1,1,8
K2(4i2–3,il)=p2(6,il) ; K2(4i2–2,il)=p2(7,i1) ; K2(4i2–1,il)=p2(9,il)
K2(4i2,i1)=p2(11,il)
repeat
repeat
comment determination of constants of integration in cases (2) to (17)
comment due to unit applied forces
invert (K3,K1,det)
if det ≠ 0 then → 4
newline
caption Determinant $ of $ K1 $ zero. Check $ data
stop
4:     invert(K4,K2,det)
       if det≠0 then → 5
caption Determinant $ of $ K2 $ zero. Check $ data.
stop
5: comment formation of matrix Q (named Q1)
null (Q1)
cycle i=1,1,2
if il=2 then → 6
x=0 ; y=0 ; →7
6: x=lx ; y=0
7: part integ (l,x,y)
       if i1=1 then → 8 ; if i1=2 then → 9
8: i2=1 ; i3=2 ; i4=3 ; i5=4 ; → 10
9: i2=5 ; i3=6 ; i4=7 ; i5=8
10: Q1(i2,1)=PI(3); Q1(i3,1)=PI(4); Q1(i4,1)=PI(9); Q1(i5,1)=PI(10)
repeat
cycle il=1,1,8
Q1(il,il+1)=4/π
repeat
comment formation of matrix Q′ (named Q2)
cycle il=1,1,8
Q2(il,il)=4/π
repeat
matrix mult(A1,K3,Q1) ; matrix mult (A2,K4,Q2)
cycle il=1,1,9
cycle i2=1,1,8
```

```
A(i1,i2)=A1(i2,i1) ; if i1+9 > 17 then → 11
A(i1+9,i2)=A2(i2,i1)
11:     repeat
        repeat
comment formation of boundary equations for the actual shell.
comment numerical integration is performed by Simpson's Rule
null(L)
x=0 ; cfa ; conmul a
cycle i1=0,1,16
simp coef (i1,i9)
y=i1*ly/16
varmul a
cycle i2=1,1,8
cycle i3=1,1,8
L(1,i2)=L(1,i2)+(RC1(3)*RV1(3)*p1(3,i3)*A1(i3,i2+1))*i9*ly/48
L(2,i2)=L(2,i2)+(RC1(4)*RV1(4)*p1(4,i3)*A1(i3,i2+2))*i9*ly/48
L(3,i2)=L(3,i2)+(RC1(9)*RV1(9)*p1(9,i3)*A1(i3,i2+2))*i9*ly/48
L(4,i2)=L(4,i2)+(RC1(10)*RV1(10)*p1(10,i3)*A1(i3,i2+1))*i9*ly/48
repeat
repeat
repeat
conmul b ; varmul b ;
cycle i1=0,1,16
simp coef (i1,i9)
y=i1*ly/16 ; cfb
cycle i2=9,1,16
cycle i3=1,1,8
i4=i2-8
L(1,i2)=L(1,i2)+(RC2(3)*RV2(3)*p2(3,i3)*A2(i3,i4))*i9*ly/48
L(2,i2)=L(2,i2)+(RC2(4)*RV2(4)*p2(4,i3)*A2(i3,i4))*i9*ly/48
L(3,i2)=L(3,i2)+(RC2(9)*RV2(9)*p2(9,i3)*A2(i3,i4))*i9*ly/48
L(4,i2)=L(4,i2)+(RC2(10)*RV2(10)*p2(10,i3)*A2(i3,i4))*i9*ly/48
repeat
repeat
repeat
y=0 ; conmul a ; varmul a
cycle i1=0,1,16
simp coef (i1,i9)
x=i1*lx/16 ; cfa
cycle i2=1,1,8
```

```
cycle i3=1,1,8
L(5,i2)=L(5,i2)+(RC1(6)*RV1(6)*p1(6,i3)*A1(i3,i2+1))*i9*lx/48
i4=i2+1
L(6,i2)=L(6,i2)+(RC1(7)*RV1(7)*p1(7,i3)*A1(i3,i4))*i9*lx/48
L(7,i2)=L(7,i2)+(RC1(9)*RV1(9)*p1(9,i3)*A1(i3,i4))*i9*lx/48
L(8,i2)=L(8,i2)+(RC1(11)*RV1(11)*p1(11,i3)*A1(i3,i4))*i9*lx/48
repeat
repeat
repeat
conmul b ; cfb
cycle i1=0,1,16
simp coef (i1,i9) ; x=i1*lx/16 ; varmul b
cycle i2=9,1,16
cycle i3=1,1,8
L(5,i2)=L(5,i2)+(RC2(6)*RV2(6)*p2(6,i3)*A2(i3,i2-8))*i9*lx/48
i4=i2-8
L(6,i2)=L(6,i2)+(RC2(7)*RV2(7)*p2(7,i3)*A2(i3,i4))*i9*lx/48
L(7,i2)=L(7,i2)+(RC2(9)*RV2(9)*p2(9,i3)*A2(i3,i4))*i9*lx/48
L(8,i2)=L(8,i2)+(RC2(11)*RV2(11)*p2(11,i3)*A2(i3,i4))*i9*lx/48
repeat
repeat
repeat
x=lx ; cfa ; conmul a
cycle i1=0,1,16
simp coef(i1,i9) ; y=i1*ly/16 ; varmul a
cycle i2=1,1,8
cycle i3=1,1,8
i4=i2+1
L(9,i2)=L(9,i2)+(RC1(4)*RV1(4)*p1(4,i3)*A1(i3,i4))*i9*ly/48
L(10,i2)=L(10,i2)+(RC1(9)*RV1(9)*p1(9,i3)*A1(i3,i4))*i9*ly/48
L(11, i2)=L(11,i2)+(RC1(2)*RV1(2)*p1(2,i3)*A1(i3,i4))*i9*ly/48
repeat
repeat
repeat
conmul b ; varmul b
cycle i1=0,1,16
simp coef(i1,i9) ; y=i1*ly/16 ; cfb
cycle i2=9,1,16
cycle i3=1,1,8
L(9,i2)= L(9,i2)+(RC2(4)*RV2(4)*p2(4,i3)*A2(i3,i2-8))*i9*ly/48
```

```
L(10,i2)=L(10,i2)+(RC2(9)*RV2(9)*p2(9,i3)*A2(i3,i2–8))*i9*ly/48
L(11,i2)=L(11,i2)+(RC2(2)*RV2(2)*p2(2,i3)*A2(i3,i2–8))*i9*ly/48
repeat
repeat
repeat
y=ly ; conmul a ; varmul a
cycle i1=0,1,16
simp coef (i1,i9) ; x=i1*lx/16 ; cfa
cycle i2=1,1,8
cycle i3=1,1,8
i4=i2+1
L(12,i3)=L(12,i3)+(RC1(7)*RV1(7)*p1(7,i3)*A1(i3,i4))*i9*ly/48
L(13,i3)=L(13,i3)+(RC1(9)*RV1(9)*p1(9,i3)*A1(i3,i4))*i9*ly/48
L(14,i3)=L(14,i3)+(RC1(5)*RV1(5)*p1(5,i3)*A1(i3,i4))*i9*ly/48
repeat
repeat
repeat

begin
array  D,F(1:16,1:16),D1,F1(1:16)
null(D) ; null(F)
cycle i1=1,1,16
D1(i1)=0 ; F1(i1)=0
repeat
cycle i1=0,1,16
y=i1*ly/16 ; conmul a ; varmul a
cycle i2=0,1,16
simp coef (i2,i5) ; x=i2*lx/16 ; cfa
cycle i3=1,1,8
cycle i4=1,1,8
D(i1,i3)=D(i1,i3)+(RC1(10)*RV1(10)*p1(10,i4)*A1(i3,i4+1))*i5*lx/48
repeat
D1(i1)=D1(i1)+(RC1(10)*RV1(10)*p1(10,i3)*A1(i3,1))*i5*lx/48
repeat
repeat
cfb ; conmul b
cycle i2=0,1,16
simp coef (i2,i5) ; x=i2*lx/16 ; varmul b
cycle i3=0,1,16
cycle i4=1,1,8
```

```
D(i1,i3)=D(i1,i3)+(RC2(10)*RV(10)*p2(10,i4)*A2(i4,i3–8))*i5*lx/48
repeat
repeat
repeat
x=lx ; conmul a ; varmul a ; cfa
cycle i2=1,1,8
cycle i3=1,1,8
D(i1,i2)=D(i1,i2)–k*y*(RC1(1)*RV1(1)*p1(1,i3)*A1(i3,i2+1))
repeat
D1(i1) = D1(i1)–k*y*(RC1(1)*RV1(1)*p1(1,i2)*A1(i2,1))
repeat
conmul b ; varmul b ; cfb
cycle i2=0,1,16
cycle i3=1,1,8
D(i1,i2)=D(i1,i2)–k*y*(RC2(1)*RV2(1)*p2(1,i3)*A2(i3,i2–8))
repeat
repeat
repeat
cycle i1=0,1,06
x=i1*lx/16 ; conmul a ; cfa
cycle i2=0,01,16
simp coef (i2,i5) ; y=i2*ly/16
varmul a
cycle i3=1,1,8
cycle i4=1,1,8
F(i1,i3)=F(i1,i3)+(RC1(11)*RV1(11)*p1(11,i4)*A1(i4,i3+1))*i5*ly/48
repeat
F1(i1)=F1(i1)+(RC1(11)*RV1(11)*p1(11,i3)*A1(i3,1))* i5*ly/48
repeat
repeat
conmul b ; varmul b
cycle i2=0,1,16
simp coef (i2,i5) ; y=i2*ly/16 ; cfb
cycle i3=0,1,16
cycle i4=1,1,8
F(i1,i3)=F(i1,i3)+(RC2(11)*RV2(11)*p2(11,i4)*A2(i4,i3–8))*i5*ly/48
repeat
repeat
repeat
y=ly
```

```
cycle i2=1,1,16
if i2 ⩾ 9 then → 20
conmul a ; varmul a ; cfa
cycle i3=1,1,8
F(i1,i2)=F(i1,i2)–k*x*(RC1(1)*RV1(1)*p1(1,i3)*A1(i3,i2+1))
repeat
F1(i1)=F1(i1)–k*x*(RC1(1)*RV1(1)*p1(1,i2)*A1(i2,1))
→ 21
20:  conmul b ; varmul b ; cfb
cycle i3=1,1,8
F(i1,i2)=F(i1,i2)–k*x*(RC2(1)*RV2(21)*p2(1,i3)*A2(i3,i2–8))
repeat
21:  repeat
repeat
cycle i1=1,1,16
cycle i2=0,1,16
simp coef (i2,i5)
L(15,i1)=L(15,i1)+i5*D(i2,i1)*ly/48
L(16,i1)=L(16,i1)+i5*F(i2,i1)*lx/48
repeat
repeat
cycle i1=0,1,16
simp coef (i1,i5)
S(15)=S(15)+i5*D1(il)*ly/48 ; S(16)=S(16)+i5*F1(il)*lx/48
end
```

```
x=0 ; cfa ; conmul a
cycle i1=0,1,16
simp coef (i1,i9) ; y=i1*ly/16 ; varmul a
cycle i1=1,1,14
S(i1)=0
repeat
x=0 ; cfa ; conmul a
cycle i1=0,1,16
simp coef (i1,i9) ; y=i1*ly/16 ; varmul a
cycle i3=1,1,8
1=A1(i3,1) ; q=i9*ly/48
S(1)=S(1)+8(RC1(3)*RV1(3)*p1(3,i3)*1)*q
```

```
S(2)=S(2)+(RC1(4)*RV1(4)*p1(4,i3)*1)*q
S(3)=S(3)+(RC1(9)*RV1(9)*p1(9,i3)*1)*q
S(4)=S(4)+(RC1(10)*RV1(10)*p1(10,i3)*1)*q
repeat
repeat
y=0 ; conmul a ; varmul a
cycle i1=0,1,16
simp coef(i1,i9) ; x=i1*1x/16 ; cfa
cycle i2=1,1,8
1=A1(i2,1) ; q=i9*1x/48
S(5)=S(5)+8(RC1(6)*RV1(6)*p1(6,i2)*1)*q
S(6)=S(6)+(RC1(7)*RV1(7)*p1(9,i2)*1)*q
S(7)=S(7)+(RC1(9)*RV1(9)*p1(9,i2)*1)*q
S(8)=S(8)+(RC1(11)*RV1(11)*p1(11,i2)*1)*q
repeat
repeat
x=1x ; cfa ; conmul a
cycle i1=0,1,16
simp coef (i1,i9) ; y=i9*1y/16 ; varmul a
cycle i2=1,1,8
S(9)=S(9)+(RC1(4)*RV1(4)*p1(4,i2)*A1(i2,1))*i9*ly/48
S(10)=S(10)+(RC1(9)*RV1(9)*p1(9,i2)*A1(i2,1))*i9*ly/48
S(11)=S(11)+(RC1(2)*RV1(2)*p1(2,i2)*A1(i2,1))*i9*ly/48
repeat
repeat
y=ly
conmul a ; varmul a
cycle i1=0,1,16
simp coef (i1,i9) ; x=i1*lx/16
cfa
cycle i2=1,1,8
S(12)=S(12)+(RC1(7)*RV1(7)*p1(7,i2)*A1(i2,1))*i9*lx/48
S(13)=S(13)+(RC1(9)*RV1(9)*p1(9,i2)*A1(i2,1))*i9*lx/48
S(14)=S(14)+(RC1(5)*RV1(5)*p1(5,i2)*A1(i2,1))*i9*lx/48
repeat
repeat
cycle i1=1,1,16
S(i1)=-S(i1)
repeat
begin
```

```
integer u,v,p
array   pb(1:2, 1:2), sb(1:2, 1:2, 1:2), area (1:4, 0:16), d1,d2,d3, c
        d4(0:16, 1:16)
read array(pb) ; read array(sb)
cycle u  =0,1,16
area(1,u)  =pb(1,1)*pb(1,2)+½*pb(1,1)*u*c*pb(1,1)/(161x)
area(2,u)  =pb(2,1)*pb(2,2)+½*(pb(2,1))*u*c*pb(2,1)/(161y)
area(3,u)  =((sb(1,1,1)+(sb(1,1,2)—sb(1,1,1))*u*ly/16)*(sb(1,2,1)+c
           (sb(1,2,2)—sb(1,2,1))*u*ly/16)+½*(sb(1,1,1)+(sb(1,1,2)—c
           sb(1,1,1))*u*ly/16)*u*(sb(1,1,1)+(sb(1,1,2)—sb(1,1,1)))*c
           c/(161x))/2
area(4,u)  =((sb(2,1,1)+sb(2,1,2)—sb(2,1,1))*u*lx/16)*(sb(2,2,1)+(sb(c
           2,2,2)—sb(2,2,1))*u*lx/16)+½*(sb(2,2,1)+(sb(2,1,2)—c
           sb(2,1,1))*u*lx/16)*u*(sb(2,1,1)+(sb(2,1,2)—sb(2,1,1))*c
           *c/(161y))/2
repeat
cycle u=1,1,16
L(3,u)=0 ; L(7,u)=0 ; L(10,u)=0 ; L(13,u)=0
repeat
S(3)=0 ; S(7)=0 ; S(10)=0 ; S(13)=0
cycle u=1,1,2
if u=1 then x=0 ; if u=2 then x=lx ; cfa ; conmul a
cycle i1=0,1,16
simp coef (i1,i9) ; y=i1*ly/16 ; varmul a
cycle i2=1,1,8
cycle i3=1,1,8
→ 1 if u=2
L(3,i2)=L(3,i2)+(RC1(11)*RV1(11)*p1(11,i3)*A1(i3,i2+1))*i9*ly/(48h)
→ 2

1: L(7,i2)=L(7,i2)+(RC1(10)*RV1(10)*p1(10,i3)*A1(i3,i2+1))*i9*ly/(48h)
2: repeat
if u=1 then S(3)=S(3)—(RC1(11)*RV1(11)*p1(11,i2)*A1(i2,1))*c
             i9*ly/(48h)
if u=2 then S(7)=S(7)—(RC1(10)*RV1(10)*p1(10,i2)*A1(i2,1))*ly/(48h)
repeat ; repeat ; repeat
cycle u=1,1,2
if u=1 then x=0  ; if u=2 then x=lx ; conmul b ; varmul b
cycle i1=0,1,16
simp coef (i1,i9) ; y=i1*ly/16 ; cfb
```

```
cycle i2=9,1,16
cycle i3=1,1,8
i4=i2*-8
→ 3 if u=2
L(3,i2)=L(3,i2)+(RC2(11)*RV2(11)*p2(11,i3)*A2(i3,i4))*i9*ly/(48h)
→ 4
3: L(7,i2)=L(7,i2)+(RC2(10)*RV2(10)*p2(10,i3)*A2(i3,i4))*i9*ly/(48h)
4: repeat ; repeat ; repeat ; repeat
cycle u=1,1,2
cycle v=1,1,2
if v=1 then y=0 ; if v=2 then y=ly ; if u=2 then → 5
cfa ; conmul a ; → 6
5: conmul b ; varmul b
6: cycle i1=0,1,16
simp coef(i1,i9) ; y=i1*lx/16 ; if u=1 then varmul a
if u=2 then cfb
cycle i2=1,1,8
cycle i3=1,1,8
→ 7 if v=2
if u=1 then L(10,i2)=L(10,i2)+(RC1(10)*RV1(10)*p1(10,i3)*A1(i3,i2+
                1))*ci9*lx/(48h)
L(10,i2+8)=L(10,i2+8)+(RC2(11)*RV2(11)*p2(11,i3)*A2(i3,i2))*i9*lx/
                (48h)c
if u=2 → 8
7: if u=1 then L(13,i2)=L(13,i2)+(RC1(10)*RV1(10)*p1(10,i3)*A1
        (i3,i2+1)c)*i9*lx/(48h)
8: repeat
if u=1 and v=1 then S(10)=S(10)–(RC1(10)*RV1(10)*p1(10,i2)*c
            A1(i2,1))*i9*lx/(48h)
if u=1 and v=2 then S(13)=S(13)–(RC1(11)*RV1(11)*p1(11,i2)*c
            A1(i2,1))*i9*lx/(48h)
repeat ; repeat ; repeat ; repeat
```

```
null(d1) ; null (d2) ; null(d3) ; null(d4)
cycle u=1,1,2
cycle p=0,1,16
cycle v=1,1,2
if v=1 then x=0 ; if v=2 then x=lx; y=p*ly/16 ; →9 if v=2
```

```
C cfa ; conmul a ; → 10
9: conmul b ; varmul b
10: cycle i1=0,1,16
simp coef (i1,i9); y=i1*y ; if  v=1 then varmul a; if v=2 then cfb
cycle i2=1,1,8
cycle i3=1,1,8
→ 11 if v=2
if u=1 then d1(p,i2)=d1(p,i2)+(RC1(9)*RV1(9)*p1(9,i3)*A1(i3,i2+1))*c
            i9*ly/(48area(1,p))
if u=2 then d1(p,i2+8)=d1(p,i2+8)+(RC2(9)*RV2(9)*p2(9,i3)*c
            A2(i3,i2))*i9*ly/(48area(1,p))
→ 12
11: if u=1 then d3(p,i2)=d3(p,i2)+(RC1(9)*RV1(9)*p1(9,i3)*c
            A1(i3,i2+1))*i9*ly/(48area(3,p))
if u=2 then d3(p,i2+8)=d3(p,i2+8)+(RC2(9)*RV2(9)*p2(9,i3)*c
            A2(i3,i2))*i9*ly/(48area(3,p))
12: repeat ; repeat ; repeat ; repeat ; repeat ; repeat

cycle u=1,1,2
cycle p=0,1,16
cycle v=1,1,2
if v=1 then y=0 if v=2 then y=ly ; x=p*lx/16
→ 13 if v=2
conmul a ; varmul a ; → 14
13: cfb ; conmul b
14: cycle i1=0,1,16
simp coef(i1,i9) ; x=i1*x ; if v=1 then cfa ; if v=2 then varmul b
cycle i2=1,1,8
cycle i3=1,1,8
→ 15 if v=2
if u=1 then d2(p,i2)=d2(p,i2)+(RC1(9)*RV1(9)*p1(9,i3)*c
                        A1(i3,i2+1))*i9*lx/(48area(2,p))
if u=2 then d2(p,i2+8)=d2(p,i2+8)+(RC2(9)*RV2(9)*p2(9,i3)*c
                        A2(i3,i2))*i9*lx/(48area(2,p))
→ 16
15: if u=1 then d4(p,i2)=d4(p,i2)+(RC1(9)*RV1(9)*p1(9,i3)*c
                        A1(i3,i2+1))*i9*lx/(48area(4,p))
if u=2 then d4(p,i2+8)=d4(p,i2+8)+(RC2(9)*RV2(9)*p2(9,i3)*c
                        A2(i3,i2))*i9*lx/(48area(4,p))
16: repeat ; repeat ; repeat ; repeat ; repeat ; repeat
```

```
cycle u=1,1,16
cycle v=0,1,16
L(3,u)=L(3,u)–d1(v,u)*i9*ly/48
L(7,u)=L(7,u)–d3(v,u)*i9*lx/48
L(10,u)=L(10,u)–d2(v,u)*i9*ly/48
L(13,u)=L(13,u)–d4(v,u)*i9*lx/48
repeat
repeat
end
```

```
eqn solve(1,S,det)
if det≠0 then → 50
newline ; caption Determinant of $
caption final $ equations $ zero. Check $ data.
stop
50: comment determination on forces and moments in shell
newpage ; newlines(3) ; caption RESULTS:
newlines(2) ; spaces(6); caption Y $$$$$$$ w $$$$$$$$$
Nx $$$$$$$$ CA caption Ny $$$$$$$ Nxy $$$$$$$$ Mx $$$$$$$$
My $$$$$$$$ Mxy $$$$$$$$ caption Vx $$$$$$$$ Vy
newline
cycle i1=0,1,8
x=i1*lx/8 ; newline ; caption x $=$ ; print (x,2,2) ; newline

cycle i2=0,1,8
y=i2*ly/8 ; newline ; spaces (4) ; print (y,2,2) ; spaces (4)
w=0 ; mx=0 ; nx=0 ;
ny=0 ; nxy=0 ; mx=0 ; my=0 ; mxy=0 ; Vx=0;
Vy=0
conmul a ; varmul a ; cfa

cycle i3=1,1,8
w=w+fora(1,i3) ; nx=nx+fora(10,i3) ; ny=ny+fora(11,i3)
nxy=nxy+fora(9,i3) ; mx=mx+fora(3,i3) ; my=my+fora(6,i3)
mxy=mxy+fora(8,i3) ; Vx=Vx+fora(4,i3) ; Vy=Vy+fora(7,i3)
repeat
cycle i3=2,1,9
```

```
cycle i4=1,1,8
w=w+forb(1,i3,i4) ; nx=nx+forb(10,i3,i4) ; ny=ny+forb(11,i3,i4)
nxy=nxy+forb(9,i3,i4) ; mx=mx+forb(3,i3,i4); my=my+forb(6,i3,i4)
mxy=mxy+forb(8,i3,i4); Vx=Vx+forb(4,i3,i4); Vy=Vy+forb(7,i3,i4)
repeat
repeat
conmul b ; varmul b ; cfb
cycle i3=10,1,17
cycle i4=1,1,8
w=w+forc(1,i3,i4) ; nx=nx+forc(11,i3,i4)
ny=ny+forc(10,i3,i4); nxy=nxy+forc(9,i3,i4); mx=mx+forc(6,i3,i4)
mxy=mxy+forc(8,i3,i4) ; my=my+forc(3,i3,i4)
Vx=Vx+forc(7,i3,i4) ; Vy=Vy+forc(4,i3,i4)
repeat
repeat
print(12w,2,4) ; spaces(2) ; print(nx,4,2) ; spaces(2) ; print(ny,4,2)
spaces(2) ; print (nxy,4,2) ; spaces(2) ; print(mx,4,2) ; spaces(2)
print(my,4,2) ; spaces(2) ; print(mxy,4,2) ; spaces(2)
print(Vx,4,2) ; spaces(2) ; print(Vy,4,2)
repeat
repeat
stop
routine const mult (integer i, arrayname R, real name m)
integer u
integer array a(1:11)
cycle u=1,1,11
a(u)=u
repeat
if i=1 then → 1
cycle u=2,1,4
a(u)=u+3 ; a(u+3)=u
repeat
a(10)=11 ; a(11)=10
1:  R(a(1))=1 ; R(a(2))=1 ; R(a(3))=-EI
    R(a(4))=-EI ; R(a(5))=-m ; R(a(6))=EI*m²
    R(a(7))=-EI*m ; R(a(8))=EI*m ; R(a(9))=½EI/k
    R(a(10))=-EI*m/(2k) ; R(a(11))=EI/(2m*k)
return
end
```

```
routine var mult (integer i, arrayname R, real name m, y)
real d,r
integer u
integer array a(1:11)
cycle u=1,1,11
a(u)=u
repeat
if i=1 then → 1
cycle u=2,1,4
a(u)=u+3 ; a(u+3)=u
repeat
a(10)=11 ; a(11)=10
1:    d=cos(m*y) ; r=sin(m*y)
cycle u=1,1,4
R(a(u))=d ; R(a(u+4))=r
repeat
R(a(6))=d ; (R(a)9))=d ; R(a(10))=r ; R(a(11))=r
return
end
```

```
routine cf coef(integer i, arrayname C, real name m,y)
integer u,v,z
integer array a(1:11)
real p,q,r,s,t,l,e
array b(1:4,0:5)
if i=1 then → 1 ; if i=2 then → 2
1: p=J1 ; q=J2 ; r=J3 ; s=K3*x ; t=C1 ; 1=wom1 → 3
2: p=J4 ; q=J5 ; r=J6 ; s=K6*x ; t=C4 ; 1=wom2
3: cycle u=1,1,11
a(u)=u
repeat
if i=1 then → >
cycle u=2,1,4
a(u)=u+3 ; a(u+3)=u
repeat
a(10)=11; a(11)=10
4:cycle u=1,1,4
C(a(1),u)=1 ; C(a(5),u)=1 ; C(a(6),u)=1
```

repeat
cycle u=2,6,8
C(a(u),1)=p ; C(a(u),2)=−p ; C(a(u),3)=q ; C(a(u),4)=−q
repeat
cycle u=1,1,2
C(a(3),u)=p^2 ; C(a(3),u+2)=q^2
repeat
C(a(4),1)=p*(p^2+2m^2) ; C(a(4),2)=−C(a(4),1)
C(a(4),3)=q*(q^2−2m^2) ; C(a(4),4)=−C(a(4),3)
cycle u=1,1,2
C(a(7),u)=m^2−2p^2 ; C(a(7),u+2)=m^2+2q^2 ;
C(a(9),u)=(p^2−m^2)*2 ; C(a(9),u+2)=(q^2−m^2)*2
C(a(10),u)=C(a(9),u)/p ; C(a(10),u+2)=C(a(9),u+2)/q
C(a(11),u)=p*C(a(9),u) ; C(a(11),u+2)=q*C(a(9),u)
repeat
cycle u=9,1,10
C(a(u),2)=−C(a(u),2) ; C(a(u),4)=−C(a(u),4)
repeat
cycle u=1,1,2
cycle v=0,1,5
if u=1 *then* z=1 ; *if* u=2 *then* z=−1
b(u,v)=(t*v)*cos(s+z*v*1) ; b(u+2,v)=(t*v)*sin(s+z*v*1)
repeat
repeat
cycle u=1,1,3
cycle v=5,1,8
C(a(u),v)=b(v−4,u−1)
repeat
repeat
cycle u=5,1,8
C(a(4),u)=b(u−4,3)−2m^2*b(u−4,1)
repeat
cycle u=5,1,6
cycle v=5,1,8
C(a(u),v)=b(v−4,0)
repeat
repeat
cycle u=5,1,8
C(a(7),u)=m^2*b(u−4,0)−2b(u−4,2) ; C(a(8),u)=b(u−4,1)
C(a(9),u)=b(u−4,4)−2m^2*b(u−4,2)+(m*4)*b(u−4,0)

```
C(a(10),u)=b(u–4,3)–2m²*b(u–4,1)
C(a(11),u)=b(u–4,5)–2m²*b(u–4,3)+(m*4)*b(u–4,1)
repeat
C(a(10),5)=C(a(10),5)+(m*4)*cos(s–1)/t
C(a(10),6)=C(a(10),6)+(m*4)*cos(s–1)/t
C(a(10),7)=C(a(10),7)+(m*4)*sin(s–1)/t
C(a(10),8)=C(a(10),8)+(m*4)*sin(s–1)/t
cycle u=1,1,11
cycle v=1,1,8
if v=1 then e=exp(p*x) ; if v=2 then e=exp(–p*x)
if v=3 then e=exp(q*x) ; if v=4 then e=exp(–q*x)
if v=5 or v=7 then e=exp(r*x)
if v=6 or v=8 then e=exp(–r*x)
C(a(u),v)=e*C(a(u),v)
repeat
repeat
return
end
```

```
routine part integ(integer i, realname x,y)
real B,H1,snx,smy,cnx,cmy
integer u
B=8Z*((m²+n²)*2)/(π²*EI*(((m²+n²)*4)+4b4*m²*n²))
snx=sin(n*x) ; cnx=cos(n*x)
PI(1)=B*snx ; PI(2)=n*B*cnx ; PI(3)=EI*n²* B*snx
PI(4)=EI*n*(n²+2m²)*B*cnx ; PI(5)=–m*b*snx ; PI(7)=–EI*m*
(2n²c+m²)*B*snx ; PI(6)=EI*m²*B*snx ; PI(8)=EI*m*n*B*cnx
H1=EI*((m²+n²)*2)*B/(2k)–8Z/(2k*π²))
PI(9)=H1*snx ; PI(10)=H1*m*cnx/n ; PI(11)=H1*n*cnx/m
if i=1 then → 1
smy=sin(m*y) ; cmy=cos(m*y)
cycle u=1,1,4
PI(u)=PI(u)–*cmy
repeat
cycle u=5,1,11
PI(u)=PI(u)*smy
repeat
PI(6)=PI(6)*cmy/smy ; PI(9)=PI(9)*cmy/smy
```

```
1: return
end

routine simp coef (integername i,j)
j=2 tracpt(i/2)
if i=0 or i=16 then j=1
return
end

routine conmul a
const mult (1,RC1,m)
return
end
routine conmul b
const mult (2,RC2,n)
return
end

routine varmul a
var mult (1,RV1,m,y)
return
end

routine varmul b
var mult (2,RV2,n,x)
return
end
routine cfa
cf coef (1,p1,m,x)
return
end
routine cfb
cf coef (2,p2,n,y)
return
end
real fn fora(integer i, integername j)
result=RC1(i)*RV1(i)*A1(j,1)*p1(i,j)
end

real fn forb(integer i, integername j,p)
```

```
result=RC1(i)*RV1(i)*A(j,p)*S(j–1)*p1(i,p)
end

real fn forc(integer i, integername j,p)
result=RC2(i)*RV2(i)*A(j,p)*S(j–1)*p2(i,p)
end

routine solve polynomial 2(integer n,F,array name a,u,v,CONV)
        integer L,i,j,m; array h,b,c,d,e(–4:2n+1)
        real t,K,p,p′,q,q′,ps,ps′,qs,qs′,pt,pt′,qt,qt′,s,s′,rev,rec,rec′,r,r′

    array fn g(addr(q),1)
          fault 1,2→ 99; → 98
99: caption $EXP$OR$DIV$OVERFLOW$$$$$$H$$$$$$BBB1$$$$$$
            $$$$$$$$$$c D$$$$$$E$$$$$$T$ETC
            cycle i=–4,1,41
    newline; print f1(h(i),8); print f1(b(i),8)
    print f1(c(i),8); print f1(d(i),8); print f1(e(i),8); print f1(g(i),8)
    repeat
    return
98: L=20
    i=addr(h(0)); 101,50,–,αi
    i=addr(b(0)); 101,51,–,αi
    i=addr(c(0)); 101,52,–,αi
    i=addr(d(0)); 101,53,–,αi
    i=addr(e(0)); 101,54,–,αi
    101,56,–,αn; 104,56,–,αn
    cycle j=–4,1,–1
    b(j)=0; c(j)=0; d(j)=0; e(j)=0; h(j)=0
    repeat
    cycle j=0,1,n
    h(2j)=a(j); h(2j+1)=0
    repeat
    t=1; K=.1↑F
 1: → 2 unless h(2n)=0
    u(n)=0; v(n)=0; CONV(n)=K
    n=n–1; 122,56,0,2; → 1
 2: return if n=0
    ps=0; qs=0; pt=0; qt=0; s=0
    ps′=0; qs′=0; pt′=0; qt′=0
```

```
          rev=1; K=.1†F
          → 3 unless n=1
comment r=–h(1)/h(0)
          1505,50,0,2; 1576,50,0,0; 1556,0,–,r; → 4
comment Scale coefficients.
      3:  cycle j=0,2,2n
          s=s+log(|h(j)|) unless h(j)=0
          repeat
          s=exp(–s/(n+1))
          cycle j=0,2,2n
comment h(j)=h(j)*s
           101,55,–,αj; 1504,0,–,s
          1542,50,55,0; 1556,50,55,0
          repeat
          → 5 unless|h(2)*h(2n)| < |h(2n–2)*h(0)|
comment Reverse order of coefficients.
     15:  t=–t; m=2int pt(½n)
          cycle j=0,2,m
          s=h(j); h(j)=h(2n–j); h(2n–j)=s
          s=h(j+1); h(j+1)=h(2n–j+1); h(2n–j+1)=s
          repeat
      5:  → 6 if qs=0
          p=ps; q=qs; p′=ps′; q′=qs′; → 7
      6:  → 8 unless|h(2n–4)| < K
          q=1; p=–2; q′=0; p′=0; → 9
comment rec=1/h(n–2), q=h(n)*rec, p=(h(n–1)–q*h(n–3))*rec
      8:  1524,0,0,1; 1576,50,56,–4; 1556,0,–,rec
          1542,50,56,0; 1556,0,–,q; 1543,50,56,–6
          1500,50,56,–2; 1542,0,–,rec; 1556,0,–,p
      9:  → 10 if n=2
          r=0; r′=0
comment Iterate.
      7:  cycle i=1,1,L
comment Bairstow method.
            cycle j=0,2,2n
comment   b(j)=h(j)–p*b(j–1)–q*b(j–2), c(j)=b(j)–p*c(j–1)–
          q*c(j–2)
            101,55,–,αj; 1504,51,55,–4; 1542,0,–p
            1556,90,0,0; 1505,51,55,–4; 1542,0,–,q
            1501,90,0,0; 1500,50,55,0; 1556,51,55,0
```

```
            1504,52,55,–2; 1542,0,–,p; 1556,90,0,0
            1505,52,55,–4; 1542,0,–,q; 1501,90,0,0
            1500,51,55,0; 1556,52,55,0
            repeat
            → 11 if |h(2n–2)|< K or|b(2n–2)|< K
            → 12 if |h(2n–2)*K| ⩽|b(2n–2)|
comment     b(n)=h(n)–q*b(n–2)
            1505,0,–,q; 1542,51,56,–4
            1500,50,56,0; 1556,51,56,0
       11:  → 10 if|b(2n)|⩽|h(2n)*K|
comment     Newton-Raphson method.
       12:  cycle j=0,2,2n
comment     d(j)=h(j)+r*d(j–1), e(j)=d(j)+r*e(j–1)
            101,55,–,αj; 1504,0,–,r; 1542,53,55,–2
            1500,50,55,0; 1556,53,55,0; 1504,0,–,r
            1542,54,55,–2; 1500,53,55,0; 1556,54,55,0
            repeat
            → 4 if |d(n)|<|K*h(n)|
comment     c(n–1)=–p*c(n–2)–q*c(n–3)
            1504,0,–,p; 1542,52,56,–4; 1556,90,0,0
            1505,0,–,q; 1542,52,56,–6
            1501,90,0,0; 1556,52,56,–2
comment     s=c(n–2)²–c(n–1)*c(n–3)
            1542,52,56,–6; 1556,90,0,0; 1504,52,56,–4
            1542,52,56,–4; 1501,90,0,0; 1556,0,–,s
            → 13 if | s | < K
comment     rec=1/s
            1524,0,0,1; 1576,0,–,s; 1556,0,–,rec
comment     p=p+(b(n–1)*c(n–2)–b(n)*c(n–3))*rec
            1504,51,56,0; 1542,52,56,–6; 1556,90,0,0
            1504,51,56,–2; 1542,52,56,–4; 1501,90,0,0
            1542,0,–,rec; 1500,0,–,p; 1556,0,–,p
comment     q=q–(b(n–1)*c(n–1)–b(n)*c(n–2))*rec
            1504,51,56,–2; 1542,52,56,–2; 1556,90,0,0
            1504,51,56,0; 1542,52,56,–4; 1501,90,0,0
            1542,0,–,rec; 1500,0,–,q; 1556,0,–,q; → 14
comment     p=p–2, q=q*(q+1)
       13:  1525,0,0,2; 1500,0,–,p; 1556,0,–,p
            1524,0,0,1; 1500,0,–,q
            1542,0,–,q; 1556,0,–,q
```

```
comment  r=r−1, r=r+1−d(n)/e(n−1) unless e(n−1)=0
    14:  →23 if |e(2n − 2)| < K
         1505,53,56,0; 1576,54,56,−2
         1500,0,−,r; 1556,0,−,r; → 24
    23:  1525,0,0,1; 1500,0,−,r; 1556,0,−,r
    24:  repeat
         ps=pt; qs=qt; pt=p; qt=q
         ps′=pt′; qs′=qt′; pt′=p′; qt′=q′
         K=10K if rev < 0
         rev= − rev; → 15
comment  Linear.
comment  r=1/r if t < 0
         4: 25 if t > 0
         1524,0,0,1; 1576,0,−,r; 1556,0,−,r
    25:  u(n)=r; v(n)=0
         CONV(n)=K; n=n−1; 122,56,0,2
         cycle j=0,1,2n+1
         h(j)=d(j)
         repeat
         → 2
comment  Quadratic.
    10:  → 17 unless t < 0
comment  q=1/q, p=p*q
         1524,0,0,1; 1576,0,−,q; 1556,0,−,q
         1542,0,−,p; 1556,0,−,p
comment  p=−½p, s=q−p²
    17:  1525,0,0,½; 1542,0,−,p; 1556,0,−,p
         1543,0,−,p; 1500,0,−,q; 1556,0,−,s
         →18 if s ⩽ 0
         u(n)=p; u(n−1)=p

         s=sqrt(s); v(n)=s; v(n−1)=−s; → 19
    18:  s=sqrt(−s)
         → 20 if p⩾ 0
         u(n)=p−s; → 21
    20:  u(n)=s+p
    21:  u(n−1)=q/u(n)
         v(n)=0; v(n−1)=0
    19:  CONV(n)=K; CONV(n−1)=K; n=n−2; 122,56,0,4
```

```
cycle j=0,1,2n+1
h(j)=b(j)
repeat
→ 2
end
end of program
* * * T
```

SELECTED BIBLIOGRAPHY

ABOUHARB (1967). Ph.D. Thesis, University of Manchester.

ABU SITTA (1963). Ph.D. Thesis, London University.

ALMOND, F. (1936). "Etude statique des Voiles Minces en Paraboloide Hyperbolique Travaillant Sans Flexion", Zurich: Publications, I.A.B.S.E. Vol. 4.

AMBARTSUMYAN, S.A. (1947). "K raschotu pologikh obolochek", (On Calculation of Shallow Shells), *Prik. Mat. i Mekh.*, **11**, (English translation published by Nat. Advisory Committee for Aeronautics (U.S.A.); Tech. Memo. 1425.)

ANON (1958). "Plywood Roof Sustains Heavy Test Load", *Engineering News Record*, **160**, Jan. 2.

ANON (1963). *Concrete and Constructional Engineering*, Vol. **LVIII**, London: The Commonwealth Institute.

BAN, S. (1953). "Formanderungen der hyperbolischen Paraboloidschale", Proceedings, Int. Assn. of Bridge and Struct. Eng., Vol. **13**.

BANERJEE, S.P. (1962). "Analysis of Hyperbolic Paraboloidal Shells", *Jnl. Int. Assn. Shell Struct.*, **5**.

BANERJEE, S.P. (1964). "Foundations of Engineering Structures", *Jnl. Inst. of Engineers* (*India*), **XLIV**, No. 11, July.

BOOTH, L.G. (1959). "Hyperbolic Paraboloid Timber Shell Roofs", *Architect and Building News*, Aug. 19.

BOUMA, A.L. (1959). "Some Applications of the Bonding Theory Regarding Doubly-curved Shells", Proc. Symposium on the Theory of Thin Elastic Shells, Delft, Int. Union of Theoretical and Applied Mech.

BOYD, R. (1953). "Engineering of Excitement", *Arch. Rev.*, **124**, No. 742.

CANDELA, F. (1955). "Structural Applications of Hyperbolic Paraboloidal Shells", *Jnl. Am. Conc. Inst.*, **26**, No. 5, Jan.

CANDELA, F. (1956). "Shell Concrete in Mexico", *Municipal Jnl. and Public Works Eng.*, **64**, March 9.

CANDELA, F. (1958). "Shells as Space Enclosure", Proceedings of a Symposium on Shell Structures, M.I.T., Cambridge, Mass., Sept.

CANDELA, F. (1960). "General Formulas for Membrane Stresses in Hyperbolic Paraboloidal Shells", *Jnl. Am. Conc. Inst.*, **32**, No. 4, Oct.

CHAKROBARTI, S. (1967). "Analysis of thin Elastic Cylindrical Shells by Finite Element Method", *Proc. Inst. of Eng.* (*India*), Jan.

CICALA, P. (1960). "Sulla Teoria Elastica della Parete Sottile con Superficio Media Rigata", *Atti della Accademia della Scienze* (*Turin*), **95**, No. 1 (In Italian).

CICALA, P. (1961). *Controlled Approximation Theory for Elastic Thin Shells—Part I* (monograph published by Institute di Scienza delle Costruzioni, Politecnico di Torino, Turin, Italy, May.

CICALA, P. (1962). "Elastic Theory of Hypar Shells", *Jnl. Am. Conc. Inst.*, January.

CSONKA, P. (1954). "A Saddle-Shaped Shell", *Eng. News Record*, **152**, May 27.

CSONKA, P. (1957). "Shell Roof Construction in Timber", *Engineering*, **154**, July 26.

CSONKA, P. (1958). "On Shells Curved in Two Directions", Second Symposium on Shell Roof Construction, Oslo.

DAS GUPTA, N.C. (1963). Ph.D. Thesis, Leeds University.

DAS GUPTA, N.C. (1963). "Edge Disturbances in a Hyperbolic Paraboloid", *Civil Eng. and Public Works Review*, Feb.

DARYARATNAM, P. AND GERSTLE, K.H. (1963). "Model Study of a Hyperbolic Paraboloidal Shell Supported on Four Elastic Edge Beams", *Ind. Conc. Jnl.,* **37**, April.

FLINT, A.R. AND LOW, A.E. (1959). "The Construction of Hyperbolic Paraboloid Type Shells without Temporary Formwork", *Proceedings of International Colloquium on Constructional Processes of Shell Structures*, Madrid, Sept.

FLUGGE, W. (1947). "Zur membrantheorie der Drehschalen negativer Krummung", *Z. fur ang. Math. und Mech.*, No. 25.

GALERKIN, B.G. (1918). "Tables of Moments & Deflections in Plates", *Bull. Polytech. Inst., St. Petersburg*, **26** and **27**.

GERARD, G. (1956). Ph.D. Thesis, London University.

GERARD, G. (1959). "Analysis of Hyperbolic Paraboloid Shells", *Trans. Canadian Eng. Inst.*, **3**, No. 1, April.

GOLDENVEIZER, A.L. AND LURYE, A.I. (1947). *Prik. Mat. i Mekh,* **11**, No. 5.

HARRENSTIEN, H.P. (1960). "Hyperbolic Paraboloidal Umbrella Shells under Working Loads", *Jnl. Am. Conc. Inst.*, **32**, No. 4, Oct.

HERRMANN, L.R. (1967). "Finite Element Bending Analysis for Plates", *Jnl. Eng. Mech. Div., A.S.C.E.*, Oct.

HOSSLI, R.I. (1965). "Corrugated Sheet Shells", *Building Research*, **2**, No. 3, May-June.

INTERNATIONAL ASSOCIATION OF SHELL STRUCTURES. (1959).

Proceedings of the International Colloquium on Constructional Processes for Shell Structures, Madrid.

JONES, L.L. (1960). "Tests on a One-tenth Scale Model of a Hyperbolic Paraboloid Shell Roof", Cement and Conc. Assn., Tech. Report No. TRA/334, August.

JONES, L.L. (1961). "Tests on a One-sixth Scale Model of a Hyperbolic Paraboloid Umbrella Shell Roof", Cement and Conc. Assn., Tech. Report No. TRA/347, January.

KONYI, K. HAJNAL. (1960). "Recent Developments in Shell Concrete Construction", *Architects' Year Book*, Vol. 9.

KONYI, K. HAJNAL. (1963). "Hyperbolic Paraboloid Shell over Lincolnshire Motor Co. Garage", *Proc. Int. Assoc. Shell Struct.*, Delft.

KHWAJA, I. (1965). Discussion to Banerjee (1964), *Jnl. Inst. of Engineers (India)*, **XLV**, No. 11, July.

LAFAILLE, B. (1934). "Thin Shells in the Shape of Hyperbolic Paraboloid", *Le Genie Civil*, **104**, No. 18.

LAFAILLE, B. (1935). "Memoire sur l'etude generale de surfaces gauches minces", Int. Assoc. of Bridge and Struct. Eng. Publications, Vol. 3.

LEE, I.D.G., *et al.* (1962). "Loading Tests on a 1/4 Scale Model Asymmetric Hyperbolic Paraboloid Timber Shell Roof", *Civ. Eng. and P.W. Rev.*, July-August.

LOVE, A. (1934). *A Treatise on Mathematical Theory of Elasticity*, London: Cambridge University Press.

MAINSTONE, R.J. (1963). "The Springs of Structural Invention", *RIBA Journal*, February.

MARGUERRE, K. (1938). "K. Zur Theorie der gekrummten Platte grosser Formanderung", Proc. 5th Int. Cong. of Applied Mech.

MIHAILESCU, M. (1952). "Asupra unor formule practice pentru calculul invelitorilor cilindrice subciri", (About Some Practical Formulae for the Calculus of Thin Cylindrical Shells), *Buletin Stiint. Acad. R.P.R.*, Sectiunea de Stiinte tehnice sichimice, **4**, No. 3-4, (In Roumanian).

MIHAILESCU, M. (1957). "On the Hyperbolic Paraboloids Bending Theory", *Revue de Mechanique Appliquee*, **2**, No. 1, (Academie de la Republique Populaire Roumaine.)

MORICE, P.B. (1952). "Research on Concrete Shell Structures", Proc. Symposium on Conc. Shell Roof Construction, Cement and Conc. Ass., London.

MUNRO, J. (1961). "Linear Analysis of Thin Elastic Shells", *Proc. Inst. C.E.*, **19**, July.

MURPHY, G. (1950). *Similitude in Engineering*, New York: Ronald Press Co.

MUSHTARI, KH. M. (1938). "Nekotorye obobshcheniia teorii tonkikh obolochek", *Izu, fiz.-mat. ob-va pri Kaz. un-te*, **XI.**

NAZAROV, A.A. (1949). "On the Theory of Thin Shallow Shells", *Prik. Mat. i Mekh.*, **13**. (English Translation published by Nat. Advisory Committee for Aeronautics (U.S.A.), Tech. Memo. 1426.)

NILSON, A.H. (1962). "Testing of Light Gage Steel Hyperbolic Paraboloidal Shell", *Jnl. Struct. Div., A.S.C.E.*, Oct.

PARME, A.L. (1956). "Shells of Double Curvature", *Trans. Am. Soc. C.E.*, paper No. 2951, **82**, Sept.

PELIKAN, J. (1958). "Membrane Structures", Second Symposium on Shell Roof Construction, Oslo.

PFLUGER, A. (1957). *Elementare Schalenstatik*, Berlin: Springer-Verlag.

"Plastics in Speciality Formwork" (1965). *Plastics in Building*, **2**, No. 1, Feb.

POPOV, E.P., PENZION, J. AND ZUNG-AN LU (1964). "Finite Element Solution for Axisymmetrical Shells", *Jnl. Eng. Mech. Div., A.S.C.E.*, Oct.

PUCHER, A. (1934). "Ueber den Spannungszustand in doppelt gekrummten Flachen", *Beton Und Eisen*, **33**, No. 19, Oct.

RAMASWAMY, G.S. *et al.* (1958). "Research on Doubly Curved Shells", *Civil Eng. and Public Works Review*, **53**, Aug., Oct. and Nov.

ROISSNER, E. (1955). "On Some Aspects of the Theory of Thin Elastic Shells", *Jnl. Boston Soc. C.E.*, **42**, No. 2, April.

RUSSEL, R.R. and GERSTLE, K.H. (1967). "Bonding of Hyperbolic Paraboloid Structures", *Jnl. Structural Div.., Proc. Am. Soc. C.E.*, **93**, No. ST3, June.

SALVADORI, M.C. (1956). "Analysis and Testing Translational Shells", *Jnl. Am. Conc. Inst.*, **27**, June.

SANDHI, J.R. AND PATEL, M.N. (1961). "Hyperbolic Paraboloid Shell Footings for a Building in Mombasa, Kenya", *Indian Conc. Jnl.*, **35**, June.

SHAH, H.C. (1961). "Analysis of Shell Roofs with Paraboloidal Surface", *Ind. Conc. Jnl.*, **35**, No. 2, Feb.

SOARO, M. (1966). "A Numerical Approach to the Bending Theory of Hypar Shells", *Ind. Conc. Jnl.*, Feb.-March.

STITCH, R.J. (1955). Discussion to Candela (1955), *Jnl. Am. Conc. Inst.*, **51**, Jan.

Technical Bulletin on Foil Strain Gauges by Saunders-Roe Division of

Westland Aircraft Ltd., Isle of Wight.
TEDESKO, A. (1960). "Shell at Denver", *Jnl. Am. Conc. Inst.*, **32**, Oct.
TESTER, K.G. (1947). "Beitrag Zur Berechnung der hyperbolischen Paraboloidschale", *Ingenieur-Archive*, **16.**
"The Construction of a Hyperbolic Paraboloid Shell without Temporary Formwork" (1959). *Architecture and Building*, **34**, No. 9, Sept.
TIMOSHENKO, S. AND WOINOWSKY-KRIEGER (1959). *Theory of Plates and Shells*, Second Edition, New York: McGraw-Hill.
TOTTENHAM, H. (1958). "Shell Roof Construction in Timber", *Civ. Eng. and Public Works Review*, **53**, April.
TOTTENHAM, H. (1960). "Timber Shell Roofs for Storage Buildings and Transit Sheds", *Dock and Harbour Authority* (*London*), **XLI**, May.
VLASOV, V.Z. (1944). "Fundamental Differential Equations of the General Theory of Elastic Shells", *Prikladnia Matematika i Mekhanika*, **8**, No. 2. (English translation published by National Advisory Committee for Aeronautics (U.S.A.), Tech. Memo. 1241.)
WALING, J.L. AND GRESZCZUK, L.B. (1960). "Experiments with Thin-Shell Structural Models", *Jnl. Am. Conc. Inst.*, **32**, Oct.
WILBY, C.B. (1962). "Proposed 'Exact Theory' for Analysing Shells and its Solution with Analogue Computer", *Proc. I.C.E.*, July.
WILBY, C.B. AND BELLAMY, N.W. (1961). *Elastic Analysis of Shells by Electronic Analogy*, London: Arnold.
YU, C.W. AND KRITZ, L.B. (1963). "Tests on a Hyperbolic Paraboloid Umbrella Shell", Proc. World Conference on Shell Structures, Cambridge, Mass. (U.S.A.).

DESIGN TABLES FOR NORTH-LIGHT CYLINDRICAL SHELL ROOFS

L = 20 FT. TABLE NO.N- 1 W = 12 FT.

SHELL DATA:

L = 20.00 FT.
W = 12.00 FT.
T = 2.50 IN.
G = 50.00 LB/FT*2
QA= 150.00 LB/FT
QV= 150.00 LB/FT

FOR 60DEG.GLAZING:
K = 6.00 FT.
R = 12.50 FT.
AL= 10.72 FT.
θ = 49.13 DEG.

FOR 90DEG.GLAZING:
K = 6.00 FT.
R = 22.50 FT.
AL= 13.62 FT.
θ = 34.69 DEG.

BEAM DATA:

FOR 60DEG.GLAZING:

B1= 0.75 FT.
D1= 1.50 FT.
B2= 0.75 FT.
D2= 1.25 FT.

EXT.NORTH BEAM:

Z'= 0.78 FT.
Y'= 0.76 FT.
IZ= 0.27 FT*4
IY= 0.20 FT*4

FOR 90DEG.GLAZING:

ALL BEAMS:

B = 1.00 FT.
D = 1.50 FT.

BEAM RESULTS:

FOR 60DEG.GLAZING:

EXT.SOUTH BEAM
F = -2139 LB.
M1= 11450 LB.FT
M2= 2471 LB.FT
S1= -1999 LB.
S2= 500 LB.
INTERNAL BEAMS
F = -2038 LB.
P = 261 LB/FT
M1= 23330 LB.FT
M2= 26 LB.FT
S1= -3889 LB.
S2= 117 LB.
EXT.NORTH BEAM
P = 276 LB/FT
M1= 26400 LB.FT
M2= 7129 LB.FT
S1= -4147 LB.
S2= -1120 LB.

FOR 90DEG.GLAZING:

EXT.SOUTH BEAM
F = -2938 LB.
M1= 20149 LB.FT
M2= 3634 LB.FT
S1= -3507 LB.
S2= 797 LB.
INTERNAL BEAMS
F = -8265 LB.
P = 311 LB/FT
M1= 27609 LB.FT
M2= 5369 LB.FT
S1= -5305 LB.
S2= 1487 LB.
EXT.NORTH BEAM
P = 314 LB/FT
M1= 29684 LB.FT
S1= -4663 LB.

SHELL RESULTS:

	SHELLS WITH 60-DEGREE GLAZING				SHELLS WITH 90-DEGREE GLAZING			
EXTERNAL SOUTH SHELL								
POINT	T1	S	T2	M2	T1	S	T2	M2
0	16131	336	91	83	13814	462	39	-181
1	5824	-1904	-45	87	5121	-1997	-146	-22
2	-899	-2366	-458	35	-679	-2535	-720	18
3	-4729	-1728	-850	-30	-4183	-1838	-1267	7
4	-6161	-543	-1047	-66	-5680	-475	-1533	4
5	-5421	713	-988	-57	-5169	1023	-1412	36
6	-2425	1581	-705	-14	-2359	2086	-943	87
7	3209	1550	-334	21	3299	2031	-329	101
8	12133	0	-125	0	12566	-0	19	-0
INTERNAL SHELLS								
POINT	T1	S	T2	M2	T1	S	T2	M2
0	16943	320	238	40	18665	1298	170	-183
1	6214	-2045	84	80	7214	-2061	72	1
2	-840	-2552	-366	45	-244	-2915	-572	44
3	-4897	-1901	-795	-15	-4600	-2206	-1224	21
4	-6441	-667	-1025	-54	-6384	-683	-1569	4
5	-5705	650	-985	-48	-5789	999	-1478	27
6	-2620	1570	-710	-10	-2659	2189	-996	77
7	3175	1564	-337	23	3472	2154	-349	95
8	12307	0	-126	0	13332	0	19	-0
EXTERNAL NORTH SHELL								
POINT	T1	S	T2	M2	T1	S	T2	M2
0	16974	399	239	29	18648	1306	169	-185
1	6245	-1972	100	82	7205	-2050	74	1
2	-777	-2489	-334	58	-246	-2903	-567	45
3	-4769	-1857	-752	6	-4590	-2194	-1215	24
4	-6229	-659	-976	-29	-6357	-676	-1558	8
5	-5438	607	-940	-25	-5745	996	-1466	31
6	-2424	1475	-680	8	-2619	2174	-987	80
7	3028	1457	-330	33	3460	2134	-345	97
8	11341	0	-133	0	13171	-0	20	-0

L = 22 FT. TABLE NO.N- 2 W = 12 FT.

SHELL DATA:

L = 22.00 FT.
W = 12.00 FT.
T = 2.50 IN.
G = 50.00 LB/FT*2
QA= 150.00 LB/FT
QV= 150.00 LB/FT

FOR 60DEG.GLAZING:
K = 6.00 FT.
R = 10.00 FT.
AL= 10.93 FT.
θ = 62.61 DEG.

FOR 90DEG.GLAZING:
K = 6.00 FT.
R = 17.50 FT.
AL= 13.77 FT.
θ = 45.08 DEG.

BEAM DATA:

FOR 60DEG.GLAZING:

B1= 0.75 FT.
D1= 1.50 FT.
B2= 0.75 FT.
D2= 1.25 FT.

EXT.NORTH BEAM:

Z'= 0.78 FT.
Y'= 0.76 FT.
IZ= 0.27 FT*4
IY= 0.20 FT*4

FOR 90DEG.GLAZING:

ALL BEAMS:

B = 1.00 FT.
D = 1.50 FT.

BEAM RESULTS:

FOR 60DEG.GLAZING:

EXT.SOUTH BEAM
F = 1316 LB.
M1= 10999 LB.FT
M2= 2741 LB.FT
S1= -1441 LB.
S2= 298 LB.
INTERNAL BEAMS
F = 523 LB.
P = 235 LB/FT
M1= 23875 LB.FT
M2= 418 LB.FT
S1= -3330 LB.
S2= 17 LB.
EXT.NORTH BEAM
P = 244 LB/FT
M1= 30214 LB.FT
M2= 7628 LB.FT
S1= -4315 LB.
S2= -1089 LB.

FOR 90DEG.GLAZING:

EXT.SOUTH BEAM
F = -2451 LB.
M1= 21394 LB.FT
M2= 4643 LB.FT
S1= -3311 LB.
S2= 830 LB.
INTERNAL BEAMS
F = -7684 LB.
P = 279 LB/FT
M1= 29357 LB.FT
M2= 6746 LB.FT
S1= -5007 LB.
S2= 1502 LB.
EXT.NORTH BEAM
P = 278 LB/FT
M1= 33631 LB.FT
S1= -4803 LB.

SHELL RESULTS:

SHELLS WITH 60-DEGREE GLAZING / SHELLS WITH 90-DEGREE GLAZING

EXTERNAL SOUTH SHELL

POINT	T1	S	T2	M2	T1	S	T2	M2
0	16897	-188	165	275	15414	350	81	-33
1	5409	-2287	-40	181	5449	-2131	-109	46
2	-1846	-2573	-471	43	-986	-2617	-653	19
3	-5753	-1784	-850	-84	-4697	-1869	-1160	-41
4	-7025	-499	-1029	-153	-6195	-488	-1404	-69
5	-6020	809	-957	-147	-5603	1006	-1293	-36
6	-2634	1695	-675	-83	-2603	2070	-866	34
7	3697	1646	-313	-14	3521	2030	-308	76
8	13928	0	-116	0	13793	0	7	0

INTERNAL SHELLS

POINT	T1	S	T2	M2	T1	S	T2	M2
0	18401	-75	272	239	20864	1097	210	-5
1	6139	-2389	66	180	7620	-2289	77	87
2	-1683	-2758	-394	57	-727	-3049	-550	51
3	-5949	-1963	-811	-68	-5343	-2236	-1160	-31
4	-7386	-621	-1019	-142	-7064	-658	-1470	-79
5	-6377	761	-964	-140	-6282	1033	-1376	-58
6	-2854	1705	-685	-80	-2864	2215	-926	14
7	3711	1677	-318	-13	3802	2175	-329	65
8	14237	0	-118	-0	14706	0	6	0

EXTERNAL NORTH SHELL

POINT	T1	S	T2	M2	T1	S	T2	M2
0	18471	-27	271	232	20775	1080	209	-4
1	6169	-2350	74	181	7587	-2291	73	87
2	-1655	-2725	-380	66	-729	-3047	-554	49
3	-5887	-1938	-790	-54	-5336	-2236	-1163	-33
4	-7264	-614	-995	-126	-7063	-659	-1474	-82
5	-6208	739	-941	-125	-6295	1033	-1380	-60
6	-2718	1651	-670	-69	-2885	2220	-929	12
7	3628	1614	-316	-7	3803	2184	-330	64
8	13607	0	-123	-0	14786	0	6	0

L = 22 FT. TABLE NO.N- 3 W = 14 FT.

SHELL DATA:

```
L  =  22.00  FT.
W  =  14.00  FT.
T  =   2.50  IN.
G  =  50.00  LB/FT*2
QA= 150.00  LB/FT
QV= 150.00  LB/FT
```

FOR 60DEG.GLAZING:

```
K  =   6.50  FT.
R  =  15.00  FT.
AL=  12.49  FT.
θ  =  47.72  DEG.
```

FOR 90DEG.GLAZING:

```
K  =   6.50  FT.
R  =  25.00  FT.
AL=  15.69  FT.
θ  =  35.96  DEG.
```

BEAM DATA:

FOR 60DEG.GLAZING:

```
B1= 0.75 FT.
D1= 1.50 FT.
B2= 0.75 FT.
D2= 1.25 FT.
```

EXT.NORTH BEAM:

```
Z'= 0.78 FT.
Y'= 0.76 FT.
IZ= 0.27 FT*4
IY= 0.20 FT*4
```

FOR 90DEG.GLAZING:

ALL BEAMS:

```
B = 1.00 FT.
D = 1.50 FT.
```

BEAM RESULTS:

FOR 60DEG.GLAZING:

```
EXT.SOUTH BEAM
  F  =    -2381 LB.
  M1=    13351 LB.FT
  M2=     2804 LB.FT
  S1=    -2109 LB.
  S2=      515 LB.
INTERNAL BEAMS
  F  =    -2636 LB.
  P  =      256 LB/FT
  M1=    27633 LB.FT
  M2=      -22 LB.FT
  S1=    -4212 LB.
  S2=      132 LB.
EXT.NORTH BEAM
  P  =      272 LB/FT
  M1=    31947 LB.FT
  M2=     8484 LB.FT
  S1=    -4562 LB.
  S2=    -1212 LB.
```

FOR 90DEG.GLAZING:

```
EXT.SOUTH BEAM
  F  =    -3757 LB.
  M1=    23281 LB.FT
  M2=     4406 LB.FT
  S1=    -3722 LB.
  S2=      893 LB.
INTERNAL BEAMS
  F  =   -10461 LB.
  P  =      304 LB/FT
  M1=    32047 LB.FT
  M2=     6458 LB.FT
  S1=    -5691 LB.
  S2=     1663 LB.
EXT.NORTH BEAM
  P  =      305 LB/FT
  M1=    35576 LB.FT
  S1=    -5080 LB.
```

SHELL RESULTS:

EXTERNAL SOUTH SHELL

	SHELLS WITH 60-DEGREE GLAZING				SHELLS WITH 90-DEGREE GLAZING			
POINT	T1	S	T2	M2	T1	S	T2	M2
0	18697	340	100	160	16084	537	57	-123
1	6338	-2354	-102	132	5658	-2405	-192	26
2	-1401	-2830	-644	34	-968	-2986	-916	26
3	-5603	-1992	-1140	-73	-4772	-2125	-1592	-28
4	-7070	-533	-1377	-130	-6340	-520	-1915	-54
5	-6169	987	-1281	-115	-5794	1231	-1764	-15
6	-2749	2032	-899	-46	-2736	2492	-1184	66
7	3849	1978	-402	17	3689	2450	-421	107
8	14709	0	-125	0	14683	0	14	0

INTERNAL SHELLS

	SHELLS WITH 60-DEGREE GLAZING				SHELLS WITH 90-DEGREE GLAZING			
POINT	T1	S	T2	M2	T1	S	T2	M2
0	19948	376	248	126	21733	1494	190	-112
1	6875	-2513	30	132	7918	-2516	38	59
2	-1359	-3049	-557	47	-636	-3431	-777	55
3	-5856	-2184	-1100	-60	-5338	-2518	-1575	-18
4	-7438	-653	-1372	-122	-7128	-711	-1982	-63
5	-6504	948	-1294	-112	-6413	1244	-1854	-34
6	-2939	2054	-912	-46	-2984	2630	-1251	47
7	3886	2017	-408	16	3920	2592	-445	97
8	15034	0	-126	0	15484	0	14	-0

EXTERNAL NORTH SHELL

	SHELLS WITH 60-DEGREE GLAZING				SHELLS WITH 90-DEGREE GLAZING			
POINT	T1	S	T2	M2	T1	S	T2	M2
0	19990	445	248	111	21668	1485	189	-113
1	6887	-2449	45	132	7891	-2512	36	59
2	-1337	-2988	-528	59	-642	-3423	-777	55
3	-5772	-2134	-1058	-38	-5331	-2510	-1573	-17
4	-7252	-633	-1322	-95	-7114	-706	-1978	-62
5	-6230	915	-1245	-85	-6393	1244	-1850	-33
6	-2703	1960	-878	-26	-2967	2624	-1247	48
7	3772	1902	-399	27	3916	2584	-443	98
8	13998	0	-133	0	15423	0	14	0

L = 24 FT. TABLE NO.N- 4 W = 12 FT.

SHELL DATA:

```
L =  24.00 FT.
W =  12.00 FT.
T =   2.50 IN.
G =  50.00 LB/FT*2
QA= 150.00 LB/FT
QV= 150.00 LB/FT

FOR 60DEG.GLAZING:
K =   6.00 FT.
R =  12.50 FT.
AL=  10.72 FT.
θ =  49.13 DEG.

FOR 90DEG.GLAZING:
K =   6.00 FT.
R =  17.50 FT.
AL=  13.77 FT.
θ =  45.08 DEG.
```

BEAM DATA:

```
FOR 60DEG.GLAZING:

B1= 0.75 FT.
D1= 1.50 FT.
B2= 0.75 FT.
D2= 1.25 FT.

EXT.NORTH BEAM:

Z'= 0.78 FT.
Y'= 0.76 FT.
IZ= 0.27 FT*4
IY= 0.20 FT*4

FOR 90DEG.GLAZING:

ALL BEAMS:

B = 1.00 FT.
D = 1.50 FT.
```

BEAM RESULTS:

```
FOR 60DEG.GLAZING:

EXT.SOUTH BEAM
F  =   -1406 LB.
M1=    16025 LB.FT
M2=     3429 LB.FT
S1=    -2196 LB.
S2=      495 LB.
INTERNAL BEAMS
F  =   -1580 LB.
P  =     254 LB/FT
M1=    32763 LB.FT
M2=       37 LB.FT
S1=    -4417 LB.
S2=       70 LB.
EXT.NORTH BEAM
P  =     274 LB/FT
M1=    37862 LB.FT
M2=    10178 LB.FT
S1=    -4956 LB.
S2=    -1332 LB.

FOR 90DEG.GLAZING:

EXT.SOUTH BEAM
F  =   -2566 LB.
M1=    25293 LB.FT
M2=     5528 LB.FT
S1=    -3554 LB.
S2=      882 LB.
INTERNAL BEAMS
F  =   -8553 LB.
P  =     275 LB/FT
M1=    34528 LB.FT
M2=     7967 LB.FT
S1=    -5348 LB.
S2=     1590 LB.
EXT.NORTH BEAM
P  =     273 LB/FT
M1=    39657 LB.FT
S1=    -5191 LB.
```

SHELL RESULTS:

EXTERNAL SOUTH SHELL

POINT	T1 (60-DEGREE GLAZING)	S	T2	M2	T1 (90-DEGREE GLAZING)	S	T2	M2
0	22877	184	105	172	18302	336	85	-11
1	8240	-2469	-66	139	6518	-2375	-115	56
2	-1608	-2986	-509	55	-1224	-2904	-671	18
3	-7348	-2144	-920	-35	-5740	-2065	-1187	-50
4	-9473	-619	-1121	-86	-7544	-520	-1432	-81
5	-8211	981	-1047	-80	-6758	1141	-1315	-48
6	-3480	2059	-739	-32	-3048	2306	-877	27
7	5115	1976	-341	12	4329	2240	-310	73
8	18247	0	-122	0	16451	0	6	-0

INTERNAL SHELLS

POINT	T1 (60-DEGREE GLAZING)	S	T2	M2	T1 (90-DEGREE GLAZING)	S	T2	M2
0	24031	207	240	126	24633	1120	214	37
1	8832	-2597	58	131	9072	-2560	66	110
2	-1453	-3178	-414	66	-918	-3386	-575	57
3	-7498	-2335	-859	-17	-6529	-2474	-1196	-38
4	-9789	-767	-1090	-69	-8624	-707	-1509	-94
5	-8561	893	-1038	-67	-7614	1180	-1408	-72
6	-3749	2027	-739	-24	-3378	2483	-943	3
7	5014	1978	-344	16	4702	2414	-334	60
8	18374	0	-124	-0	17645	0	6	-0

EXTERNAL NORTH SHELL

POINT	T1 (60-DEGREE GLAZING)	S	T2	M2	T1 (90-DEGREE GLAZING)	S	T2	M2
0	24054	334	241	119	24523	1097	213	39
1	8908	-2478	81	141	9029	-2566	61	109
2	-1286	-3080	-372	90	-924	-3387	-580	54
3	-7210	-2277	-803	17	-6525	-2475	-1201	-41
4	-9385	-770	-1029	-31	-8630	-708	-1515	-97
5	-8124	815	-983	-32	-7637	1182	-1413	-76
6	-3488	1883	-704	1	-3408	2491	-948	1
7	4708	1830	-336	29	4707	2427	-336	58
8	16872	0	-132	0	17768	0	6	-0

TABLE NO.N- 5

L = 24 FT. W = 14 FT.

SHELL DATA:

L = 24.00 FT.
W = 14.00 FT.
T = 2.50 IN.
G = 50.00 LB/FT*2
QA= 150.00 LB/FT
QV= 150.00 LB/FT

FOR 60DEG.GLAZING:
K = 6.50 FT.
R = 12.50 FT.
AL= 12.67 FT.
θ = 58.08 DEG.

FOR 90DEG.GLAZING:
K = 6.50 FT.
R = 22.50 FT.
AL= 15.76 FT.
θ = 40.12 DEG.

BEAM DATA:

FOR 60DEG.GLAZING:

B1= 0.75 FT.
D1= 1.50 FT.
B2= 0.75 FT.
D2= 1.25 FT.

EXT.NORTH BEAM:

Z'= 0.78 FT.
Y'= 0.76 FT.
IZ= 0.27 FT*4
IY= 0.20 FT*4

FOR 90DEG.GLAZING:

ALL BEAMS:

B = 1.00 FT.
D = 1.50 FT.

BEAM RESULTS:

FOR 60DEG.GLAZING:

EXT.SOUTH BEAM
F = 801 LB.
M1= 13040 LB.FT
M2= 3091 LB.FT
S1= -1629 LB.
S2= 346 LB.
INTERNAL BEAMS
F = -325 LB.
P = 233 LB/FT
M1= 28408 LB.FT
M2= 361 LB.FT
S1= -3728 LB.
S2= 52 LB.
EXT.NORTH BEAM
P = 243 LB/FT
M1= 36144 LB.FT
M2= 9014 LB.FT
S1= -4731 LB.
S2= -1180 LB.

FOR 90DEG.GLAZING:

EXT.SOUTH BEAM
F = -3932 LB.
M1= 26064 LB.FT
M2= 5362 LB.FT
S1= -3791 LB.
S2= 953 LB.
INTERNAL BEAMS
F = -11193 LB.
P = 288 LB/FT
M1= 35856 LB.FT
M2= 7784 LB.FT
S1= -5784 LB.
S2= 1743 LB.
EXT.NORTH BEAM
P = 287 LB/FT
M1= 41021 LB.FT
S1= -5370 LB.

SHELL RESULTS:

SHELLS WITH 60-DEGREE GLAZING / SHELLS WITH 90-DEGREE GLAZING

EXTERNAL SOUTH SHELL

POINT	T1	S	T2	M2	T1	S	T2	M2
0	19583	-105	156	344	18402	515	75	-44
1	6020	-2655	-98	219	6391	-2572	-172	61
2	-2225	-2969	-638	37	-1206	-3159	-876	25
3	-6470	-2009	-1107	-129	-5502	-2233	-1528	-56
4	-7777	-488	-1318	-217	-7217	-542	-1837	-95
5	-6666	1051	-1214	-203	-6547	1285	-1690	-55
6	-2952	2098	-846	-113	-3073	2594	-1133	38
7	4241	2036	-375	-17	4174	2546	-404	95
8	16365	0	-117	-0	16576	-0	9	0

INTERNAL SHELLS

POINT	T1	S	T2	M2	T1	S	T2	M2
0	21517	42	273	319	24852	1465	208	-9
1	6861	-2789	17	226	8907	-2735	41	109
2	-2116	-3195	-564	54	-925	-3651	-761	59
3	-6775	-2210	-1078	-115	-6237	-2645	-1534	-48
4	-8233	-606	-1322	-211	-8172	-724	-1921	-111
5	-7069	1024	-1234	-203	-7265	1326	-1791	-82
6	-3160	2137	-863	-115	-3328	2763	-1205	12
7	4325	2089	-382	-19	4489	2710	-429	81
8	16802	0	-117	-0	17545	-0	8	0

EXTERNAL NORTH SHELL

POINT	T1	S	T2	M2	T1	S	T2	M2
0	21563	83	272	309	24732	1443	206	-8
1	6870	-2754	23	226	8863	-2737	36	108
2	-2111	-3161	-550	63	-927	-3648	-765	58
3	-6733	-2180	-1057	-101	-6224	-2644	-1538	-50
4	-8119	-591	-1297	-193	-8166	-725	-1925	-113
5	-6888	1008	-1208	-185	-7276	1325	-1795	-84
6	-2997	2082	-845	-102	-3351	2768	-1208	11
7	4258	2019	-377	-12	4486	2718	-431	80
8	16113	0	-122	-0	17628	-0	8	0

L = 26 FT. TABLE NO.N- 6 W = 12 FT.

```
SHELL DATA:

   L =  26.00 FT.
   W =  12.00 FT.
   T =   2.50 IN.
   G =  50.00 LB/FT*2
   QA= 150.00 LB/FT
   QV= 150.00 LB/FT

 FOR 60DEG.GLAZING:
   K =   6.25 FT.
   R =  10.00 FT.
   AL=  10.93 FT.
   θ =  62.63 DEG.

 FOR 90DEG.GLAZING:
   K =   6.25 FT.
   R =  15.00 FT.
   AL=  14.04 FT.
   θ =  53.62 DEG.

BEAM DATA:

 FOR 60DEG.GLAZING:

   B1= 0.75 FT.
   D1= 1.75 FT.
   B2= 0.75 FT.
   D2= 1.50 FT.

   EXT.NORTH BEAM:

   Z'= 0.91 FT.
   Y'= 0.78 FT.
   IZ= 0.43 FT*4
   IY= 0.31 FT*4

 FOR 90DEG.GLAZING:

   ALL BEAMS:

   B = 1.00 FT.
   D = 1.75 FT.
```

```
BEAM RESULTS:

FOR 60DEG.GLAZING:

EXT.SOUTH BEAM
  F =      1829 LB.
  M1=     20515 LB.FT
  M2=      3433 LB.FT
  S1=     -2297 LB.
  S2=       301 LB.
INTERNAL BEAMS
  F =      1281 LB.
  P =       259 LB/FT
  M1=     39173 LB.FT
  M2=       343 LB.FT
  S1=     -4567 LB.
  S2=       -36 LB.
EXT.NORTH BEAM
  P =       270 LB/FT
  M1=     49076 LB.FT
  M2=     11792 LB.FT
  S1=     -5930 LB.
  S2=     -1425 LB.

FOR 90DEG.GLAZING:

 EXT.SOUTH BEAM
  F =     -1184 LB.
  M1=     33258 LB.FT
  M2=      5808 LB.FT
  S1=     -4135 LB.
  S2=       761 LB.
INTERNAL BEAMS
  F =     -6634 LB.
  P =       276 LB/FT
  M1=     45034 LB.FT
  M2=      8318 LB.FT
  S1=     -6131 LB.
  S2=      1389 LB.
 EXT.NORTH BEAM
  P =       276 LB/FT
  M1=     50109 LB.FT
  S1=     -6055 LB.
```

SHELL RESULTS:

SHELLS WITH 60-DEGREE GLAZING / SHELLS WITH 90-DEGREE GLAZING

EXTERNAL SOUTH SHELL

	SHELLS WITH 60-DEGREE GLAZING				SHELLS WITH 90-DEGREE GLAZING			
POINT	T1	S	T2	M2	T1	S .	T2	M2
0	20621	-221	131	260	17583	143	74	37
1	7049	-2439	-45	186	6153	-2296	-110	76
2	-2024	-2796	-431	68	-1366	-2742	-594	20
3	-7219	-1984	-778	-46	-5729	-1939	-1039	-59
4	-9039	-598	-945	-109	-7444	-499	-1246	-93
5	-7745	830	-883	-105	-6638	1040	-1140	-56
6	-3284	1786	-627	-52	-2986	2116	-758	25
7	4713	1720	-302	3	4231	2055	-269	76
8	16885	-0	-126	·0	16006	-0	1	0

INTERNAL SHELLS

POINT	T1	S	T2	M2	T1	S	T2	M2
0	21958	-155	268	185	23533	802	190	87
1	7839	-2548	87	168	8506	-2489	41	132
2	-1724	-2994	-325	81	-1150	-3184	-518	59
3	-7304	-2197	-706	-21	-6533	-2303	-1052	-49
4	-9366	-775	-906	-84	-8489	-655	-1315	-108
5	-8155	716	-868	-85	-7438	1085	-1220	-83
6	-3630	1736	-626	-39	-3271	2280	-814	1
7	4550	1714	-305	8	4603	2211	-288	62
8	16980	-0	-130	0	17126	-0	1	0

EXTERNAL NORTH SHELL

POINT	T1	S	T2	M2	T1	S	T2	M2
0	22090	-79	268	180	23411	787	188	86
1	7923	-2489	98	174	8458	-2486	38	131
2	-1636	-2949	-305	95	-1155	-3176	-520	58
3	-7169	-2170	-681	-1	-6517	-2296	-1052	-49
4	-9163	-776	-878	-62	-8470	-653	-1315	-108
5	-7918	678	-844	-65	-7426	1085	-1219	-83
6	-3472	1663	-611	-24	-3270	2277	-814	1
7	4394	1637	-303	16	4596	2210	-288	62
8	16125	-0	-135	0	17119	-0	1	0

L = 26 FT. TABLE NO.N- 7 W = 13 FT.

SHELL DATA:

L = 26.00 FT.
W = 13.00 FT.
T = 2.20 IN.
G = 50.00 LB/FT*2
QA= 150.00 LB/FT
QV= 150.00 LB/FT

FOR 60DEG.GLAZING:
K = 6.75 FT.
R = 10.00 FT.
AL= 11.96 FT.
Ø = 68.53 DEG.

FOR 90DEG.GLAZING:
K = 6.75 FT.
R = 17.50 FT.
AL= 12.11 FT.
Ø = 49.48 DEG.

BEAM DATA:

FOR 60DEG.GLAZING:

B1= 0.75 FT.
D1= 1.75 FT.
B2= 0.75 FT.
D2= 1.50 FT.

EXT.NORTH BEAM:

Z'= 0.91 FT.
Y'= 0.78 FT.
IZ= 0.43 FT*4
IY= 0.31 FT*4

FOR 90DEG.GLAZING:

ALL BEAMS:

B = 1.00 FT.
D = 1.75 FT.

BEAM RESULTS:

FOR 60DEG.GLAZING:

EXT.SOUTH BEAM
F = 3325 LB.
M1= 17792 LB.FT
M2= 3353 LB.FT
S1= -1835 LB.
S2= 226 LB.
INTERNAL BEAMS
F = 1944 LB.
P = 248 LB/FT
M1= 35746 LB.FT
M2= 670 LB.FT
S1= -4087 LB.
S2= -26 LB.
EXT.NORTH BEAM
P = 253 LB/FT
M1= 48080 LB.FT
M2= 11015 LB.FT
S1= -5809 LB.
S2= -1331 LB.

FOR 90DEG.GLAZING:

EXT.SOUTH BEAM
F = -2078 LB.
M1= 33714 LB.FT
M2= 5848 LB.FT
S1= -4286 LB.
S2= 822 LB.
INTERNAL BEAMS
F = -8327 LB.
P = 285 LB/FT
M1= 45913 LB.FT
M2= 8414 LB.FT
S1= -6418 LB.
S2= 1507 LB.
EXT.NORTH BEAM
P = 285 LB/FT
M1= 51253 LB.FT
S1= -6193 LB.

SHELL RESULTS:

	SHELLS WITH 60-DEGREE GLAZING				SHELLS WITH 90-DEGREE GLAZING			
EXTERNAL SOUTH SHELL								
POINT	T1	S	T2	M2	T1	S	T2	M2
0	18762	-402	179	343	17691	251	67	10
1	5845	-2547	-39	223	6167	-2385	-136	72
2	-2341	-2799	-469	55	-1294	-2873	-686	20
3	-6725	-1929	-842	-97	-5577	-2035	-1193	-62
4	-8100	-549	-1014	-179	-7279	-522	-1431	-98
5	-6883	842	-941	-169	-6550	1105	-1312	-55
6	-2980	1777	-664	-94	-3015	2256	-876	35
7	4205	1722	-312	-13	4150	2206	-311	90
8	15676	0	-123	-0	16073	0	5	-0
INTERNAL SHELLS								
POINT	T1	S	T2	M2	T1	S	T2	M2
0	20613	-235	301	275	23759	1006	185	50
1	6857	-2636	87	212	8547	-2561	32	124
2	-2013	-3007	-371	73	-1060	-3322	-599	57
3	-6883	-2149	-784	-72	-6340	-2405	-1203	-53
4	-8519	-714	-991	-157	-8260	-682	-1503	-113
5	-7345	761	-941	-155	-7288	1147	-1396	-83
6	-3298	1768	-671	-86	-3273	2415	-935	9
7	4167	1747	-317	-10	4487	2357	-331	76
8	15999	0	-125	-0	17080	0	4	-0
EXTERNAL NORTH SHELL								
POINT	T1	S	T2	M2	T1	S	T2	M2
0	20723	-206	301	273	23638	990	183	50
1	6899	-2621	91	214	8500	-2559	29	123
2	-2001	-2996	-365	78	-1064	-3314	-601	56
3	-6865	-2140	-776	-65	-6326	-2399	-1204	-54
4	-8471	-711	-982	-149	-8244	-679	-1503	-114
5	-7266	752	-932	-148	-7279	1147	-1396	-83
6	-3228	1745	-666	-80	-3274	2414	-934	9
7	4133	1719	-316	-7	4481	2357	-331	76
8	15687	0	-128	-0	17084	0	4	-0

L = 26 FT. TABLE NO.N- 8 W = 15 FT.

SHELL DATA:

L = 26.00 FT.
W = 15.00 FT.
T = 2.50 IN.
G = 50.00 LB/FT*2
QA= 150.00 LB/FT
QV= 150.00 LB/FT

FOR 60DEG.GLAZING:
K = 7.25 FT.
R = 15.00 FT.
AL= 13.44 FT.
θ = 51.33 DEG.

FOR 90DEG.GLAZING:
K = 7.25 FT.
R = 22.50 FT.
AL= 17.07 FT.
θ = 43.46 DEG.

BEAM DATA:

FOR 60DEG.GLAZING:

B1= 0.75 FT.
D1= 1.75 FT.
B2= 0.75 FT.
D2= 1.50 FT.

EXT.NORTH BEAM:

Z'= 0.91 FT.
Y'= 0.78 FT.
IZ= 0.43 FT*4
IY= 0.31 FT*4

FOR 90DEG.GLAZING:

ALL BEAMS:

B = 1.00 FT.
D = 1.75 FT.

BEAM RESULTS:

FOR 60DEG.GLAZING:

EXT.SOUTH BEAM
F = -1743 LB.
M1= 22221 LB.FT
M2= 3430 LB.FT
S1= -2831 LB.
S2= 477 LB.
INTERNAL BEAMS
F = -2139 LB.
P = 270 LB/FT
M1= 42069 LB.FT
M2= 119 LB.FT
S1= -5290 LB.
S2= 103 LB.
EXT.NORTH BEAM
P = 283 LB/FT
M1= 50700 LB.FT
M2= 12326 LB.FT
S1= -6126 LB.
S2= -1489 LB.

FOR 90DEG.GLAZING:

EXT.SOUTH BEAM
F = -3562 LB.
M1= 35166 LB.FT
M2= 5635 LB.FT
S1= -4620 LB.
S2= 890 LB.
INTERNAL BEAMS
F = -11178 LB.
P = 302 LB/FT
M1= 48209 LB.FT
M2= 8164 LB.FT
S1= -6999 LB.
S2= 1653 LB.
EXT.NORTH BEAM
P = 303 LB/FT
M1= 53146 LB.FT
S1= -6422 LB.

SHELL RESULTS:

SHELLS WITH 60-DEGREE GLAZING / SHELLS WITH 90-DEGREE GLAZING

EXTERNAL SOUTH SHELL

POINT	T1	S	T2	M2	T1	S	T2	M2
0	21219	211	85	195	17954	430	48	-43
1	7212	-2579	-120	156	6164	-2570	-197	70
2	-1723	-3060	-647	41	-1245	-3123	-885	26
3	-6642	-2151	-1126	-80	-5406	-2206	-1519	-64
4	-8355	-580	-1352	-144	-7081	-548	-1819	-105
5	-7242	1051	-1254	-124	-6468	1250	-1672	-54
6	-3159	2160	-879	-45	-3092	2551	-1122	53
7	4545	2089	-398	25	4084	2518	-401	117
8	16923	-0	-131	0	16450	-0	8	0

INTERNAL SHELLS

POINT	T1	S	T2	M2	T1	S	T2	M2
0	22706	258	262	129	24178	1351	168	-17
1	7943	-2754	42	147	8556	-2719	1	115
2	-1583	-3320	-532	58	-1005	-3580	-778	60
3	-6901	-2395	-1063	-58	-6111	-2582	-1524	-57
4	-8808	-748	-1332	-126	-7958	-708	-1894	-121
5	-7695	977	-1259	-113	-7098	1290	-1761	-83
6	-3448	2164	-890	-40	-3298	2701	-1184	27
7	4541	2124	-404	26	4368	2660	-422	103
8	17285	-0	-133	0	17272	-0	8	0

EXTERNAL NORTH SHELL

POINT	T1	S	T2	M2	T1	S	T2	M2
0	22784	324	262	116	24070	1336	165	-19
1	7973	-2698	54	147	8513	-2714	-3	114
2	-1555	-3269	-509	70	-1012	-3570	-779	60
3	-6822	-2355	-1031	-38	-6100	-2573	-1523	-56
4	-8637	-733	-1294	-101	-7939	-703	-1891	-120
5	-7443	948	-1222	-89	-7077	1290	-1757	-81
6	-3234	2086	-864	-21	-3281	2696	-1181	29
7	4433	2028	-397	36	4363	2652	-421	103
8	16339	-0	-138	-0	17213	-0	8	-0

L = 28 FT. TABLE NO.N- 9 W = 12 FT.

SHELL DATA:

L = 28.00 FT.
W = 12.00 FT.
T = 2.50 IN.
G = 50.00 LB/FT*2
QA= 150.00 LB/FT
QV= 150.00 LB/FT

FOR 60DEG.GLAZING:
K = 6.25 FT.
R = 10.00 FT.
AL= 10.93 FT.
θ = 62.63 DEG.

FOR 90DEG.GLAZING:
K = 6.25 FT.
R = 15.00 FT.
AL= 14.04 FT.
θ = 53.62 DEG.

BEAM DATA:

FOR 60DEG.GLAZING:

B1= 0.75 FT.
D1= 1.75 FT.
B2= 0.75 FT.
D2= 1.50 FT.

EXT.NORTH BEAM:

Z'= 0.91 FT.
Y'= 0.78 FT.
IZ= 0.43 FT*4
IY= 0.31 FT*4

FOR 90DEG.GLAZING:

ALL BEAMS:

B = 1.00 FT.
D = 1.75 FT.

BEAM RESULTS:

FOR 60DEG.GLAZING:

EXT.SOUTH BEAM
F = 3223 LB.
M1= 23396 LB.FT
M2= 3871 LB.FT
S1= -2334 LB.
S2= 260 LB.
INTERNAL BEAMS
F = 2358 LB.
P = 257 LB/FT
M1= 44877 LB.FT
M2= 375 LB.FT
S1= -4766 LB.
S2= -82 LB.
EXT.NORTH BEAM
P = 269 LB/FT
M1= 56768 LB.FT
M2= 13590 LB.FT
S1= -6369 LB.
S2= -1525 LB.

FOR 90DEG.GLAZING:

EXT.SOUTH BEAM
F = -795 LB.
M1= 38332 LB.FT
M2= 6704 LB.FT
S1= -4368 LB.
S2= 781 LB.
INTERNAL BEAMS
F = -6793 LB.
P = 272 LB/FT
M1= 51650 LB.FT
M2= 9534 LB.FT
S1= -6447 LB.
S2= 1431 LB.
EXT.NORTH BEAM
P = 271 LB/FT
M1= 57621 LB.FT
S1= -6465 LB.

SHELL RESULTS:

	SHELLS WITH 60-DEGREE GLAZING				SHELLS WITH 90-DEGREE GLAZING			
EXTERNAL SOUTH SHELL								
POINT	T1	S	T2	M2	T1	S	T2	M2
0	23662	-362	141	316	20339	89	80	68
1	8027	-2722	-51	223	7121	-2534	-117	92
2	-2548	-3082	-449	86	-1682	-3004	-613	21
3	-8657	-2169	-805	-42	-6826	-2112	-1065	-68
4	-10785	-630	-973	-115	-8823	-522	-1273	-107
5	-9187	948	-905	-114	-7797	1165	-1161	-70
6	-3809	1994	-640	-59	-3417	2329	-769	16
7	5709	1905	-305	-1	5059	2242	-271	71
8	20001	-0	-125	0	18672	-0	1	0
INTERNAL SHELLS								
POINT	T1	S	T2	M2	T1	S	T2	M2
0	25273	-265	272	242	27083	762	196	140
1	8976	-2821	77	205	9802	-2760	31	162
2	-2171	-3280	-344	100	-1451	-3493	-543	67
3	-8720	-2390	-732	-16	-7790	-2511	-1086	-56
4	-11125	-818	-933	-88	-10084	-690	-1352	-123
5	-9630	823	-889	-92	-8773	1224	-1249	-100
6	-4199	1935	-638	-45	-3764	2523	-831	-12
7	5501	1893	-308	5	5534	2426	-293	56
8	20058	-0	-128	-0	20085	-0	1	-0
EXTERNAL NORTH SHELL								
POINT	T1	S	T2	M2	T1	S	T2	M2
0	25427	-178	271	241	26933	744	194	139
1	9083	-2754	89	214	9743	-2758	27	160
2	-2053	-3229	-324	118	-1457	-3484	-546	66
3	-8545	-2361	-706	8	-7772	-2504	-1087	-57
4	-10879	-821	-904	-62	-10064	-687	-1352	-124
5	-9357	778	-864	-70	-8764	1224	-1249	-100
6	-4031	1854	-623	-29	-3768	2522	-831	-13
7	5306	1810	-306	14	5527	2426	-293	55
8	19087	-0	-134	0	20098	-0	1	0

L = 28 FT. TABLE NO.N- 10 W = 14 FT.

SHELL DATA:

L = 28.00 FT.
W = 14.00 FT.
T = 2.50 IN.
G = 50.00 LB/FT*2
QA= 150.00 LB/FT
QV= 150.00 LB/FT

FOR 60DEG.GLAZING:
K = 6.75 FT.
R = 12.50 FT.
AL= 12.66 FT.
θ = 58.04 DEG.

FOR 90DEG.GLAZING:
K = 6.75 FT.
R = 20.00 FT.
AL= 15.96 FT.
θ = 45.73 DEG.

BEAM DATA:

FOR 60DEG.GLAZING:

B1= 0.75 FT.
D1= 1.75 FT.
B2= 0.75 FT.
D2= 1.50 FT.

EXT.NORTH BEAM:

Z'= 0.91 FT.
Y'= 0.78 FT.
IZ= 0.43 FT*4
IY= 0.31 FT*4

FOR 90DEG.GLAZING:

ALL BEAMS:

B = 1.00 FT.
D = 1.75 FT.

BEAM RESULTS:

FOR 60DEG.GLAZING:

EXT.SOUTH BEAM
F = 1704 LB.
M1= 23720 LB.FT
M2= 3750 LB.FT
S1= -2502 LB.
S2= 323 LB.
INTERNAL BEAMS
F = 934 LB.
P = 255 LB/FT
M1= 45455 LB.FT
M2= 228 LB.FT
S1= -4980 LB.
S2= -29 LB.
EXT.NORTH BEAM
P = 267 LB/FT
M1= 56986 LB.FT
M2= 13482 LB.FT
S1= -6394 LB.
S2= -1513 LB.

FOR 90DEG.GLAZING:

EXT.SOUTH BEAM
F = -2864 LB.
M1= 40520 LB.FT
M2= 6450 LB.FT
S1= -4819 LB.
S2= 875 LB.
INTERNAL BEAMS
F = -10685 LB.
P = 290 LB/FT
M1= 54994 LB.FT
M2= 9248 LB.FT
S1= -7208 LB.
S2= 1624 LB.
EXT.NORTH BEAM
P = 292 LB/FT
M1= 50108 LB.FT
S1= -6744 LB.

SHELL RESULTS:

EXTERNAL SOUTH SHELL

POINT	T1 (60°)	S (60°)	T2 (60°)	M2 (60°)	T1 (90°)	S (90°)	T2 (90°)	M2 (90°)
	SHELLS WITH 60-DEGREE GLAZING				SHELLS WITH 90-DEGREE GLAZING			
0	23441	-191	125	327	20804	321	53	-1
1	7695	-2866	-106	221	7276	-2723	-181	77
2	-2519	-3251	-602	56	-1542	-3287	-809	22
3	-8176	-2241	-1040	-98	-6632	-2310	-1385	-70
4	-10065	-570	-1241	-181	-8647	-546	-1652	-110
5	-8576	1135	-1144	-168	-7740	1345	-1510	-63
6	-3586	2270	-798	-86	-3487	2671	-1005	39
7	5531	2165	-361	-3	5042	2588	-356	100
8	19817	-0	-126	0	19125	-0	5	-0

INTERNAL SHELLS

POINT	T1	S	T2	M2	T1	S	T2	M2
0	25175	-105	270	264	27780	1199	169	52
1	8581	-3010	32	212	10046	-2900	-4	137
2	-2296	-3491	-500	73	-1234	-3780	-717	63
3	-8409	-2476	-982	-75	-7484	-2722	-1400	-60
4	-10531	-740	-1219	-161	-9768	-724	-1734	-128
5	-9064	1052	-1144	-155	-8595	1395	-1605	-94
6	-3916	2262	-805	-79	-3790	2855	-1072	10
7	5496	2192	-367	0	5436	2763	-379	84
8	20165	-0	-128	0	20314	-0	5	-0

EXTERNAL NORTH SHELL

POINT	T1	S	T2	M2	T1	S	T2	M2
0	25288	-32	270	255	27646	1186	166	49
1	8640	-2931	43	215	9991	-2892	-7	135
2	-2238	-3442	-480	87	-1242	-3765	-717	63
3	-8296	-2442	-954	-53	-7464	-2709	-1396	-58
4	-10333	-733	-1187	-136	-9733	-717	-1729	-125
5	-8804	1018	-1114	-131	-8554	1393	-1599	-91
6	-3719	2185	-786	-61	-3759	2845	-1067	13
7	5355	2103	-363	9	5424	2750	-377	86
8	19215	-0	-133	-0	20194	-0	5	-0

L = 28 FT. TABLE NO.N- 11 W = 16 FT.

SHELL DATA:

L = 28.00 FT.
W = 16.00 FT.
T = 2.50 IN.
G = 50.00 LB/FT*2
QA= 150.00 LB/FT
QV= 150.00 LB/FT

FOR 60DEG.GLAZING:
K = 7.75 FT.
R = 15.00 FT.
AL= 14.41 FT.
θ = 55.03 DEG.

FOR 90DEG.GLAZING:
K = 7.75 FT.
R = 25.00 FT.
AL= 18.18 FT.
θ = 41.66 DEG.

BEAM DATA:

FOR 60DEG.GLAZING:

B1= 0.75 FT.
D1= 1.75 FT.
B2= 0.75 FT.
D2= 1.50 FT.

EXT.NORTH BEAM:

Z'= 0.91 FT.
Y'= 0.78 FT.
IZ= 0.43 FT*4
IY= 0.31 FT*4

FOR 90DEG.GLAZING:

ALL BEAMS:

B = 1.00 FT.
D = 1.75 FT.

BEAM RESULTS:

FOR 60DEG.GLAZING:

EXT.SOUTH BEAM
F = -110 LB.
M1= 23112 LB.FT
M2= 3878 LB.FT
S1= -2590 LB.
S2= 421 LB.
INTERNAL BEAMS
F = -1532 LB.
P = 259 LB/FT
M1= 45225 LB.FT
M2= 444 LB.FT
S1= -5203 LB.
S2= 104 LB.
EXT.NORTH BEAM
P = 267 LB/FT
M1= 57841 LB.FT
M2= 13508 LB.FT
S1= -6490 LB.
S2= -1516 LB.

FOR 90DEG.GLAZING:

EXT.SOUTH BEAM
F = -4571 LB.
M1= 40475 LB.FT
M2= 6619 LB.FT
S1= -4984 LB.
S2= 992 LB.
INTERNAL BEAMS
F = -13895 LB.
P = 305 LB/FT
M1= 55609 LB.FT
M2= 9572 LB.FT
S1= -7595 LB.
S2= 1845 LB.
EXT.NORTH BEAM
P = 305 LB/FT
M1= 62246 LB.FT
S1= -6984 LB.

SHELL RESULTS:

EXTERNAL SOUTH SHELL

	SHELLS WITH 60-DEGREE GLAZING				SHELLS WITH 90-DEGREE GLAZING			
POINT	T1	S	T2	M2	T1	S	T2	M2
0	22845	12	122	328	20704	513	52	-31
1	7262	-2917	-133	224	7068	-2902	-226	88
2	-2347	-3325	-713	42	-1456	-3524	-1000	30
3	-7390	-2275	-1223	-132	-6206	-2479	-1711	-80
4	-9000	-568	-1455	-221	-8098	-601	-2045	-130
5	-7734	1170	-1340	-197	-7387	1431	-1878	-74
6	-3398	2352	-933	-93	-3531	2900	-1261	51
7	4915	2276	-414	7	4689	2861	-450	127
8	18671	0	-128	0	18914	0	11	0

INTERNAL SHELLS

	SHELLS WITH 60-DEGREE GLAZING				SHELLS WITH 90-DEGREE GLAZING			
POINT	T1	S	T2	M2	T1	S	T2	M2
0	24933	172	284	262	27880	1559	176	6
1	8289	-3067	26	219	9800	-3071	-14	141
2	-2118	-3598	-600	63	-1203	-4038	-889	68
3	-7678	-2537	-1166	-108	-7027	-2896	-1724	-74
4	-9527	-744	-1442	-204	-9100	-771	-2133	-152
5	-8251	1104	-1352	-189	-8094	1485	-1980	-108
6	-3709	2372	-949	-90	-3748	3074	-1331	20
7	4951	2327	-421	8	5023	3021	-474	110
8	19144	0	-128	0	19846	0	10	0

EXTERNAL NORTH SHELL

	SHELLS WITH 60-DEGREE GLAZING				SHELLS WITH 90-DEGREE GLAZING			
POINT	T1	S	T2	M2	T1	S	T2	M2
0	25018	215	283	252	27736	1537	174	5
1	8310	-3034	33	218	9747	-3068	-18	140
2	-2117	-3566	-587	71	-1206	-4029	-891	67
3	-7645	-2509	-1147	-93	-7010	-2889	-1724	-74
4	-9422	-729	-1418	-186	-9083	-770	-2133	-152
5	-8073	1090	-1328	-171	-8087	1483	-1979	-108
6	-3541	2321	-931	-76	-3752	3072	-1330	20
7	4890	2259	-417	15	5016	3021	-474	110
8	18452	0	-132	0	19856	0	10	0

L = 30 FT. TABLE NO.N- 12 W = 13 FT.

```
SHELL DATA:                 BEAM RESULTS:

  L =  30.00 FT.            FOR 60DEG.GLAZING:
  W =  13.00 FT.
  T =   2.50 IN.            EXT.SOUTH BEAM
  G =  50.00 LB/FT*2          F =      4467 LB.
  QA= 150.00 LB/FT            M1=     30157 LB.FT
  QV= 150.00 LB/FT            M2=      4039 LB.FT
                              S1=     -2733 LB.
 FOR 60DEG.GLAZING:           S2=       210 LB.
  K =   7.00 FT.            INTERNAL BEAMS
  R =  10.00 FT.              F =      3289 LB.
  AL=  11.97 FT.              P =       269 LB/FT
  θ =  68.59 DEG.             M1=     54668 LB.FT
                              M2=       610 LB.FT
 FOR 90DEG.GLAZING:           S1=     -5347 LB.
  K =   7.00 FT.              S2=       -91 LB.
  R =  15.00 FT.            EXT.NORTH BEAM
  AL=  15.44 FT.              P =       275 LB/FT
  θ =  58.97 DEG.             M1=     72766 LB.FT
                              M2=     15948 LB.FT
BEAM DATA:                    S1=     -7620 LB.
                              S2=     -1670 LB.
 FOR 60DEG.GLAZING:

  B1= 0.75 FT.             FOR 90DEG.GLAZING:
  D1= 2.00 FT.
  B2= 0.75 FT.              EXT.SOUTH BEAM
  D2= 1.75 FT.                F =      -193 LB.
                              M1=     48316 LB.FT
  EXT.NORTH BEAM:             M2=      7146 LB.FT
                              S1=     -5070 LB.
  Z'= 1.04 FT.                S2=       743 LB.
  Y'= 0.80 FT.             INTERNAL BEAMS
  IZ= 0.64 FT*4               F =     -6390 LB.
  IY= 0.45 FT*4               P =       281 LB/FT
                              M1=     65076 LB.FT
 FOR 90DEG.GLAZING:           M2=     10147 LB.FT
                              S1=     -7471 LB.
  ALL BEAMS:                  S2=      1377 LB.
                            EXT.NORTH BEAM
  B = 1.00 FT.                P =       281 LB/FT
  D = 2.00 FT.                M1=     72067 LB.FT
                              S1=     -7547 LB.
```

SHELL RESULTS:

EXTERNAL SOUTH SHELL

	SHELLS WITH 60-DEGREE GLAZING				SHELLS WITH 90-DEGREE GLAZING			
POINT	T1	S	T2	M2	T1	S	T2	M2
0	22250	-468	152	338	19545	20	77	103
1	7335	-2717	-44	234	6649	-2542	-122	115
2	-2599	-3027	-437	82	-1740	-2970	-618	23
3	-8221	-2127	-785	-59	-6521	-2082	-1067	-85
4	-10111	-645	-949	-136	-8347	-533	-1272	-131
5	-8590	864	-884	-130	-7412	1107	-1160	-84
6	-3618	1867	-627	-64	-3340	2252	-771	20
7	5204	1797	-304	3	4719	2189	-273	87
8	18561	-0	-131	-0	17909	-0	-2	-0

INTERNAL SHELLS

POINT	T1	S	T2	M2	T1	S	T2	M2
0	23913	-344	299	226	26012	669	188	169
1	8416	-2809	105	203	9114	-2765	18	182
2	-2108	-3242	-314	97	-1613	-3430	-555	66
3	-8236	-2378	-699	-25	-7458	-2441	-1090	-78
4	-10471	-864	-901	-100	-9481	-672	-1347	-153
5	-9095	716	-864	-101	-8231	1171	-1241	-119
6	-4071	1797	-625	-45	-3592	2427	-825	-11
7	4961	1783	-308	11	5143	2348	-292	70
8	18631	-0	-136	0	19066	-0	-2	-0

EXTERNAL NORTH SHELL

POINT	T1	S	T2	M2	T1	S	T2	M2
0	24116	-297	299	225	25847	651	185	167
1	8511	-2784	110	207	9052	-2761	14	181
2	-2063	-3227	-305	106	-1615	-3420	-557	65
3	-8191	-2369	-689	-13	-7435	-2434	-1090	-79
4	-10394	-864	-889	-88	-9457	-670	-1346	-153
5	-8987	700	-854	-90	-8219	1170	-1240	-119
6	-3987	1765	-619	-37	-3595	2425	-825	-11
7	4896	1748	-307	15	5134	2347	-292	70
8	18202	-0	-139	0	19071	-0	-2	-0

L = 30 FT. TABLE NO.N- 13 W = 15 FT.

SHELL DATA:

L = 30.00 FT.
W = 15.00 FT.
T = 2.50 IN.
G = 50.00 LB/FT*2
QA= 150.00 LB/FT
QV= 150.00 LB/FT

FOR 60DEG.GLAZING:
K = 7.50 FT.
R = 12.50 FT.
AL= 13.66 FT.
θ = 62.61 DEG.

FOR 90DEG.GLAZING:
K = 7.50 FT.
R = 22.50 FT.
AL= 17.19 FT.
θ = 43.76 DEG.

BEAM DATA:

FOR 60DEG.GLAZING:

B1= 0.75 FT.
D1= 2.00 FT.
B2= 0.75 FT.
D2= 1.75 FT.

EXT.NORTH BEAM:

Z'= 1.04 FT.
Y'= 0.80 FT.
IZ= 0.64 FT*4
IY= 0.45 FT*4

FOR 90DEG.GLAZING:

ALL BEAMS:

B = 1.00 FT.
D = 2.00 FT.

BEAM RESULTS:

FOR 60DEG.GLAZING:

EXT.SOUTH BEAM
F = 1929 LB.
M1= 31492 LB.FT
M2= 3978 LB.FT
S1= -3107 LB.
S2= 315 LB.
INTERNAL BEAMS
F = 1026 LB.
P = 274 LB/FT
M1= 56732 LB.FT
M2= 444 LB.FT
S1= -5807 LB.
S2= -10 LB.
EXT.NORTH BEAM
P = 280 LB/FT
M1= 73799 LB.FT
M2= 16277 LB.FT
S1= -7728 LB.
S2= -1704 LB.

FOR 90DEG.GLAZING:

EXT.SOUTH BEAM
F = -2859 LB.
M1= 55175 LB.FT
M2= 6545 LB.FT
S1= -6070 LB.
S2= 826 LB.
INTERNAL BEAMS
F = -11759 LB.
P = 316 LB/FT
M1= 74438 LB.FT
M2= 9399 LB.FT
S1= -9017 LB.
S2= 1589 LB.
EXT.NORTH BEAM
P = 320 LB/FT
M1= 77093 LB.FT
S1= -8073 LB.

SHELL RESULTS:

EXTERNAL SOUTH SHELL

	SHELLS WITH 60-DEGREE GLAZING				SHELLS WITH 90-DEGREE GLAZING			
POINT	T1	S	T2	M2	T1	S	T2	M2
0	22497	-202	118	320	21035	299	11	-78
1	7324	-2780	-98	222	7485	-2813	-231	48
2	-2406	-3146	-571	56	-1517	-3408	-881	21
3	-7716	-2182	-989	-101	-6840	-2406	-1479	-54
4	-9461	-598	-1181	-182	-9013	-566	-1756	-82
5	-8080	1016	-1092	-163	-8098	1419	-1601	-24
6	-3457	2100	-768	-74	-3621	2811	-1063	85
7	5050	2022	-357	11	5316	2714	-373	138
8	18492	-0	-135	-0	19854	0	11	-0

INTERNAL SHELLS

POINT	T1	S	T2	M2	T1	S	T2	M2
0	24274	-107	291	220	27912	1231	115	-45
1	8329	-2933	68	201	10309	-2938	-53	99
2	-2092	-3415	-445	76	-1099	-3870	-779	60
3	-7928	-2458	-913	-68	-7587	-2811	-1479	-42
4	-9972	-806	-1150	-153	-10068	-756	-1824	-96
5	-8645	908	-1089	-143	-8931	1449	-1686	-51
6	-3858	2081	-775	-63	-3941	2975	-1125	59
7	4981	2047	-363	15	5662	2877	-396	123
8	18856	-0	-138	-0	20971	0	10	0

EXTERNAL NORTH SHELL

POINT	T1	S	T2	M2	T1	S	T2	M2
0	24408	-62	290	214	27820	1234	112	-52
1	8383	-2904	74	203	10262	-2919	-53	97
2	-2070	-3392	-434	84	-1115	-3845	-774	63
3	-7888	-2440	-899	-56	-7569	-2786	-1469	-35
4	-9879	-799	-1133	-138	-10007	-740	-1809	-86
5	-8502	893	-1073	-129	-8829	1446	-1669	-40
6	-3733	2040	-764	-52	-3843	2948	-1111	67
7	4918	1998	-360	21	5641	2839	-389	128
8	18302	-0	-141	-0	20605	0	11	0

L = 30 FT. TABLE NO.N- 14 W = 17 FT.

SHELL DATA:

L = 30.00 FT.
W = 17.00 FT.
T = 2.50 IN.
G = 50.00 LB/FT*2
QA= 150.00 LB/FT
QV= 150.00 LB/FT

FOR 60DEG.GLAZING
K = 8.00 FT.
R = 15.00 FT.
AL= 15.40 FT.
θ = 58.82 DEG.

FOR 90DEG.GLAZING:
K = 8.00 FT.
R = 25.00 FT.
AL= 19.26 FT.
θ = 44.14 DEG.

BEAM DATA:

FOR 60DEG.GLAZING:

B1= 0.75 FT.
D1= 2.00 FT.
B2= 0.75 FT.
D2= 1.75 FT.

EXT.NORTH BEAM:

Z'= 1.04 FT.
Y'= 0.80 FT.
IZ= 0.64 FT*4
IY= 0.45 FT*4

FOR 90DEG.GLAZING:

ALL BEAMS:

B = 1.00 FT.
D = 2.00 FT.

BEAM RESULTS:

FOR 60DEG.GLAZING:

EXT.SOUTH BEAM
F = -561 LB.
M1= 32247 LB.FT
M2= 3946 LB.FT
S1= -3416 LB.
S2= 416 LB.
INTERNAL BEAMS
F = -1603 LB.
P = 283 LB/FT
M1= 58409 LB.FT
M2= 360 LB.FT
S1= -6263 LB.
S2= 90 LB.
EXT.NORTH BEAM
P = 290 LB/FT
M1= 75190 LB.FT
M2= 16812 LB.FT
S1= -7874 LB.
S2= -1761 LB.

FOR 90DEG.GLAZING:

EXT.SOUTH BEAM
F = -4065 LB.
M1= 53723 LB.FT
M2= 6551 LB.FT
S1= -6045 LB.
S2= 892 LB.
INTERNAL BEAMS
F = -13723 LB.
P = 319 LB/FT
M1= 73225 LB.FT
M2= 9436 LB.FT
S1= -9097 LB.
S2= 1698 LB.
EXT.NORTH BEAM
P = 322 LB/FT
M1= 77777 LB.FT
S1= -8145 LB.

SHELL RESULTS:

SHELLS WITH 60-DEGREE GLAZING / SHELLS WITH 90-DEGREE GLAZING

EXTERNAL SOUTH SHELL

POINT	T1	S	T2	M2	T1	S	T2	M2
0	22471	59	94	302	20459	426	19	-46
1	7158	-2816	-137	218	6956	-2909	-258	89
2	-2229	-3225	-688	44	-1511	-3504	-1015	31
3	-7114	-2219	-1176	-128	-6243	-2460	-1707	-83
4	-8675	-579	-1399	-215	-8153	-591	-2030	-133
5	-7489	1095	-1293	-185	-7468	1436	-1861	-66
6	-3351	2243	-906	-74	-3585	2908	-1249	72
7	4674	2186	-412	25	4759	2869	-446	152
8	18057	-0	-142	-0	19154	0	7	-0

INTERNAL SHELLS

POINT	T1	S	T2	M2	T1	S	T2	M2
0	24515	168	290	211	27351	1437	126	-18
1	8181	-3013	46	208	9580	-3052	-64	139
2	-2024	-3542	-560	68	-1263	-3975	-910	68
3	-7461	-2517	-1112	-99	-7014	-2842	-1715	-76
4	-9278	-776	-1385	-194	-9081	-746	-2107	-153
5	-8074	1022	-1306	-175	-8107	1485	-1950	-99
6	-3699	2267	-924	-71	-3771	3062	-1309	42
7	4717	2243	-421	25	5061	3010	-467	136
8	18583	-0	-143	-0	19970	0	7	-0

EXTERNAL NORTH SHELL

POINT	T1	S	T2	M2	T1	S	T2	M2
0	24590	198	289	202	27241	1424	124	-22
1	8195	-2991	51	207	9528	-3045	-67	137
2	-2033	-3520	-551	75	-1283	-3959	-910	69
3	-7449	-2494	-1098	-87	-7010	-2824	-1711	-73
4	-9204	-762	-1368	-179	-9051	-732	-2098	-148
5	-7931	1014	-1288	-160	-8051	1487	-1939	-93
6	-3554	2228	-910	-60	-3708	3049	-1300	48
7	4680	2189	-417	31	5060	2986	-463	139
8	18012	-0	-146	-0	19754	0	7	-0

L = 32 FT. TABLE NO.N- 15 W = 14 FT.

SHELL DATA:

L = 32.00 FT.
W = 14.00 FT.
T = 2.50 IN.
G = 50.00 LB/FT*2
QA= 150.00 LB/FT
QV= 150.00 LB/FT

FOR 60DEG.GLAZING:
K = 7.00 FT.
R = 10.00 FT.
AL= 13.03 FT.
Θ = 74.63 DEG.

FOR 90DEG.GLAZING:
K = 7.00 FT.
R = 17.50 FT.
AL= 16.23 FT.
Θ = 53.13 DEG.

BEAM DATA:

FOR 60DEG.GLAZING:

B1= 0.75 FT.
D1= 2.00 FT.
B2= 0.75 FT.
D2= 1.75 FT.

EXT.NORTH BEAM:

Z'= 1.04 FT.
Y'= 0.80 FT.
IZ= 0.64 FT*4
IY= 0.45 FT*4

FOR 90DEG.GLAZING:

ALL BEAMS:

B = 1.00 FT.
D = 2.00 FT.

BEAM RESULTS:

FOR 60DEG.GLAZING:

EXT.SOUTH BEAM
F = 8099 LB.
M1= 30487 LB.FT
M2= 4144 LB.FT
S1= -2280 LB.
S2= 68 LB.
INTERNAL BEAMS
F = 5947 LB.
P = 258 LB/FT
M1= 57511 LB.FT
M2= 745 LB.FT
S1= -5025 LB.
S2= -174 LB.
EXT.NORTH BEAM
P = 258 LB/FT
M1= 80833 LB.FT
M2= 17015 LB.FT
S1= -7936 LB.
S2= -1670 LB.

FOR 90DEG.GLAZING:

EXT.SOUTH BEAM
F = -1022 LB.
M1= 57956 LB.FT
M2= 7638 LB.FT
S1= -5779 LB.
S2= 785 LB.
INTERNAL BEAMS
F = -8933 LB.
P = 289 LB/FT
M1= 77771 LB.FT
M2= 10823 LB.FT
S1= -8498 LB.
S2= 1482 LB.
EXT.NORTH BEAM
P = 291 LB/FT
M1= 83328 LB.FT
S1= -8181 LB.

SHELL RESULTS:

EXTERNAL SOUTH SHELL

	SHELLS WITH 60-DEGREE GLAZING				SHELLS WITH 90-DEGREE GLAZING			
POINT	T1	S	T2	M2	T1	S	T2	M2
0	23139	-795	204	492	22778	100	51	76
1	6896	-3113	-51	314	7861	-2857	-174	111
2	-3540	-3310	-504	82	-1971	-3368	-738	23
3	-9152	-2238	-887	-124	-7656	-2347	-1249	-89
4	-10834	-592	-1058	-233	-9851	-549	-1481	-136
5	-9074	1044	-974	-224	-8718	1357	-1346	-83
6	-3775	2121	-681	-128	-3833	2675	-891	30
7	5744	2027	-317	-24	5730	2573	-315	100
8	20618	0	-129	0	21219	0	-0	0

INTERNAL SHELLS

POINT	T1	S	T2	M2	T1	S	T2	M2
0	25617	-584	328	408	30137	877	156	151
1	8270	-3207	80	301	10750	-3068	-27	185
2	-3074	-3548	-401	105	-1727	-3863	-670	70
3	-9336	-2495	-825	-90	-8656	-2748	-1274	-80
4	-11373	-788	-1033	-204	-11124	-711	-1562	-158
5	-9678	943	-973	-204	-9670	1424	-1436	-120
6	-4197	2106	-689	-117	-4147	2872	-953	-4
7	5680	2052	-324	-20	6211	2756	-336	81
8	21007	0	-132	0	22583	0	-0	0

EXTERNAL NORTH SHELL

POINT	T1	S	T2	M2	T1	S	T2	M2
0	25761	-573	329	412	29973	869	153	146
1	8322	-3211	81	303	10682	-3053	-30	182
2	-3076	-3556	-401	107	-1734	-3842	-669	71
3	-9362	-2500	-826	-89	-8625	-2729	-1269	-76
4	-11401	-789	-1035	-204	-11066	-701	-1554	-152
5	-9695	946	-974	-204	-9599	1421	-1427	-114
6	-4198	2110	-690	-118	-4091	2855	-946	1
7	5694	2055	-324	-20	6188	2734	-333	84
8	21027	0	-132	0	22364	0	-0	0

L = 32 FT. TABLE NO.N- 16 W = 16 FT.

SHELL DATA:

```
   L =  32.00 FT.
   W =  16.00 FT.
   T =   2.50 IN.
   G =  50.00 LB/FT*2
  QA= 150.00 LB/FT
  QV= 150.00 LB/FT

 FOR 60DEG.GLAZING:
   K =   8.00 FT.
   R =  12.50 FT.
  AL=  14.69 FT.
   θ =  67.32 DEG.

 FOR 90DEG.GLAZING:
   K =   8.00 FT.
   R =  22.50 FT.
  AL=  18.40 FT.
   θ =  46.85 DEG.
```

BEAM DATA:

```
 FOR 60DEG.GLAZING:

  B1= 0.75 FT.
  D1= 2.00 FT.
  B2= 0.75 FT.
  D2= 1.75 FT.

  EXT.NORTH BEAM:

  Z'= 1.04 FT.
  Y'= 0.80 FT.
  IZ= 0.64 FT*4
  IY= 0.45 FT*4

 FOR 90DEG.GLAZING:

  ALL BEAMS:

  B = 1.00 FT.
  D = 2.00 FT.
```

BEAM RESULTS:

```
FOR 60DEG.GLAZING:

EXT.SOUTH BEAM
 F =     4617 LB.
 M1=    31599 LB.FT
 M2=     4398 LB.FT
 S1=    -2691 LB.
 S2=      233 LB.
INTERNAL BEAMS
 F =     2247 LB.
 P =      264 LB/FT
 M1=    59494 LB.FT
 M2=      855 LB.FT
 S1=    -5592 LB.
 S2=      -18 LB.
EXT.NORTH BEAM
 P =      265 LB/FT
 M1=    82731 LB.FT
 M2=    17501 LB.FT
 S1=    -8122 LB.
 S2=    -1718 LB.

FOR 90DEG.GLAZING:

EXT.SOUTH BEAM
 F =    -3272 LB.
 M1=    59050 LB.FT
 M2=     7775 LB.FT
 S1=    -6110 LB.
 S2=      914 LB.
INTERNAL BEAMS
 F =   -13178 LB.
 P =      307 LB/FT
 M1=    80150 LB.FT
 M2=    11117 LB.FT
 S1=    -9152 LB.
 S2=     1725 LB.
EXT.NORTH BEAM
 P =      308 LB/FT
 M1=    86697 LB.FT
 S1=    -8511 LB.
```

SHELL RESULTS:

EXTERNAL SOUTH SHELL

	SHELLS WITH 60-DEGREE GLAZING				SHELLS WITH 90-DEGREE GLAZING			
POINT	T1	S	T2	M2	T1	S	T2	M2
0	23514	-453	169	460	22972	321	37	21
1	7046	-3103	-95	293	7877	-3046	-227	107
2	-3107	-3373	-615	57	-1793	-3643	-927	26
3	-8343	-2278	-1060	-156	-7276	-2551	-1565	-97
4	-9884	-586	-1256	-264	-9438	-606	-1858	-149
5	-8367	1103	-1155	-241	-8503	1480	-1697	-85
6	-3627	2237	-806	-125	-3921	2960	-1132	51
7	5266	2162	-367	-8	5494	2887	-402	133
8	19819	-0	-132	-0	21290	-0	7	-0

INTERNAL SHELLS

POINT	T1	S	T2	M2	T1	S	T2	M2
0	26052	-221	321	362	30622	1294	148	75
1	8432	-3221	64	279	10830	-3228	-47	171
2	-2675	-3652	-495	82	-1525	-4155	-837	69
3	-8590	-2568	-992	-121	-8204	-2965	-1584	-90
4	-10485	-796	-1235	-237	-10584	-773	-1944	-173
5	-9005	1010	-1161	-224	-9323	1542	-1793	-124
6	-4041	2242	-819	-117	-4173	3146	-1197	16
7	5263	2207	-375	-5	5897	3057	-424	114
8	20328	-0	-134	0	22404	0	6	-0

EXTERNAL NORTH SHELL

POINT	T1	S	T2	M2	T1	S	T2	M2
0	26152	-208	322	363	30464	1278	145	72
1	8467	-3220	65	280	10764	-3219	-51	170
2	-2676	-3654	-493	84	-1537	-4139	-837	69
3	-8600	-2569	-991	-120	-8186	-2950	-1581	-88
4	-10488	-795	-1233	-235	-10547	-763	-1938	-170
5	-8993	1010	-1159	-223	-9274	1542	-1786	-119
6	-4021	2239	-818	-116	-4132	3135	-1192	20
7	5263	2202	-375	-5	5887	3041	-422	116
8	20258	-0	-135	-0	22252	-0	6	-0

TABLE NO.N- 17

L = 32 FT. W = 18 FT.

SHELL DATA:

L = 32.00 FT.
W = 18.00 FT.
T = 2.50 IN.
G = 50.00 LB/FT*2
QA= 150.00 LB/FT
QV= 150.00 LB/FT

FOR 60DEG.GLAZING:
K = 8.50 FT.
R = 20.00 FT.
AL= 16.02 FT.
θ = 45.90 DEG.

FOR 90DEG.GLAZING:
K = 8.50 FT.
R = 30.00 FT.
AL= 20.29 FT.
θ = 38.75 DEG.

BEAM DATA:

FOR 60DEG.GLAZING:

B1= 0.75 FT.
D1= 2.00 FT.
B2= 0.75 FT.
D2= 1.75 FT.

EXT.NORTH BEAM:

Z'= 1.04 FT.
Y'= 0.80 FT.
IZ= 0.64 FT*4
IY= 0.45 FT*4

FOR 90DEG.GLAZING:

ALL BEAMS:

B = 1.00 FT.
D = 2.00 FT.

BEAM RESULTS:

FOR 60DEG.GLAZING:

EXT.SOUTH BEAM
F = -3748 LB.
M1= 42037 LB.FT
M2= 4456 LB.FT
S1= -4436 LB.
S2= 560 LB.
INTERNAL BEAMS
F = -4313 LB.
P = 297 LB/FT
M1= 73886 LB.FT
M2= -137 LB.FT
S1= -7657 LB.
S2= 138 LB.
EXT.NORTH BEAM
P = 310 LB/FT
M1= 88449 LB.FT
M2= 20497 LB.FT
S1= -8683 LB.
S2= -2012 LB.

FOR 90DEG.GLAZING:

EXT.SOUTH BEAM
F = -5109 LB.
M1= 63527 LB.FT
M2= 7373 LB.FT
S1= -6732 LB.
S2= 968 LB.
INTERNAL BEAMS
F = -17305 LB.
P = 332 LB/FT
M1= 86429 LB.FT
M2= 10619 LB.FT
S1= -10176 LB.
S2= 1884 LB.
EXT.NORTH BEAM
P = 336 LB/FT
M1= 90829 LB.FT
S1= -8917 LB.

SHELL RESULTS:

EXTERNAL SOUTH SHELL

	SHELLS WITH 60-DEGREE GLAZING				SHELLS WITH 90-DEGREE GLAZING			
POINT	T1	S	T2	M2	T1	S	T2	M2
0	27941	368	39	168	23928	502	6	-102
1	9679	-3214	-204	165	8283	-3376	-324	77
2	-2158	-3859	-851	47	-1736	-4089	-1207	34
3	-8803	-2705	-1442	-90	-7472	-2865	-2016	-80
4	-11179	-676	-1716	-159	-9808	-648	-2390	-128
5	-9693	1439	-1582	-124	-8921	1754	-2185	-53
6	-4134	2870	-1098	-19	-4153	3474	-1460	96
7	6287	2747	-482	61	5829	3390	-517	177
8	22786	0	-141	0	22702	-0	15	-0

INTERNAL SHELLS

POINT	T1	S	T2	M2	T1	S	T2	M2
0	30034	423	271	79	31844	1699	117	-71
1	10687	-3460	5	154	11403	-3503	-105	131
2	-1988	-4216	-704	70	-1341	-4617	-1083	75
3	-9190	-3036	-1364	-61	-8301	-3314	-2020	-71
4	-11828	-900	-1693	-136	-10886	-844	-2476	-148
5	-10330	1345	-1590	-111	-9712	1799	-2288	-88
6	-4531	2881	-1113	-13	-4418	3653	-1532	64
7	6298	2798	-490	63	6177	3560	-543	159
8	23316	-0	-142	0	23732	-0	15	0

EXTERNAL NORTH SHELL

POINT	T1	S	T2	M2	T1	S	T2	M2
0	30121	502	271	61	31729	1689	114	-77
1	10720	-3392	19	154	11344	-3492	-107	129
2	-1956	-4154	-677	85	-1367	-4596	-1081	77
3	-9093	-2986	-1326	-34	-8297	-3290	-2012	-65
4	-11614	-880	-1648	-104	-10844	-826	-2463	-140
5	-10014	1313	-1546	-79	-9629	1801	-2272	-79
6	-4259	2787	-1082	11	-4327	3633	-1518	71
7	6165	2682	-480	76	6172	3526	-536	164
8	22128	0	-147	0	23414	-0	15	-0

L = 34 FT. TABLE NO.N- 18 W = 15 FT.

SHELL DATA:

L = 34.00 FT.
W = 15.00 FT.
T = 2.50 IN.
G = 50.00 LB/FT*2
QA= 150.00 LB/FT
QV= 150.00 LB/FT

FOR 60DEG.GLAZING:
K = 7.75 FT.
R = 10.00 FT.
AL= 14.14 FT.
θ = 81.03 DEG.

FOR 90DEG.GLAZING:
K = 7.75 FT.
R = 20.00 FT.
AL= 17.43 FT.
θ = 49.93 DEG.

BEAM DATA:

FOR 60DEG.GLAZING:

B1= 0.75 FT.
D1= 2.25 FT.
B2= 0.75 FT.
D2= 2.00 FT.

EXT.NORTH BEAM:

Z'= 1.17 FT.
Y'= 0.83 FT.
IZ= 0.91 FT*4
IY= 0.61 FT*4

FOR 90DEG.GLAZING:

ALL BEAMS:

B = 1.00 FT.
D = 2.25 FT.

BEAM RESULTS:

FOR 60DEG.GLAZING:

EXT.SOUTH BEAM
F = 9010 LB.
M1= 37930 LB.FT
M2= 4452 LB.FT
S1= -2659 LB.
S2= 59 LB.
INTERNAL BEAMS
F = 6209 LB.
P = 277 LB/FT
M1= 69161 LB.FT
M2= 1057 LB.FT
S1= -5712 LB.
S2= -147 LB.
EXT.NORTH BEAM
P = 272 LB/FT
M1= 102148 LB.FT
M2= 20311 LB.FT
S1= -9438 LB.
S2= -1877 LB.

FOR 90DEG.GLAZING:

EXT.SOUTH BEAM
F = -1404 LB.
M1= 76846 LB.FT
M2= 7700 LB.FT
S1= -7237 LB.
S2= 764 LB.
INTERNAL BEAMS
F = -10655 LB.
P = 315 LB/FT
M1= 102655 LB.FT
M2= 10939 LB.FT
S1= -10580 LB.
S2= 1487 LB.
EXT.NORTH BEAM
P = 320 LB/FT
M1= 104603 LB.FT
S1= -9665 LB.

SHELL RESULTS:

	SHELLS WITH 60-DEGREE GLAZING				SHELLS WITH 90-DEGREE GLAZING			
EXTERNAL SOUTH SHELL								
POINT	T1	S	T2	M2	T1	S	T2	M2
0	22002	-833	222	497	23122	130	5	-18
1	6340	-3061	-28	319	8177	-2931	-222	72
2	-3466	-3223	-469	79	-1867	-3491	-808	19
3	-8550	-2186	-841	-135	-7828	-2452	-1344	-70
4	-9983	-628	-1008	-246	-10222	-578	-1586	-101
5	-8364	911	-932	-230	-9101	1429	-1440	-36
6	-3603	1935	-658	-124	-3988	2820	-952	82
7	5063	1876	-316	-14	6007	2707	-334	141
8	18892	0	-141	0	21985	-0	5	0
INTERNAL SHELLS								
POINT	T1	S	T2	M2	T1	S	T2	M2
0	24579	-574	360	370	30438	985	101	35
1	7904	-3140	122	293	11169	-3081	-71	134
2	-2853	-3479	-349	105	-1475	-3957	-727	64
3	-8685	-2476	-767	-92	-8699	-2850	-1354	-58
4	-10560	-856	-977	-207	-11418	-755	-1656	-118
5	-9046	790	-929	-203	-10033	1473	-1522	-68
6	-4092	1913	-667	-109	-4330	2996	-1010	52
7	4979	1903	-324	-8	6427	2878	-355	124
8	19317	0	-145	0	23271	-0	5	0
EXTERNAL NORTH SHELL								
POINT	T1	S	T2	M2	T1	S	T2	M2
0	24708	-585	362	380	30327	993	97	25
1	7952	-3166	122	296	11113	-3057	-71	132
2	-2865	-3507	-353	102	-1489	-3925	-722	68
3	-8747	-2499	-776	-101	-8668	-2820	-1342	-49
4	-10669	-864	-988	-219	-11331	-737	-1639	-105
5	-9184	803	-941	-215	-9897	1468	-1504	-54
6	-4197	1947	-674	-118	-4203	2963	-996	63
7	5044	1943	-325	-13	6393	2832	-348	131
8	19789	0	-142	-0	22792	-0	5	0

L = 34 FT. TABLE NO.N- 19 W = 17 FT.

SHELL DATA:

```
L =  34.00 FT.
W =  17.00 FT.
T =   2.50 IN.
G =  50.00 LB/FT*2
QA= 150.00 LB/FT
QV= 150.00 LB/FT

FOR 60DEG.GLAZING:
K =   8.25 FT.
R =  15.00 FT.
AL=  15.39 FT.
θ =  58.79 DEG.

FOR 90DEG.GLAZING:
K =   8.25 FT.
R =  22.50 FT.
AL=  19.50 FT.
θ =  49.66 DEG.
```

BEAM DATA:

```
FOR 60DEG.GLAZING:

B1= 0.75 FT.
D1= 2.25 FT.
B2= 0.75 FT.
D2= 2.00 FT.

EXT.NORTH BEAM:

Z'= 1.17 FT.
Y'= 0.83 FT.
IZ= 0.91 FT*4
IY= 0.61 FT*4

FOR 90DEG.GLAZING:

ALL BEAMS:

B = 1.00 FT.
D = 2.25 FT.
```

BEAM RESULTS:

```
FOR 60DEG.GLAZING:

EXT.SOUTH BEAM
 F =      568 LB.
 M1=    49768 LB.FT
 M2=     4491 LB.FT
 S1=    -4531 LB.
 S2=      372 LB.
INTERNAL BEAMS
 F =      154 LB.
 P =      298 LB/FT
 M1=    83693 LB.FT
 M2=      211 LB.FT
 S1=    -7692 LB.
 S2=        0 LB.
EXT.NORTH BEAM
 P =      306 LB/FT
 M1=   107040 LB.FT
 M2=    22791 LB.FT
 S1=    -9890 LB.
 S2=    -2106 LB.

FOR 90DEG.GLAZING:

EXT.SOUTH BEAM
 F =    -2818 LB.
 M1=    75273 LB.FT
 M2=     7711 LB.FT
 S1=    -7240 LB.
 S2=      833 LB.
INTERNAL BEAMS
 F =   -13057 LB.
 P =      320 LB/FT
 M1=   101526 LB.FT
 M2=    10986 LB.FT
 S1=   -10728 LB.
 S2=     1605 LB.
EXT.NORTH BEAM
 P =      323 LB/FT
 M1=   105709 LB.FT
 S1=    -9767 LB.
```

SHELL RESULTS:

EXTERNAL SOUTH SHELL

	SHELLS WITH 60-DEGREE GLAZING				SHELLS WITH 90-DEGREE GLAZING			
POINT	T1	S	T2	M2	T1	S	T2	M2
0	25806	-53	68	284	22609	260	11	12
1	8632	-3016	-150	215	7697	-3039	-250	111
2	-2486	-3480	-666	56	-1830	-3611	-936	27
3	-8641	-2425	-1129	-103	-7219	-2524	-1560	-102
4	-10737	-648	-1341	-182	-9364	-600	-1846	-153
5	-9216	1179	-1238	-152	-8478	1469	-1685	-79
6	-3924	2406	-867	-48	-3950	2947	-1124	72
7	5842	2310	-400	39	5472	2881	-400	158
8	21179	-0	-147	-0	21318	-0	3	0

INTERNAL SHELLS

POINT	T1	S	T2	M2	T1	S	T2	M2
0	27781	-14	291	160	29962	1206	108	57
1	9709	-3248	53	191	10516	-3197	-83	171
2	-2215	-3830	-517	82	-1583	-4082	-851	69
3	-8979	-2766	-1043	-64	-8092	-2902	-1576	-95
4	-11407	-894	-1308	-148	-10416	-749	-1922	-178
5	-9921	1059	-1237	-129	-9207	1527	-1769	-117
6	-4401	2395	-878	-36	-4157	3112	-1181	37
7	5789	2347	-407	44	5835	3029	-419	139
8	21670	-0	-150	-0	22277	-0	2	-0

EXTERNAL NORTH SHELL

POINT	T1	S	T2	M2	T1	S	T2	M2
0	27914	37	290	150	29837	1197	105	51
1	9761	-3213	60	192	10453	-3186	-86	168
2	-2192	-3801	-505	91	-1608	-4061	-851	71
3	-8931	-2742	-1025	-48	-8086	-2880	-1570	-90
4	-11290	-884	-1288	-129	-10371	-732	-1911	-169
5	-9738	1041	-1217	-111	-9122	1529	-1756	-108
6	-4240	2345	-864	-21	-4064	3093	-1170	45
7	5713	2286	-404	52	5829	2998	-414	144
8	20969	-0	-153	-0	21954	-0	2	-0

L = 34 FT. TABLE NO.N- 20 W = 19 FT.

SHELL DATA:

L = 34.00 FT.
W = 19.00 FT.
T = 2.50 IN.
G = 50.00 LB/FT*2
QA= 150.00 LB/FT
QV= 150.00 LB/FT

FOR 60DEG.GLAZING:
K = 9.25 FT.
R = 20.00 FT.
AL= 16.96 FT.
θ = 48.59 DEG.

FOR 90DEG.GLAZING:
K = 9.25 FT.
R = 30.00 FT.
AL= 21.60 FT.
θ = 41.24 DEG.

BEAM DATA:

FOR 60DEG.GLAZING:

B1= 0.75 FT.
D1= 2.25 FT.
B2= 0.75 FT.
D2= 2.00 FT.

EXT.NORTH BEAM:

Z'= 1.17 FT.
Y'= 0.83 FT.
IZ= 0.91 FT*4
IY= 0.61 FT*4

FOR 90DEG.GLAZING:

ALL BEAMS:

B = 1.00 FT.
D = 2.25 FT.

BEAM RESULTS:

FOR 60DEG.GLAZING:

EXT.SOUTH BEAM
F = -3537 LB.
M1= 53994 LB.FT
M2= 4589 LB.FT
S1= -5304 LB.
S2= 531 LB.
INTERNAL BEAMS
F = -3803 LB.
P = 314 LB/FT
M1= 90106 LB.FT
M2= -0 LB.FT
S1= -8702 LB.
S2= 124 LB.
EXT.NORTH BEAM
P = 323 LB/FT
M1= 110485 LB.FT
M2= 24092 LB.FT
S1= -10209 LB.
S2= -2226 LB.

FOR 90DEG.GLAZING:

EXT.SOUTH BEAM
F = -4506 LB.
M1= 79375 LB.FT
M2= 7598 LB.FT
S1= -7796 LB.
S2= 904 LB.
INTERNAL BEAMS
F = -17066 LB.
P = 345 LB/FT
M1= 107610 LB.FT
M2= 10916 LB.FT
S1= -11709 LB.
S2= 1789 LB.
EXT.NORTH BEAM
P = 348 LB/FT
M1= 110653 LB.FT
S1= -10224 LB.

SHELL RESULTS:

EXTERNAL SOUTH SHELL

	SHELLS WITH 60-DEGREE GLAZING				SHELLS WITH 90-DEGREE GLAZING			
POINT	T1	S	T2	M2	T1	S	T2	M2
0	27105	327	19	144	23403	416	-15	-101
1	9400	-3136	-209	158	8035	-3372	-344	88
2	-2073	-3763	-827	47	-1753	-4055	-1213	35
3	-8511	-2653	-1392	-87	-7333	-2846	-2007	-90
4	-10824	-697	-1656	-151	-9642	-667	-2375	-140
5	-9415	1346	-1528	-108	-8850	1708	-2173	-54
6	-4061	2735	-1065	4	-4210	3430	-1455	115
7	6008	2630	-475	82	5715	3370	-516	205
8	21933	-0	-148	-0	22607	-0	15	-0

INTERNAL SHELLS

POINT	T1	S	T2	M2	T1	S	T2	M2
0	29195	351	284	26	31054	1577	86	-79
1	10452	-3417	28	140	11014	-3489	-137	138
2	-1886	-4161	-660	74	-1405	-4550	-1096	76
3	-8933	-3023	-1303	-50	-8133	-3260	-2010	-81
4	-11546	-950	-1628	-122	-10642	-841	-2452	-161
5	-10131	1238	-1536	-91	-9553	1753	-2264	-89
6	-4512	2746	-1082	11	-4425	3591	-1518	83
7	6013	2686	-484	85	6032	3518	-538	187
8	22515	-0	-150	0	23488	-0	14	0

EXTERNAL NORTH SHELL

POINT	T1	S	T2	M2	T1	S	T2	M2
0	29287	407	283	12	30957	1567	83	-86
1	10480	-3372	36	139	10957	-3480	-140	136
2	-1876	-4119	-643	85	-1439	-4530	-1095	78
3	-8882	-2986	-1277	-30	-8141	-3235	-2003	-75
4	-11404	-932	-1597	-97	-10605	-819	-2439	-152
5	-9897	1219	-1505	-67	-9462	1759	-2248	-79
6	-4295	2679	-1059	31	-4314	3570	-1503	92
7	5931	2601	-477	95	6036	3479	-531	192
8	21611	-0	-153	-0	23114	-0	15	0

L = 36 FT. TABLE NO.N- 21 W = 16 FT.

SHELL DATA:

L = 36.00 FT.
W = 16.00 FT.
T = 2.50 IN.
G = 50.00 LB/FT*2
QA= 150.00 LB/FT
QV= 150.00 LB/FT

FOR 60DEG.GLAZING:
K = 8.25 FT.
R = 12.50 FT.
AL= 14.69 FT.
Θ = 67.33 DEG.

FOR 90DEG.GLAZING:
K = 8.25 FT.
R = 20.00 FT.
AL= 18.67 FT.
Θ = 53.49 DEG.

BEAM DATA:

FOR 60DEG.GLAZING:

B1= 0.75 FT.
D1= 2.25 FT.
B2= 0.75 FT.
D2= 2.00 FT.

EXT.NORTH BEAM:

Z'= 1.17 FT.
Y'= 0.83 FT.
IZ= 0.91 FT*4
IY= 0.61 FT*4

FOR 90DEG.GLAZING:

ALL BEAMS:

B = 1.00 FT.
D = 2.25 FT.

BEAM RESULTS:

FOR 60DEG.GLAZING:

EXT.SOUTH BEAM
F = 6497 LB.
M1= 49015 LB.FT
M2= 5040 LB.FT
S1= -3693 LB.
S2= 191 LB.
INTERNAL BEAMS
F = 4486 LB.
P = 281 LB/FT
M1= 84924 LB.FT
M2= 751 LB.FT
S1= -6937 LB.
S2= -105 LB.
EXT.NORTH BEAM
P = 283 LB/FT
M1= 116743 LB.FT
M2= 23669 LB.FT
S1= -10188 LB.
S2= -2065 LB.

FOR 90DEG.GLAZING:

EXT.SOUTH BEAM
F = -1310 LB.
M1= 80895 LB.FT
M2= 9052 LB.FT
S1= -7178 LB.
S2= 833 LB.
INTERNAL BEAMS
F = -11367 LB.
P = 306 LB/FT
M1= 108710 LB.FT
M2= 12805 LB.FT
S1= -10589 LB.
S2= 1595 LB.
EXT.NORTH BEAM
P = 308 LB/FT
M1= 115897 LB.FT
S1= -10114 LB.

SHELL RESULTS:

SHELLS WITH 60-DEGREE GLAZING / SHELLS WITH 90-DEGREE GLAZING

EXTERNAL SOUTH SHELL

POINT	T1	S	T2	M2	T1	S	T2	M2
0	26925	-567	147	454	24882	114	38	100
1	8484	-3311	-101	301	8432	-3169	-215	140
2	-3464	-3633	-591	79	-2200	-3715	-850	25
3	-9995	-2490	-1015	-124	-8231	-2584	-1422	-119
4	-12062	-668	-1205	-228	-10554	-613	-1680	-177
5	-10174	1166	-1108	-206	-9411	1482	-1527	-106
6	-4233	2377	-775	-98	-4251	2950	-1014	43
7	6444	2270	-359	6	6119	2860	-359	135
8	22976	-0	-139	-0	23226	-0	1	-0

INTERNAL SHELLS

POINT	T1	S	T2	M2	T1	S	T2	M2
0	29254	-391	323	312	32904	992	141	177
1	9928	-3438	76	266	11492	-3380	-62	218
2	-2886	-3921	-448	102	-1996	-4228	-780	74
3	-10128	-2810	-923	-79	-9294	-2987	-1450	-113
4	-12640	-927	-1160	-184	-11829	-763	-1765	-207
5	-10886	1014	-1097	-174	-10306	1559	-1618	-151
6	-4787	2330	-779	-78	-4505	3147	-1075	3
7	6275	2283	-365	15	6595	3035	-380	113
8	23326	-0	-143	-0	24475	-0	1	0

EXTERNAL NORTH SHELL

POINT	T1	S	T2	M2	T1	S	T2	M2
0	29463	-362	323	313	32719	978	137	171
1	10010	-3431	80	269	11415	-3368	-65	216
2	-2874	-3921	-445	107	-2009	-4207	-780	74
3	-10135	-2810	-919	-74	-9269	-2967	-1445	-109
4	-12631	-927	-1156	-179	-11776	-752	-1757	-200
5	-10849	1010	-1093	-169	-10236	1557	-1609	-144
6	-4743	2320	-776	-75	-4444	3131	-1068	8
7	6262	2269	-365	17	6579	3014	-377	116
8	23138	-0	-144	-0	24251	-0	1	-0

L = 36 FT. TABLE NO.N- 22 W = 18 FT.

SHELL DATA:

L = 36.00 FT.
W = 18.00 FT.
T = 2.50 IN.
G = 50.00 LB/FT*2
QA= 150.00 LB/FT
QV= 150.00 LB/FT

FOR 60DEG.GLAZING:
K = 8.75 FT.
R = 15.00 FT.
AL= 16.39 FT.
θ = 62.62 DEG.

FOR 90DEG.GLAZING:
K = 8.75 FT.
R = 25.00 FT.
AL= 20.59 FT.
θ = 47.19 DEG.

BEAM DATA:

FOR 60DEG.GLAZING:

B1= 0.75 FT.
D1= 2.25 FT.
B2= 0.75 FT.
D2= 2.00 FT.

EXT.NORTH BEAM:

Z'= 1.17 FT.
Y'= 0.83 FT.
IZ= 0.91 FT*4
IY= 0.61 FT*4

FOR 90DEG.GLAZING:

ALL BEAMS:

B = 1.00 FT.
D = 2.25 FT.

BEAM RESULTS:

FOR 60DEG.GLAZING:

EXT.SOUTH BEAM
F = 2960 LB.
M1= 50891 LB.FT
M2= 5018 LB.FT
S1= -4168 LB.
S2= 314 LB.
INTERNAL BEAMS
F = 1085 LB.
P = 289 LB/FT
M1= 88053 LB.FT
M2= 618 LB.FT
S1= -7550 LB.
S2= 2 LB.
EXT.NORTH BEAM
P = 293 LB/FT
M1= 118798 LB.FT
M2= 24469 LB.FT
S1= -10367 LB.
S2= -2135 LB.

FOR 90DEG.GLAZING:

EXT.SOUTH BEAM
F = -3705 LB.
M1= 84584 LB.FT
M2= 8706 LB.FT
S1= -7737 LB.
S2= 912 LB.
INTERNAL BEAMS
F = -15918 LB.
P = 325 LB/FT
M1= 114345 LB.FT
M2= 12398 LB.FT
S1= -11530 LB.
S2= 1764 LB.
EXT.NORTH BEAM
P = 328 LB/FT
M1= 119978 LB.FT
S1= -10470 LB.

SHELL RESULTS:

SHELLS WITH 60-DEGREE GLAZING / SHELLS WITH 90-DEGREE GLAZING

EXTERNAL SOUTH SHELL

	SHELLS WITH 60-DEGREE GLAZING				SHELLS WITH 90-DEGREE GLAZING			
POINT	T1	S	T2	M2	T1	S	T2	M2
0	27191	-258	112	430	25378	323	10	17
1	8512	-3336	-150	290	8618	-3366	-283	126
2	-3161	-3721	-714	58	-2056	-4003	-1049	30
3	-9316	-2534	-1205	-159	-8070	-2791	-1743	-118
4	-11235	-640	-1422	-267	-10452	-648	-2061	-178
5	-9567	1272	-1306	-232	-9457	1654	-1879	-95
6	-4117	2558	-910	-100	-4402	3298	-1254	73
7	6101	2462	-413	21	6136	3222	-446	173
8	22633	-0	-145	-0	23913	0	6	-0

INTERNAL SHELLS

	SHELLS WITH 60-DEGREE GLAZING				SHELLS WITH 90-DEGREE GLAZING			
POINT	T1	S	T2	M2	T1	S	T2	M2
0	29733	-95	315	299	33647	1389	111	72
1	9921	-3524	47	267	11777	-3537	-103	195
2	-2751	-4069	-567	87	-1784	-4523	-959	76
3	-9644	-2885	-1122	-117	-9047	-3206	-1764	-112
4	-11957	-891	-1395	-232	-11621	-808	-2146	-206
5	-10330	1159	-1311	-211	-10261	1723	-1974	-139
6	-4615	2561	-925	-89	-4623	3484	-1317	34
7	6089	2513	-421	25	6545	3387	-467	151
8	23233	-0	-147	-0	24977	0	5	-0

EXTERNAL NORTH SHELL

	SHELLS WITH 60-DEGREE GLAZING				SHELLS WITH 90-DEGREE GLAZING			
POINT	T1	S	T2	M2	T1	S	T2	M2
0	29857	-68	315	295	33496	1376	108	65
1	9962	-3512	50	268	11705	-3526	-106	192
2	-2750	-4060	-563	92	-1810	-4502	-958	77
3	-9642	-2875	-1116	-110	-9038	-3183	-1758	-106
4	-11920	-885	-1387	-224	-11574	-791	-2136	-198
5	-10250	1154	-1303	-202	-10174	1725	-1961	-130
6	-4532	2541	-919	-83	-4531	3464	-1306	42
7	6066	2486	-420	28	6538	3355	-462	156
8	22905	-0	-149	-0	24656	0	6	-0

L = 36 FT.   TABLE NO.N- 23   W = 20 FT.

SHELL DATA:

L = 36.00 FT.
W = 20.00 FT.
T = 2.50 IN.
G = 50.00 LB/FT*2
QA= 150.00 LB/FT
QV= 150.00 LB/FT

FOR 60DEG.GLAZING:
K = 9.25 FT.
R = 20.00 FT.
AL= 17.93 FT.
θ = 51.37 DEG.

FOR 90DEG.GLAZING:
K = 9.25 FT.
R = 30.00 FT.
AL= 22.56 FT.
θ = 43.09 DEG.

BEAM DATA:

FOR 60DEG.GLAZING:

B1= 0.75 FT.
D1= 2.25 FT.
B2= 0.75 FT.
D2= 2.00 FT.

EXT.NORTH BEAM:

Z'= 1.17 FT.
Y'= 0.83 FT.
IZ= 0.91 FT*4
IY= 0.61 FT*4

FOR 90DEG.GLAZING:

ALL BEAMS:

B = 1.00 FT.
D = 2.25 FT.

BEAM RESULTS:

FOR 60DEG.GLAZING:

EXT.SOUTH BEAM
F = -2553 LB.
M1= 57747 LB.FT
M2= 4930 LB.FT
S1= -5249 LB.
S2= 496 LB.
INTERNAL BEAMS
F = -3677 LB.
P = 309 LB/FT
M1= 97836 LB.FT
M2= 49 LB.FT
S1= -8877 LB.
S2= 115 LB.
EXT.NORTH BEAM
P = 317 LB/FT
M1= 123045 LB.FT
M2= 26535 LB.FT
S1= -10738 LB.
S2= -2316 LB.

FOR 90DEG.GLAZING:

EXT.SOUTH BEAM
F = -5713 LB.
M1= 86878 LB.FT
M2= 8465 LB.FT
S1= -8135 LB.
S2= 981 LB.
INTERNAL BEAMS
F = -19729 LB.
P = 342 LB/FT
M1= 118163 LB.FT
M2= 12125 LB.FT
S1= -12240 LB.
S2= 1909 LB.
EXT.NORTH BEAM
P = 345 LB/FT
M1= 123451 LB.FT
S1= -10773 LB.

SHELL RESULTS:

EXTERNAL SOUTH SHELL

POINT	T1	S	T2	M2	T1	S	T2	M2
	SHELLS WITH 60-DEGREE GLAZING				SHELLS WITH 90-DEGREE GLAZING			
0	29184	223	41	269	25625	499	-6	-47
1	9660	-3445	-227	226	8607	-3563	-345	128
2	-2641	-4027	-898	50	-1982	-4265	-1250	41
3	-9290	-2780	-1499	-139	-7837	-2975	-2073	-120
4	-11547	-678	-1772	-231	-10197	-692	-2453	-189
5	-9992	1491	-1628	-184	-9378	1787	-2247	-96
6	-4344	2966	-1131	-44	-4546	3598	-1508	96
7	6459	2855	-499	67	6012	3555	-540	208
8	24011	0	-151	-0	24350	-0	9	0

INTERNAL SHELLS

POINT	T1	S	T2	M2	T1	S	T2	M2
0	31803	321	293	156	34085	1722	96	-9
1	10929	-3723	3	214	11778	-3716	-138	189
2	-2433	-4444	-740	81	-1720	-4798	-1143	85
3	-9783	-3163	-1421	-102	-8776	-3397	-2089	-115
4	-12355	-927	-1756	-205	-11275	-849	-2541	-217
5	-10763	1401	-1646	-172	-10078	1854	-2344	-140
6	-4796	3000	-1153	-41	-4713	3773	-1573	58
7	6525	2929	-509	67	6370	3706	-561	187
8	24715	0	-152	-0	25225	0	9	-0

EXTERNAL NORTH SHELL

POINT	T1	S	T2	M2	T1	S	T2	M2
0	31890	365	292	142	33954	1705	93	-15
1	10943	-3689	10	213	11711	-3709	-142	187
2	-2443	-4409	-727	91	-1754	-4779	-1143	86
3	-9757	-3128	-1401	-84	-8781	-3373	-2084	-109
4	-12238	-906	-1730	-182	-11241	-829	-2530	-209
5	-10547	1389	-1619	-149	-9995	1860	-2330	-130
6	-4581	2943	-1133	-23	-4612	3755	-1560	66
7	6465	2851	-503	77	6375	3672	-555	192
8	23863	0	-156	-0	24886	-0	9	0

L = 38 FT. TABLE NO.N- 24 W = 17 FT.

```
SHELL DATA:                   BEAM RESULTS:

  L =  38.00 FT.              FOR 60DEG.GLAZING:
  W =  17.00 FT.
  T =   2.50 IN.              EXT.SOUTH BEAM
  G =  50.00 LB/FT*2            F =      854 LB.
  QA= 150.00 LB/FT              M1=    72930 LB.FT
  QV= 150.00 LB/FT              M2=     5050 LB.FT
                                S1=    -5933 LB.
 FOR 60DEG.GLAZING:             S2=      362 LB.
  K =   8.50 FT.              INTERNAL BEAMS
  R =  15.00 FT.                F =     1215 LB.
  AL=  15.39 FT.                P =      318 LB/FT
  θ =  58.78 DEG.               M1=   115854 LB.FT
                                M2=       56 LB.FT
 FOR 90DEG.GLAZING:             S1=    -9422 LB.
  K =   8.50 FT.                S2=      -50 LB.
  R =  20.00 FT.              EXT.NORTH BEAM
  AL=  19.81 FT.                P =      326 LB/FT
  θ =  56.74 DEG.               M1=   147445 LB.FT
                                M2=    30351 LB.FT
BEAM DATA:                      S1=   -12190 LB.
                                S2=    -2509 LB.
 FOR 60DEG.GLAZING:

  B1= 0.75 FT.                FOR 90DEG.GLAZING:
  D1= 2.50 FT.
  B2= 0.75 FT.                EXT.SOUTH BEAM
  D2= 2.25 FT.                  F =     -995 LB.
                                M1=   100007 LB.FT
  EXT.NORTH BEAM:               M2=     9048 LB.FT
                                S1=    -8361 LB.
  Z'= 1.29 FT.                  S2=      775 LB.
  Y'= 0.85 FT.                INTERNAL BEAMS
  IZ= 1.26 FT*4                 F =   -11401 LB.
  IY= 0.82 FT*4                 P =      318 LB/FT
                                M1=   133740 LB.FT
 FOR 90DEG.GLAZING:             M2=    12755 LB.FT
                                S1=   -12222 LB.
  ALL BEAMS:                    S2=     1508 LB.
                              EXT.NORTH BEAM
  B = 1.00 FT.                  P =      322 LB/FT
  D = 2.50 FT.                  M1=   138711 LB.FT
                                S1=   -11468 LB.
```

SHELL RESULTS:

EXTERNAL SOUTH SHELL

POINT	T1 (60-DEGREE GLAZING)	S	T2	M2	T1 (90-DEGREE GLAZING)	S	T2	M2
0	29103	-71	35	235	24442	82	18	98
1	10179	-3107	-157	197	8202	-3148	-229	146
2	-2565	-3636	-628	72	-2211	-3673	-853	26
3	-9932	-2576	-1057	-62	-8072	-2553	-1416	-126
4	-12571	-727	-1257	-128	-10343	-612	-1669	-184
5	-10798	1189	-1163	-97	-9279	1456	-1517	-102
6	-4506	2469	-819	-5	-4259	2918	-1009	63
7	6800	2357	-386	62	5998	2842	-359	159
8	23886	0	-154	-0	23036	0	-4	0

INTERNAL SHELLS

POINT	T1 (60-DEGREE GLAZING)	S	T2	M2	T1 (90-DEGREE GLAZING)	S	T2	M2
0	31079	-100	291	71	32175	943	110	165
1	11332	-3386	73	158	11113	-3337	-86	220
2	-2227	-4033	-454	97	-2039	-4147	-788	72
3	-10262	-2970	-947	-12	-9076	-2919	-1440	-121
4	-13306	-1033	-1204	-79	-11505	-744	-1744	-214
5	-11630	1012	-1148	-60	-10059	1529	-1597	-147
6	-5139	2411	-822	18	-4455	3092	-1062	23
7	6596	2369	-393	72	6423	2993	-377	137
8	24249	0	-158	0	24091	0	-4	0

EXTERNAL NORTH SHELL

POINT	T1 (60-DEGREE GLAZING)	S	T2	M2	T1 (90-DEGREE GLAZING)	S	T2	M2
0	31271	-28	290	62	32024	933	106	157
1	11430	-3335	82	161	11041	-3325	-89	217
2	-2159	-3994	-438	111	-2064	-4124	-787	74
3	-10162	-2945	-926	8	-9064	-2896	-1434	-115
4	-13135	-1028	-1180	-55	-11451	-727	-1734	-205
5	-11401	984	-1126	-38	-9963	1530	-1585	-137
6	-4963	2349	-808	35	-4354	3072	-1052	32
7	6471	2298	-389	81	6413	2961	-372	142
8	23393	0	-162	0	23738	0	-4	0

L = 38 FT.　　TABLE NO.N- 25　　W = 19 FT.

SHELL DATA:

L = 38.00 FT.
W = 19.00 FT.
T = 2.50 IN.
G = 50.00 LB/FT*2
QA= 150.00 LB/FT
QV= 150.00 LB/FT

FOR 60DEG.GLAZING:
K = 9.50 FT.
R = 15.00 FT.
AL= 17.42 FT.
θ = 66.53 DEG.

FOR 90DEG.GLAZING:
K = 9.50 FT.
R = 30.00 FT.
AL= 21.71 FT.
θ = 41.47 DEG.

BEAM DATA:

FOR 60DEG.GLAZING:

B1= 0.75 FT.
D1= 2.50 FT.
B2= 0.75 FT.
D2= 2.25 FT.

EXT.NORTH BEAM:

Z'= 1.29 FT.
Y'= 0.85 FT.
IZ= 1.26 FT*4
IY= 0.82 FT*4

FOR 90DEG.GLAZING:

ALL BEAMS:

B = 1.00 FT.
D = 2.50 FT.

BEAM RESULTS:

FOR 60DEG.GLAZING:

EXT.SOUTH BEAM
F = 3093 LB.
M1= 63248 LB.FT
M2= 5307 LB.FT
S1= -4928 LB.
S2= 316 LB.
INTERNAL BEAMS
F = 1293 LB.
P = 308 LB/FT
M1= 105159 LB.FT
M2= 835 LB.FT
S1= -8534 LB.
S2= 12 LB.
EXT.NORTH BEAM
P = 309 LB/FT
M1= 146157 LB.FT
M2= 28747 LB.FT
S1= -12083 LB.
S2= -2377 LB.

FOR 90DEG.GLAZING:

EXT.SOUTH BEAM
F = -2986 LB.
M1= 112176 LB.FT
M2= 8425 LB.FT
S1= -9575 LB.
S2= 812 LB.
INTERNAL BEAMS
F = -17089 LB.
P = 357 LB/FT
M1= 150037 LB.FT
M2= 12023 LB.FT
S1= -14161 LB.
S2= 1690 LB.
EXT.NORTH BEAM
P = 363 LB/FT
M1= 147834 LB.FT
S1= -12222 LB.

SHELL RESULTS:

EXTERNAL SOUTH SHELL

	SHELLS WITH 60-DEGREE GLAZING				SHELLS WITH 90-DEGREE GLAZING			
POINT	T1	S	T2	M2	T1	S	T2	M2
0	26446	-256	106	418	26399	247	-45	-141
1	8199	-3258	-140	288	9302	-3640	-376	62
2	-3052	-3628	-683	57	-2089	-4354	-1210	28
3	-8878	-2484	-1155	-161	-8858	-3047	-1972	-80
4	-10661	-673	-1366	-265	-11672	-672	-2317	-115
5	-9109	1155	-1258	-222	-10538	1898	-2104	-18
6	-4019	2395	-883	-81	-4720	3708	-1395	152
7	5635	2329	-410	40	7039	3572	-489	229
8	21419	-0	-156	-0	26137	-0	17	-0

INTERNAL SHELLS

POINT	T1	S	T2	M2	T1	S	T2	M2
0	28928	-107	334	248	34595	1413	45	-113
1	9679	-3467	80	255	12672	-3730	-185	117
2	-2585	-4011	-519	91	-1561	-4848	-1097	72
3	-9220	-2874	-1063	-110	-9677	-3486	-1969	-67
4	-11457	-954	-1336	-223	-12820	-876	-2387	-133
5	-9961	1028	-1265	-196	-11420	1930	-2193	-52
6	-4577	2400	-900	-69	-5043	3879	-1459	120
7	5624	2388	-419	45	7406	3740	-511	211
8	22091	-0	-158	-0	27280	-0	17	0

EXTERNAL NORTH SHELL

POINT	T1	S	T2	M2	T1	S	T2	M2
0	29035	-94	335	248	34507	1417	41	-125
1	9716	-3466	81	257	12610	-3709	-186	113
2	-2586	-4013	-518	93	-1604	-4816	-1092	77
3	-9232	-2874	-1062	-108	-9678	-3448	-1955	-55
4	-11461	-953	-1335	-221	-12745	-845	-2366	-116
5	-9950	1029	-1263	-195	-11259	1934	-2167	-34
6	-4557	2397	-899	-67	-4861	3842	-1437	135
7	5626	2383	-419	45	7395	3678	-501	220
8	22020	-0	-158	-0	26644	-0	17	-0

L = 38 FT. TABLE NO.N- 26 W = 21 FT.

SHELL DATA:

```
L  =  38.00 FT.
W  =  21.00 FT.
T  =   2.50 IN.
G  =  50.00 LB/FT*2
QA= 150.00 LB/FT
QV= 150.00 LB/FT

FOR 60DEG.GLAZING:
K  =  10.00 FT.
R  =  20.00 FT.
AL=  18.89 FT.
θ  =  54.11 DEG.

FOR 90DEG.GLAZING:
K  =  10.00 FT.
R  =  32.50 FT.
AL=  23.79 FT.
θ  =  41.93 DEG.
```

BEAM DATA:

```
FOR 60DEG.GLAZING:

B1= 0.75 FT.
D1= 2.50 FT.
B2= 0.75 FT.
D2= 2.25 FT.

EXT.NORTH BEAM:

Z'= 1.29 FT.
Y'= 0.85 FT.
IZ= 1.26 FT*4
IY= 0.82 FT*4

FOR 90DEG.GLAZING:

ALL BEAMS:

B = 1.00 FT.
D = 2.50 FT.
```

BEAM RESULTS:

```
FOR 60DEG.GLAZING:

EXT.SOUTH BEAM
F  =    -2483 LB.
M1=     72063 LB.FT
M2=      5112 LB.FT
S1=     -6177 LB.
S2=       482 LB.
INTERNAL BEAMS
F  =    -3247 LB.
P  =      326 LB/FT
M1=    117081 LB.FT
M2=       207 LB.FT
S1=     -9995 LB.
S2=       108 LB.
EXT.NORTH BEAM
P  =      332 LB/FT
M1=    150641 LB.FT
M2=     30905 LB.FT
S1=    -12454 LB.
S2=     -2555 LB.

FOR 90DEG.GLAZING:

EXT.SOUTH BEAM
F  =    -4798 LB.
M1=    109891 LB.FT
M2=      8469 LB.FT
S1=     -9574 LB.
S2=       892 LB.
INTERNAL BEAMS
F  =   -19825 LB.
P  =      362 LB/FT
M1=    148327 LB.FT
M2=     12109 LB.FT
S1=    -14303 LB.
S2=      1812 LB.
EXT.NORTH BEAM
P  =      366 LB/FT
M1=    149264 LB.FT
S1=    -12340 LB.
```

SHELL RESULTS:

EXTERNAL SOUTH SHELL

	SHELLS WITH 60-DEGREE GLAZING				SHELLS WITH 90-DEGREE GLAZING			
POINT	T1	S	T2	M2	T1	S	T2	M2
0	28467	205	25	244	25813	397	-38	-112
1	9430	-3365	-225	219	8792	-3712	-400	108
2	-2532	-3936	-869	50	-2013	-4434	-1338	41
3	-8976	-2735	-1447	-136	-8148	-3104	-2191	-112
4	-11175	-707	-1711	-222	-10707	-719	-2585	-172
5	-9712	1391	-1576	-167	-9880	1886	-2364	-65
6	-4289	2828	-1100	-19	-4748	3788	-1584	141
7	6155	2739	-493	90	6390	3729	-563	250
8	23146	-0	-160	0	25410	0	14	0

INTERNAL SHELLS

POINT	T1	S	T2	M2	T1	S	T2	M2
0	31045	268	307	100	34068	1639	53	-91
1	10732	-3679	29	200	11976	-3823	-195	163
2	-2310	-4394	-694	86	-1663	-4941	-1223	85
3	-9508	-3157	-1360	-91	-9006	-3520	-2196	-103
4	-12060	-983	-1693	-191	-11744	-888	-2662	-195
5	-10561	1290	-1594	-152	-10579	1937	-2454	-105
6	-4789	2863	-1124	-15	-4938	3949	-1645	105
7	6224	2819	-504	90	6718	3873	-584	230
8	23911	-0	-161	-0	26262	-0	13	0

EXTERNAL NORTH SHELL

POINT	T1	S	T2	M2	T1	S	T2	M2
0	31121	298	306	89	33975	1627	51	-98
1	10743	-3658	32	199	11913	-3816	-198	160
2	-2328	-4372	-686	92	-1712	-4920	-1222	87
3	-9504	-3132	-1347	-79	-9028	-3490	-2189	-95
4	-11986	-966	-1676	-174	-11708	-858	-2648	-184
5	-10407	1285	-1576	-135	-10465	1948	-2434	-91
6	-4624	2825	-1109	-1	-4789	3926	-1626	117
7	6192	2763	-500	98	6733	3824	-575	237
8	23284	-0	-164	-0	25769	0	14	0

L = 40 FT. TABLE NO.N- 27 W = 18 FT.

SHELL DATA:

L = 40.00 FT.
W = 18.00 FT.
T = 2.50 IN.
G = 50.00 LB/FT*2
QA= 150.00 LB/FT
QV= 150.00 LB/FT

FOR 60DEG.GLAZING:
K = 9.00 FT.
R = 15.00 FT.
AL= 16.39 FT.
θ = 62.61 DEG.

FOR 90DEG.GLAZING:
K = 9.00 FT.
R = 25.00 FT.
AL= 20.71 FT.
θ = 47.47 DEG.

BEAM DATA:

FOR 60DEG.GLAZING:

B1= 0.75 FT.
D1= 2.50 FT.
B2= 0.75 FT.
D2= 2.25 FT.

EXT.NORTH BEAM:

Z'= 1.29 FT.
Y'= 0.85 FT.
IZ= 1.26 FT*4
IY= 0.82 FT*4

FOR 90DEG.GLAZING:

ALL BEAMS:

B = 1.00 FT.
D = 2.50 FT.

BEAM RESULTS:

FOR 60DEG.GLAZING:

EXT.SOUTH BEAM
F = 4419 LB.
M1= 74262 LB.FT
M2= 5598 LB.FT
S1= -5424 LB.
S2= 277 LB.
INTERNAL BEAMS
F = 3221 LB.
P = 306 LB/FT
M1= 120664 LB.FT
M2= 471 LB.FT
S1= -9128 LB.
S2= -78 LB.
EXT.NORTH BEAM
P = 310 LB/FT
M1= 161425 LB.FT
M2= 32028 LB.FT
S1= -12678 LB.
S2= -2515 LB.

FOR 90DEG.GLAZING:

EXT.SOUTH BEAM
F = -2021 LB.
M1= 118498 LB.FT
M2= 9572 LB.FT
S1= -9495 LB.
S2= 819 LB.
INTERNAL BEAMS
F = -15528 LB.
P = 337 LB/FT
M1= 158073 LB.FT
M2= 13538 LB.FT
S1= -13927 LB.
S2= 1658 LB.
EXT.NORTH BEAM
P = 342 LB/FT
M1= 158789 LB.FT
S1= -12471 LB.

SHELL RESULTS:

EXTERNAL SOUTH SHELL

	SHELLS WITH 60-DEGREE GLAZING				SHELLS WITH 90-DEGREE GLAZING			
POINT	T1	S	T2	M2	T1	S	T2	M2
0	30523	-347	87	403	28317	159	-18	-16
1	9991	-3504	-157	285	9854	-3606	-311	105
2	-3421	-3945	-686	75	-2404	-4269	-1045	24
3	-10862	-2723	-1152	-125	-9591	-2972	-1710	-107
4	-13326	-715	-1362	-224	-12468	-659	-2008	-153
5	-11319	1327	-1251	-189	-11133	1815	-1820	-62
6	-4705	2682	-873	-66	-4915	3539	-1205	107
7	7251	2556	-402	39	7434	3399	-423	195
8	25705	-0	-151	-0	27389	-0	8	-0

INTERNAL SHELLS

POINT	T1	S	T2	M2	T1	S	T2	M2
0	32954	-253	317	231	37090	1220	72	46
1	11455	-3724	61	244	13381	-3755	-146	178
2	-2917	-4322	-519	104	-1967	-4788	-959	75
3	-11145	-3115	-1048	-73	-10599	-3410	-1726	-97
4	-14084	-1018	-1314	-174	-13798	-845	-2088	-179
5	-12191	1160	-1241	-153	-12127	1874	-1912	-105
6	-5349	2641	-880	-45	-5246	3736	-1268	68
7	7098	2582	-409	49	7905	3583	-445	173
8	26197	-0	-155	-0	28744	-0	7	-0

EXTERNAL NORTH SHELL

POINT	T1	S	T2	M2	T1	S	T2	M2
0	33163	-208	317	227	36956	1224	68	34
1	11540	-3702	65	247	13305	-3730	-147	175
2	-2892	-4308	-511	112	-2000	-4753	-953	80
3	-11120	-3104	-1038	-61	-10580	-3372	-1713	-85
4	-14017	-1014	-1303	-161	-13705	-818	-2069	-162
5	-12074	1149	-1231	-141	-11957	1873	-1890	-86
6	-5241	2611	-873	-35	-5072	3698	-1250	83
7	7048	2544	-408	53	7879	3525	-437	181
8	25724	-0	-157	-0	28115	-0	7	-0

L = 40 FT. TABLE NO.N- 28 W = 20 FT.

SHELL DATA:

L = 40.00 FT.
W = 20.00 FT.
T = 2.50 IN.
G = 50.00 LB/FT*2
QA= 150.00 LB/FT
QV= 150.00 LB/FT

FOR 60DEG.GLAZING:
K = 9.50 FT.
R = 17.50 FT.
AL= 18.13 FT.
θ = 59.35 DEG.

FOR 90DEG.GLAZING:
K = 9.50 FT.
R = 27.50 FT.
AL= 22.79 FT.
θ = 47.48 DEG.

BEAM DATA:

FOR 60DEG.GLAZING:

B1= 0.75 FT.
D1= 2.50 FT.
B2= 0.75 FT.
D2= 2.25 FT.

EXT.NORTH BEAM:

Z'= 1.29 FT.
Y'= 0.85 FT.
IZ= 1.26 FT*4
IY= 0.82 FT*4

FOR 90DEG.GLAZING:

ALL BEAMS:

B = 1.00 FT.
D = 2.50 FT.

BEAM RESULTS:

FOR 60DEG.GLAZING:

EXT.SOUTH BEAM
F = 1376 LB.
M1= 75509 LB.FT
M2= 5532 LB.FT
S1= -5793 LB.
S2= 369 LB.
INTERNAL BEAMS
F = -5 LB.
P = 314 LB/FT
M1= 123499 LB.FT
M2= 375 LB.FT
S1= -9673 LB.
S2= 15 LB.
EXT.NORTH BEAM
P = 318 LB/FT
M1= 163688 LB.FT
M2= 32858 LB.FT
S1= -12856 LB.
S2= -2581 LB.

FOR 90DEG.GLAZING:

EXT.SOUTH BEAM
F = -4148 LB.
M1= 116378 LB.FT
M2= 9620 LB.FT
S1= -9539 LB.
S2= 909 LB.
INTERNAL BEAMS
F = -18860 LB.
P = 344 LB/FT
M1= 156762 LB.FT
M2= 13650 LB.FT
S1= -14153 LB.
S2= 1800 LB.
EXT.NORTH BEAM
P = 348 LB/FT
M1= 160715 LB.FT
S1= -12623 LB.

SHELL RESULTS:

EXTERNAL SOUTH SHELL

	SHELLS WITH 60-DEGREE GLAZING				SHELLS WITH 90-DEGREE GLAZING			
POINT	T1	S	T2	M2	T1	S	T2	M2
0	30493	-108	65	386	27740	326	-14	8
1	9831	-3572	-200	283	9348	-3681	-336	146
2	-3226	-4059	-808	59	-2304	-4359	-1166	35
3	-10234	-2783	-1344	-161	-8834	-3031	-1918	-139
4	-12516	-695	-1583	-267	-11443	-696	-2260	-207
5	-10722	1432	-1452	-219	-10415	1821	-2060	-105
6	-4603	2866	-1011	-69	-4907	3632	-1376	100
7	6893	2753	-456	55	6762	3557	-490	217
8	25353	-0	-157	-0	26556	-0	5	-0

INTERNAL SHELLS

POINT	T1	S	T2	M2	T1	S	T2	M2
0	33266	0	316	231	36606	1481	78	61
1	11321	-3842	33	257	12703	-3842	-155	217
2	-2866	-4490	-640	94	-2030	-4886	-1076	83
3	-10702	-3198	-1253	-112	-9857	-3448	-1938	-133
4	-13410	-984	-1556	-228	-12631	-850	-2344	-239
5	-11630	1310	-1462	-196	-11197	1893	-2153	-153
6	-5175	2880	-1030	-59	-5097	3815	-1437	57
7	6909	2819	-466	59	7173	3715	-511	194
8	26099	-0	-159	-0	27562	-0	4	-0

EXTERNAL NORTH SHELL

POINT	T1	S	T2	M2	T1	S	T2	M2
0	33387	33	315	223	36464	1468	74	53
1	11358	-3823	37	257	12626	-3831	-159	214
2	-2868	-4473	-633	101	-2071	-4863	-1075	86
3	-10691	-3182	-1243	-101	-9863	-3418	-1932	-126
4	-13345	-973	-1543	-214	-12582	-825	-2331	-227
5	-11503	1303	-1449	-183	-11083	1900	-2136	-140
6	-5046	2849	-1020	-48	-4960	3791	-1422	69
7	6873	2776	-463	65	7175	3672	-504	200
8	25587	-0	-161	-0	27099	-0	5	-0

L = 40 FT. TABLE NO.N- 29 W = 22 FT.

SHELL DATA:

L = 40.00 FT.
W = 22.00 FT.
T = 2.50 IN.
G = 50.00 LB/FT*2
QA= 150.00 LB/FT
QV= 150.00 LB/FT

FOR 60DEG.GLAZING:
K = 10.50 FT.
R = 20.00 FT.
AL= 19.87 FT.
θ = 56.91 DEG.

FOR 90DEG.GLAZING:
K = 10.50 FT.
R = 32.50 FT.
AL= 24.99 FT.
θ = 44.05 DEG.

BEAM DATA:

FOR 60DEG.GLAZING:

B1= 0.75 FT.
D1= 2.50 FT.
B2= 0.75 FT.
D2= 2.25 FT.

EXT.NORTH BEAM:

Z'= 1.29 FT.
Y'= 0.85 FT.
IZ= 1.26 FT*4
IY= 0.82 FT*4

FOR 90DEG.GLAZING:

ALL BEAMS:

B = 1.00 FT.
D = 2.50 FT.

BEAM RESULTS:

FOR 60DEG.GLAZING:

EXT.SOUTH BEAM
F = -1495 LB.
M1= 75180 LB.FT
M2= 5740 LB.FT
S1= -6024 LB.
S2= 475 LB.
INTERNAL BEAMS
F = -3699 LB.
P = 322 LB/FT
M1= 124460 LB.FT
M2= 590 LB.FT
S1= -10116 LB.
S2= 144 LB.
EXT.NORTH BEAM
P = 325 LB/FT
M1= 166555 LB.FT
M2= 33560 LB.FT
S1= -13081 LB.
S2= -2636 LB.

FOR 90DEG.GLAZING:

EXT.SOUTH BEAM
F = -6678 LB.
M1= 117336 LB.FT
M2= 9806 LB.FT
S1= -9864 LB.
S2= 1025 LB.
INTERNAL BEAMS
F = -23441 LB.
P = 360 LB/FT
M1= 159623 LB.FT
M2= 14008 LB.FT
S1= -14829 LB.
S2= 2011 LB.
EXT.NORTH BEAM
P = 363 LB/FT
M1= 165631 LB.FT
S1= -13009 LB.

SHELL RESULTS:

	SHELLS WITH 60-DEGREE GLAZING				SHELLS WITH 90-DEGREE GLAZING			
EXTERNAL SOUTH SHELL								
POINT	T1	S	T2	M2	T1	S	T2	M2
0	30235	117	54	368	27814	525	-24	-49
1	9546	-3614	-226	284	9247	-3854	-385	155
2	-3034	-4134	-907	52	-2166	-4597	-1353	47
3	-9503	-2828	-1506	-185	-8396	-3214	-2232	-147
4	-11560	-712	-1774	-296	-10942	-776	-2640	-228
5	-10016	1450	-1631	-235	-10187	1889	-2423	-114
6	-4515	2940	-1137	-60	-5088	3870	-1632	121
7	6317	2866	-507	79	6428	3861	-586	258
8	24483	-0	-160	-0	26697	-0	11	0
INTERNAL SHELLS								
POINT	T1	S	T2	M2	T1	S	T2	M2
0	33302	290	324	210	36975	1841	72	-13
1	11144	-3894	29	265	12611	-4012	-176	221
2	-2689	-4598	-729	92	-1945	-5148	-1246	95
3	-10034	-3266	-1419	-137	-9415	-3636	-2251	-142
4	-12502	-999	-1759	-264	-12047	-921	-2728	-260
5	-10925	1349	-1653	-221	-10851	1967	-2517	-162
6	-5042	2984	-1164	-58	-5204	4043	-1693	79
7	6409	2956	-519	78	6790	4002	-606	235
8	25323	-0	-161	-0	27484	-0	10	-0
EXTERNAL NORTH SHELL								
POINT	T1	S	T2	M2	T1	S	T2	M2
0	33363	306	323	204	36843	1821	70	-19
1	11151	-3884	31	264	12540	-4008	-180	218
2	-2707	-4587	-725	96	-1986	-5130	-1248	96
3	-10042	-3252	-1413	-130	-9430	-3611	-2247	-137
4	-12466	-987	-1750	-254	-12018	-897	-2718	-251
5	-10834	1349	-1643	-211	-10765	1976	-2503	-151
6	-4937	2963	-1155	-49	-5090	4026	-1680	89
7	6397	2923	-516	83	6804	3966	-599	240
8	24942	-0	-163	-0	27110	-0	10	0

L = 42 FT. TABLE NO.N- 30 W = 19 FT.

SHELL DATA:

```
L =  42.00 FT.
W =  19.00 FT.
T =   2.50 IN.
G =  50.00 LB/FT*2
QA= 150.00 LB/FT
QV= 150.00 LB/FT

FOR 60DEG.GLAZING:
K  =   9.75 FT.
R  =  12.00 FT.
AL=  17.42 FT.
Ø  =  66.23 DEG.

FOR 90DEG.GLAZING:
K  =   9.75 FT.
R  =  22.00 FT.
AL=  22.06 FT.
Ø  =  50.27 DEG.
```

BEAM DATA:

```
FOR 60DEG.GLAZING:

B1= 0.75 FT.
D1= 2.75 FT.
B2= 0.75 FT.
D2= 2.20 FT.

EXT.NORTH BEAM:

Z'= 1.41 FT.
Y'= 0.88 FT.
IZ= 1.69 FT*4
IY= 1.06 FT*4

FOR 90DEG.GLAZING:

ALL BEAMS:

B = 1.00 FT.
D = 2.75 FT.
```

BEAM RESULTS:

```
FOR 60DEG.GLAZING:

EXT.SOUTH BEAM
 F  =     5133 LB.
 M1=     89050 LB.FT
 M2=      5855 LB.FT
 S1=     -6166 LB.
 S2=       262 LB.
INTERNAL BEAMS
 F  =     4078 LB.
 P  =      320 LB/FT
 M1=    140265 LB.FT
 M2=       705 LB.FT
 S1=    -10042 LB.
 S2=       -82 LB.
EXT.NORTH BEAM
 P  =      322 LB/FT
 M1=    193968 LB.FT
 M2=     36631 LB.FT
 S1=    -14509 LB.
 S2=     -2740 LB.

FOR 90DEG.GLAZING:

EXT.SOUTH BEAM
 F  =    -1465 LB.
 M1=    140211 LB.FT
 M2=      9952 LB.FT
 S1=    -10629 LB.
 S2=       787 LB.
INTERNAL BEAMS
 F  =   -15322 LB.
 P  =      347 LB/FT
 M1=    186797 LB.FT
 M2=     14042 LB.FT
 S1=    -15536 LB.
 S2=      1608 LB.
EXT.NORTH BEAM
 P  =      352 LB/FT
 M1=    186728 LB.FT
 S1=    -13967 LB.
```

SHELL RESULTS:

EXTERNAL SOUTH SHELL

	SHELLS WITH 60-DEGREE GLAZING				SHELLS WITH 90-DEGREE GLAZING			
POINT	T1	S	T2	M2	T1	S	T2	M2
0	29521	-384	88	410	27588	110	-27	8
1	9525	-3459	-150	290	9477	-3595	-317	124
2	-3365	-3873	-666	72	-2386	-4230	-1047	25
3	-10397	-2681	-1121	-135	-9262	-2952	-1707	-125
4	-12679	-743	-1325	-234	-12040	-689	-2004	-176
5	-10788	1224	-1219	-191	-10855	1745	-1820	-72
6	-4578	2537	-856	-56	-4935	3466	-1208	121
7	6740	2438	-401	54	7119	3360	-427	221
8	24367	0	-160	-0	26857	-0	5	-0

INTERNAL SHELLS

POINT	T1	S	T2	M2	T1	S	T2	M2
0	31894	-305	339	200	36097	1146	58	61
1	11051	-3701	87	240	12817	-3739	-159	195
2	-2815	-4284	-485	105	-2032	-4722	-965	74
3	-10703	-3109	-1009	-74	-10243	-3356	-1722	-117
4	-13520	-1072	-1277	-176	-13263	-851	-2078	-204
5	-11754	1047	-1213	-150	-11718	1905	-1903	-116
6	-5279	2502	-866	-32	-5189	3644	-1264	80
7	6602	2474	-410	64	7544	3520	-446	198
8	24961	0	-164	0	27996	-0	5	-0

EXTERNAL NORTH SHELL

POINT	T1	S	T2	M2	T1	S	T2	M2
0	32087	-278	340	200	35976	1143	54	50
1	11125	-3695	90	242	12743	-3722	-161	191
2	-2807	-4284	-482	109	-2075	-4693	-962	78
3	-10714	-3108	-1007	-69	-10245	-3322	-1713	-107
4	-13515	-1071	-1274	-171	-13195	-823	-2063	-139
5	-11720	1046	-1209	-145	-11569	1810	-1884	-99
6	-5235	2493	-863	-29	-5019	3613	-1248	95
7	6596	2461	-409	66	7537	3468	-438	206
8	24787	0	-164	0	27412	-0	5	-0

L = 42 FT. TABLE NO.N- 31 W = 21 FT.

SHELL DATA:

L = 42.00 FT.
W = 21.00 FT.
T = 2.50 IN.
G = 50.00 LB/FT*2
QA= 150.00 LB/FT
QV= 150.00 LB/FT

FOR 60DEG.GLAZING:
K = 10.25 FT.
R = 20.00 FT.
AL= 18.88 FT.
θ = 54.09 DEG.

FOR 90DEG.GLAZING:
K = 10.25 FT.
R = 30.00 FT.
AL= 24.00 FT.
θ = 45.84 DEG.

BEAM DATA:

FOR 60DEG.GLAZING:

B1= 0.75 FT.
D1= 2.75 FT.
B2= 0.75 FT.
D2= 2.50 FT.

EXT.NORTH BEAM:

Z'= 1.41 FT.
Y'= 0.88 FT.
IZ= 1.69 FT*4
IY= 1.06 FT*4

FOR 90DEG.GLAZING:

ALL BEAMS:

B = 1.00 FT.
D = 2.75 FT.

BEAM RESULTS:

FOR 60DEG.GLAZING:

EXT.SOUTH BEAM
F = -1395 LB.
M1= 100654 LB.FT
M2= 5605 LB.FT
S1= -7645 LB.
S2= 437 LB.
INTERNAL BEAMS
F = -1176 LB.
P = 340 LB/FT
M1= 155907 LB.FT
M2= 26 LB.FT
S1= -11759 LB.
S2= 23 LB.
EXT.NORTH BEAM
P = 346 LB/FT
M1= 199693 LB.FT
M2= 39410 LB.FT
S1= -14937 LB.
S2= -2948 LB.

FOR 90DEG.GLAZING:

EXT.SOUTH BEAM
F = -3323 LB.
M1= 144349 LB.FT
M2= 9584 LB.FT
S1= -11131 LB.
S2= 832 LB.
INTERNAL BEAMS
F = -19217 LB.
P = 364 LB/FT
M1= 193169 LB.FT
M2= 13590 LB.FT
S1= -16416 LB.
S2= 1724 LB.
EXT.NORTH BEAM
P = 369 LB/FT
M1= 191522 LB.FT
S1= -14326 LB.

SHELL RESULTS:

SHELLS WITH 60-DEGREE GLAZING / SHELLS WITH 90-DEGREE GLAZING

EXTERNAL SOUTH SHELL

POINT	T1	S	T2	M2	T1	S	T2	M2
0	31668	104	-1	204	27964	249	-47	-66
1	10890	-3538	-238	201	9569	-3826	-388	122
2	-2738	-4162	-845	57	-2310	-4534	-1252	34
3	-10435	-2920	-1395	-107	-9141	-3166	-2036	-128
4	-13176	-768	-1646	-181	-11967	-728	-2393	-186
5	-11393	1468	-1514	-123	-10938	1920	-2180	-70
6	-4839	2975	-1056	16	-5120	3825	-1454	148
7	7304	2848	-479	109	7150	3736	-515	261
8	26190	-0	-164	0	27651	0	10	-0

INTERNAL SHELLS

POINT	T1	S	T2	M2	T1	S	T2	M2
0	34262	88	311	27	36650	1437	37	-31
1	12258	-3904	36	170	12955	-3944	-206	184
2	-2483	-4668	-650	95	-1933	-5036	-1153	82
3	-11009	-3392	-1288	-53	-10072	-3580	-2045	-119
4	-14178	-1095	-1612	-137	-13108	-894	-2468	-213
5	-12400	1324	-1522	-96	-11718	1975	-2266	-113
6	-5482	2982	-1074	30	-5337	3994	-1512	109
7	7297	2915	-489	115	7525	3887	-535	239
8	26977	0	-166	0	28628	-0	9	-0

EXTERNAL NORTH SHELL

POINT	T1	S	T2	M2	T1	S	T2	M2
0	34386	136	310	14	36549	1429	33	-41
1	12299	-3869	42	169	12883	-3933	-209	181
2	-2474	-4637	-639	104	-1989	-5011	-1152	85
3	-10972	-3365	-1272	-36	-10095	-3546	-2037	-109
4	-14063	-1080	-1591	-115	-13058	-862	-2452	-198
5	-12202	1310	-1501	-75	-11574	1986	-2245	-96
6	-5294	2932	-1059	47	-5154	3966	-1493	124
7	7230	2850	-484	123	7536	3831	-526	248
8	26204	-0	-169	-0	28020	-0	10	0

L = 42 FT. TABLE NO.N- 32 W = 23 FT.

```
SHELL DATA:

   L =  42.00 FT.
   W =  23.00 FT.
   T =   2.50 IN.
   G =  50.00 LB/FT*2
   QA= 150.00 LB/FT
   QV= 150.00 LB/FT

 FOR 60DEG.GLAZING:
   K =  11.25 FT.
   R =  20.00 FT.
   AL=  20.85 FT.
   θ =  59.74 DEG.

 FOR 90DEG.GLAZING:
   K =  11.25 FT.
   R =  35.00 FT.
   AL=  26.21 FT.
   θ =  42.91 DEG.

BEAM DATA:

 FOR 60DEG.GLAZING:

   B1= 0.75 FT.
   D1= 2.75 FT.
   B2= 0.75 FT.
   D2= 2.50 FT.

   EXT.NORTH BEAM:

   Z'= 1.41 FT.
   Y'= 0.88 FT.
   IZ= 1.69 FT*4
   IY= 1.06 FT*4

 FOR 90DEG.GLAZING:

   ALL BEAMS:

   B = 1.00 FT.
   D = 2.75 FT.
```

```
BEAM RESULTS:

FOR 60DEG.GLAZING:

EXT.SOUTH BEAM
 F =    -1492 LB.
 M1=    91808 LB.FT
 M2=     5979 LB.FT
 S1=    -6995 LB.
 S2=      469 LB.
INTERNAL BEAMS
 F =    -3232 LB.
 P =      339 LB/FT
 M1=   146601 LB.FT
 M2=      750 LB.FT
 S1=   -11277 LB.
 S2=      135 LB.
EXT.NORTH BEAM
 P =      341 LB/FT
 M1=   200421 LB.FT
 M2=    38779 LB.FT
 S1=   -14991 LB.
 S2=    -2901 LB.

FOR 90DEG.GLAZING:

EXT.SOUTH BEAM
 F =    -5575 LB.
 M1=   145082 LB.FT
 M2=     9779 LB.FT
 S1=   -11419 LB.
 S2=      933 LB.
INTERNAL BEAMS
 F =   -23379 LB.
 P =      379 LB/FT
 M1=   195872 LB.FT
 M2=    13937 LB.FT
 S1=   -17047 LB.
 S2=     1908 LB.
EXT.NORTH BEAM
 P =      383 LB/FT
 M1=   196573 LB.FT
 S1=   -14704 LB.
```

SHELL RESULTS:

EXTERNAL SOUTH SHELL

	SHELLS WITH 60-DEGREE GLAZING				SHELLS WITH 90-DEGREE GLAZING			
POINT	T1	S	T2	M2	T1	S	T2	M2
0	29631	112	40	344	28003	417	-55	-112
1	9337	-3541	-222	277	9445	-4011	-441	137
2	-2926	-4052	-878	51	-2190	-4778	-1443	48
3	-9183	-2790	-1457	-183	-8715	-3354	-2355	-140
4	-11179	-746	-1717	-288	-11481	-810	-2779	-213
5	-9738	1349	-1582	-217	-10731	1991	-2548	-85
6	-4478	2805	-1109	-34	-5320	4072	-1713	167
7	6001	2757	-504	104	6825	4048	-612	301
8	23651	0	-170	0	27852	0	16	0

INTERNAL SHELLS

POINT	T1	S	T2	M2	T1	S	T2	M2
0	32588	242	339	152	36939	1749	33	-93
1	10949	-3857	55	251	12817	-4129	-232	196
2	-2575	-4556	-685	97	-1881	-5303	-1329	95
3	-9759	-3265	-1362	-127	-9656	-3772	-2363	-132
4	-12200	-1057	-1701	-251	-12548	-966	-2858	-241
5	-10725	1240	-1607	-202	-11398	2053	-2636	-129
6	-5048	2854	-1138	-31	-5461	4233	-1772	127
7	6104	2854	-516	103	7161	4184	-631	280
8	24559	0	-171	0	28628	0	15	-0

EXTERNAL NORTH SHELL

POINT	T1	S	T2	M2	T1	S	T2	M2
0	32630	250	339	149	36844	1733	30	-100
1	10954	-3853	55	251	12750	-4126	-236	193
2	-2588	-4550	-683	99	-1938	-5285	-1330	97
3	-9768	-3257	-1359	-123	-9689	-3742	-2358	-125
4	-12184	-1049	-1697	-245	-12522	-934	-2845	-229
5	-10675	1241	-1602	-196	-11285	2068	-2617	-114
6	-4986	2844	-1134	-26	-5302	4213	-1753	140
7	6102	2836	-515	106	7186	4135	-622	287
8	24342	0	-171	0	28111	0	15	-0

L = 44 FT. TABLE NO.N- 33 W = 20 FT.

SHELL DATA:

```
L =  44.00 FT.
W =  20.00 FT.
T =   2.50 IN.
G =  50.00 LB/FT*2
QA= 150.00 LB/FT
QV= 150.00 LB/FT

FOR 60DEG.GLAZING:
 K =   9.75 FT.
 R =  15.00 FT.
 AL=  18.47 FT.
 θ =  70.54 DEG.

FOR 90DEG.GLAZING:
 K =   9.75 FT.
 R =  25.00 FT.
 AL=  23.06 FT.
 θ =  52.85 DEG.
```

BEAM DATA:

```
FOR 60DEG.GLAZING:

 B1= 0.75 FT.
 D1= 2.75 FT.
 B2= 0.75 FT.
 D2= 2.50 FT.

 EXT.NORTH BEAM:

 Z'= 1.41 FT.
 Y'= 0.88 FT.
 IZ= 1.69 FT*4
 IY= 1.06 FT*4

FOR 90DEG.GLAZING:

 ALL BEAMS:

 B = 1.00 FT.
 D = 2.75 FT.
```

BEAM RESULTS:

```
FOR 60DEG.GLAZING:

EXT.SOUTH BEAM
 F =     7626 LB.
 M1=    92015 LB.FT
 M2=     6200 LB.FT
 S1=    -5873 LB.
 S2=      204 LB.
INTERNAL BEAMS
 F =     5048 LB.
 P =      318 LB/FT
 M1=   148358 LB.FT
 M2=      911 LB.FT
 S1=   -10064 LB.
 S2=      -93 LB.
EXT.NORTH BEAM
 P =      317 LB/FT
 M1=   211768 LB.FT
 M2=    39560 LB.FT
 S1=   -15120 LB.
 S2=    -2825 LB.

FOR 90DEG.GLAZING:

EXT.SOUTH BEAM
 F =    -2173 LB.
 M1=   150056 LB.FT
 M2=    10918 LB.FT
 S1=   -10917 LB.
 S2=      844 LB.
INTERNAL BEAMS
 F =   -17231 LB.
 P =      344 LB/FT
 M1=   200512 LB.FT
 M2=    15356 LB.FT
 S1=   -15995 LB.
 S2=     1695 LB.
EXT.NORTH BEAM
 P =      348 LB/FT
 M1=   203959 LB.FT
 S1=   -14563 LB.
```

SHELL RESULTS:

EXTERNAL SOUTH SHELL

POINT	T1 (60°)	S (60°)	T2 (60°)	M2 (60°)	T1 (90°)	S (90°)	T2 (90°)	M2 (90°)
	SHELLS WITH 60-DEGREE GLAZING				SHELLS WITH 90-DEGREE GLAZING			
0	30896	-545	123	552	29601	155	-13	84
1	9361	-3740	-151	367	9902	-3773	-316	175
2	-4045	-4078	-711	74	-2662	-4410	-1079	31
3	-11012	-2760	-1192	-195	-9711	-3055	-1764	-161
4	-13084	-715	-1401	-324	-12479	-704	-2071	-234
5	-11069	1330	-1285	-276	-11268	1811	-1882	-122
6	-4757	2699	-898	-110	-5225	3604	-1253	96
7	6993	2600	-417	35	7345	3514	-446	222
8	25925	0	-164	-0	28379	-0	-1	0

INTERNAL SHELLS

POINT	T1	S	T2	M2	T1	S	T2	M2
0	33943	-360	357	347	38817	1230	73	156
1	11206	-3961	77	325	13365	-3954	-159	257
2	-3429	-4500	-539	114	-2433	-4933	-1005	83
3	-11390	-3197	-1093	-132	-10839	-3459	-1790	-158
4	-14032	-1037	-1366	-271	-13764	-846	-2154	-270
5	-12105	1179	-1289	-242	-12102	1894	-1970	-176
6	-5449	2695	-914	-93	-5411	3793	-1310	49
7	6956	2660	-427	42	7808	3675	-465	196
8	26699	0	-167	0	29474	-0	-1	0

EXTERNAL NORTH SHELL

POINT	T1	S	T2	M2	T1	S	T2	M2
0	34081	-352	358	352	38658	1219	69	146
1	11259	-3968	77	327	13279	-3941	-162	253
2	-3427	-4511	-540	113	-2478	-4907	-1004	86
3	-11417	-3207	-1095	-135	-10841	-3428	-1784	-149
4	-14073	-1040	-1370	-274	-13703	-820	-2141	-257
5	-12150	1183	-1292	-246	-11968	1900	-1953	-161
6	-5479	2706	-917	-96	-5255	3767	-1295	63
7	6976	2672	-427	40	7806	3628	-458	204
8	26835	0	-166	0	28945	-0	-1	-0

L = 44 FT. TABLE NO.N- 34 W = 22 FT.

SHELL DATA:

```
L =  44.00 FT.
W OUTH SHELL
G =  50.00 LB/FT*2
QA= 150.00 LB/FT
QV= 150.00 LB/FT

FOR 60DEG.GLAZING:
K =  10.75 FT.
R =  20.00 FT.
AL=  19.86 FT.
θ =  56.90 DEG.

FOR 90DEG.GLAZING:
K =  10.75 FT.
R =  32.50 FT.
AL=  25.11 FT.
θ =  44.26 DEG.
```

BEAM DATA:

```
FOR 60DEG.GLAZING:

B1= 0.75 FT.
D1= 2.75 FT.
B2= 0.75 FT.
D2= 2.50 FT.

EXT.NORTH BEAM:

Z'= 1.41 FT.
Y'= 0.88 FT.
IZ= 1.69 FT*4
IY= 1.06 FT*4

FOR 90DEG.GLAZING:

ALL BEAMS:

B = 1.00 FT.
D = 2.75 FT.
```

BEAM RESULTS:

```
FOR 60DEG.GLAZING:
 F =       361 LB.
 M1=    104307 LB.FT
 M2=      6251 LB.FT
 S1=     -7401 LB.
 S2=       413 LB.
INTERNAL BEAMS
 F =      -909 LB.
 P =       334 LB/FT
 M1=    164109 LB.FT
 M2=       422 LB.FT
 S1=    -11781 LB.
 S2=        41 LB.
EXT.NORTH BEAM
 P =       338 LB/FT
 M1=    218350 LB.FT
 M2=     42207 LB.FT
 S1=    -15590 LB.
 S2=     -3014 LB.

FOR 90DEG.GLAZING:

EXT.SOUTH BEAM
 F =     -4366 LB.
 M1=    158510 LB.FT
 M2=     10639 LB.FT
 S1=    -11738 LB.
 S2=       906 LB.
INTERNAL BEAMS
 F =    -22696 LB.
 P =       369 LB/FT
 M1=    212715 LB.FT
 M2=     15084 LB.FT
 S1=    -17406 LB.
 S2=      1876 LB.
EXT.NORTH BEAM
 P =       375 LB/FT
 M1=    212543 LB.FT
 S1=    -15176 LB.
```

SHELL RESULTS:

	SHELLS WITH 60-DEGREE GLAZING				SHELLS WITH 90-DEGREE GLAZING			
POINT	T1	S	T2	M2	T1	S	T2	M2
0	33412	-26	33	346	30705	312	-48	-65
1	10971	-3827	-239	276	10471	-4150	-420	139
2	-3321	-4397	-890	60	-2534	-4920	-1362	39
3	-11085	-3035	-1467	-163	-9968	-3429	-2216	-145
4	-13689	-771	-1726	-265	-13026	-777	-2604	-213
5	-11777	1553	-1583	-201	-11905	2099	-2372	-87
6	-5060	3122	-1101	-32	-5585	4171	-1583	153
7	7561	2999	-494	95	7795	4075	-561	280
8	27667	-0	-164	0	30248	-0	13	-0

INTERNAL SHELLS

POINT	T1	S	T2	M2	T1	S	T2	M2
0	36416	65	328	154	40297	1620	39	-23
1	12623	-4149	32	243	14187	-4278	-226	209
2	-2920	-4898	-695	101	-2137	-5468	-1258	90
3	-11629	-3517	-1362	-105	-10999	-3877	-2229	-137
4	-14728	-1105	-1694	-219	-14273	-954	-2687	-244
5	-12837	1411	-1594	-174	-12746	2164	-2467	-136
6	-5729	3139	-1122	-20	-5808	4357	-1646	109
7	7580	3076	-505	99	8209	4239	-582	256
8	28540	-0	-165	-0	31303	-0	12	-0

EXTERNAL NORTH SHELL

POINT	T1	S	T2	M2	T1	S	T2	M2
0	36541	97	328	146	40176	1609	36	-33
1	12660	-4131	36	243	14107	-4267	-229	205
2	-2925	-4881	-688	107	-2196	-5441	-1257	93
3	-11624	-3501	-1353	-93	-11021	-3841	-2221	-126
4	-14667	-1094	-1682	-204	-14220	-920	-2671	-228
5	-12709	1405	-1581	-160	-12596	2175	-2445	-117
6	-5596	3109	-1112	-8	-5619	4328	-1626	125
7	7548	3033	-502	105	8220	4181	-573	265
8	28022	-0	-168	-0	30674	-0	13	-0

L = 44 FT. TABLE NO.N- 35 W = 24 FT.

SHELL DATA:

L = 44.00 FT.
W = 24.00 FT.
T = 2.50 IN.
G = 50.00 LB/FT*2
QA= 150.00 LB/FT
QV= 150.00 LB/FT

FOR 60DEG.GLAZING:
K = 11.25 FT.
R = 22.50 FT.
AL= 21.62 FT.
θ = 55.06 DEG.

FOR 90DEG.GLAZING:
K = 11.25 FT.
R = 35.00 FT.
AL= 27.18 FT.
θ = 44.50 DEG.

BEAM DATA:

FOR 60DEG.GLAZING:

B1= 0.75 FT.
D1= 2.75 FT.
B2= 0.75 FT.
D2= 2.50 FT.

ÆXT.NORTH BEAM:

Z'= 1.41 FT.
Y'= 0.88 FT.
IZ= 1.69 FT*4
IY= 1.06 FT*4

FOR 90DEG.GLAZING:

ALL BEAMS:

B = 1.00 FT.
D = 2.75 FT.

BEAM RESULTS:

FOR 60DEG.GLAZING:

EXT.SOUTH BEAM
F = -2773 LB.
M1= 105139 LB.FT
M2= 6195 LB.FT
S1= -7742 LB.
S2= 498 LB.
INTERNAL BEAMS
F = -4683 LB.
P = 344 LB/FT
M1= 167097 LB.FT
M2= 371 LB.FT
S1= -12369 LB.
S2= 141 LB.
EXT.NORTH BEAM
P = 348 LB/FT
M1= 221576 LB.FT
M2= 43492 LB.FT
S1= -15820 LB.
S2= -3105 LB.

FOR 90DEG.GLAZING:

EXT.SOUTH BEAM
F = -7326 LB.
M1= 156421 LB.FT
M2= 10716 LB.FT
S1= -11881 LB.
S2= 1019 LB.
INTERNAL BEAMS
F = -26900 LB.
P = 379 LB/FT
M1= 211935 LB.FT
M2= 15246 LB.FT
S1= -17764 LB.
S2= 2039 LB.
EXT.NORTH BEAM
P = 383 LB/FT
M1= 215570 LB.FT
S1= -15392 LB.

SHELL RESULTS:

EXTERNAL SOUTH SHELL

	SHELLS WITH 60-DEGREE GLAZING				SHELLS WITH 90-DEGREE GLAZING			
POINT	T1	S	T2	M2	T1	S	T2	M2
0	33236	198	17	325	30171	523	-48	-68
1	10703	-3886	-273	282	9985	-4168	-437	176
2	-3171	-4490	-1002	55	-2392	-4953	-1468	55
3	-10443	-3086	-1649	-190	-9129	-3462	-2402	-168
4	-12849	-771	-1939	-301	-11925	-839	-2837	-260
5	-11186	1613	-1781	-224	-11188	2043	-2605	-124
6	-5022	3259	-1240	-27	-5665	4206	-1758	153
7	7111	3169	-551	118	7044	4211	-632	310
8	27253	-0	-171	-0	29485	-0	10	-0

INTERNAL SHELLS

POINT	T1	S	T2	M2	T1	S	T2	M2
0	36653	334	330	147	39923	1921	41	-36
1	12422	-4242	14	263	13542	-4319	-228	244
2	-2877	-5034	-807	102	-2175	-5513	-1361	105
3	-11122	-3585	-1558	-135	-10203	-3885	-2421	-164
4	-13961	-1088	-1928	-266	-13055	-977	-2924	-295
5	-12227	1511	-1810	-212	-11835	2126	-2696	-175
6	-5605	3319	-1272	-27	-5750	4377	-1815	108
7	7242	3275	-564	116	7410	4345	-651	286
8	28237	-0	-171	-0	30220	-0	9	0

EXTERNAL NORTH SHELL

POINT	T1	S	T2	M2	T1	S	T2	M2
0	36704	351	329	138	39803	1900	38	-43
1	12421	-4230	15	262	13467	-4317	-232	241
2	-2904	-5019	-802	107	-2231	-5495	-1363	106
3	-11133	-3566	-1550	-126	-10235	-3856	-2418	-157
4	-13913	-1072	-1916	-253	-13031	-946	-2912	-283
5	-12106	1511	-1797	-198	-11728	2141	-2679	-161
6	-5465	3292	-1261	-15	-5595	4358	-1798	121
7	7229	3232	-560	122	7438	4299	-642	294
8	27731	-0	-173	-0	29720	-0	10	0

L = 46 FT. TABLE NO.N- 36 W = 21 FT.

SHELL DATA:

L = 46.00 FT.
W = 21.00 FT.
T = 2.50 IN.
G = 50.00 LB/FT*2
QA= 150.00 LB/FT
QV= 150.00 LB/FT

FOR 60DEG.GLAZING:
K = 10.50 FT.
R = 15.00 FT.
AL= 19.54 FT.
θ = 74.63 DEG.

FOR 90DEG.GLAZING:
K = 10.50 FT.
R = 27.50 FT.
AL= 24.26 FT.
θ = 50.54 DEG.

BEAM DATA:

FOR 60DEG.GLAZING:

B1= 0.75 FT.
D1= 3.00 FT.
B2= 0.75 FT.
D2= 2.50 FT.

EXT.NORTH BEAM:

Z'= 1.48 FT.
Y'= 0.94 FT.
IZ= 1.97 FT*4
IY= 1.27 FT*4

FOR 90DEG.GLAZING:

ALL BEAMS:

B = 1.00 FT.
D = 3.00 FT.

BEAM RESULTS:

FOR 60DEG.GLAZING:

EXT.SOUTH BEAM
F = 10488 LB.
M1= 91847 LB.FT
M2= 7009 LB.FT
S1= -5362 LB.
S2= 172 LB.
INTERNAL BEAMS
F = 5804 LB.
P = 336 LB/FT
M1= 169688 LB.FT
M2= 1171 LB.FT
S1= -10964 LB.
S2= -92 LB.
EXT.NORTH BEAM
P = 327 LB/FT
M1= 242530 LB.FT
M2= 44568 LB.FT
S1= -16564 LB.
S2= -3044 LB.

FOR 90DEG.GLAZING:

EXT.SOUTH BEAM
F = -1549 LB.
M1= 183575 LB.FT
M2= 10829 LB.FT
S1= -12687 LB.
S2= 781 LB.
INTERNAL BEAMS
F = -18009 LB.
P = 365 LB/FT
M1= 243895 LB.FT
M2= 15230 LB.FT
S1= -18490 LB.
S2= 1640 LB.
EXT.NORTH BEAM
P = 370 LB/FT
M1= 240248 LB.FT
S1= -16408 LB.

SHELL RESULTS:

	SHELLS WITH 60-DEGREE GLAZING				SHELLS WITH 90-DEGREE GLAZING			
EXTERNAL SOUTH SHELL								
POINT	T1	S	T2	M2	T1	S	T2	M2
0	31888	-716	173	682	29901	106	-48	-1
1	8866	-3973	-133	442	10195	-3914	-367	144
2	-4821	-4195	-721	88	-2628	-4592	-1164	29
3	-11380	-2760	-1209	-225	-10022	-3199	-1883	-149
4	-12910	-676	-1410	-371	-13042	-741	-2206	-208
5	-10613	1336	-1284	-314	-11836	1913	-2004	-81
6	-4436	2651	-894	-125	-5458	3806	-1333	150
7	6796	2536	-417	39	7759	3702	-472	269
8	24941	-0	-171	-0	29554	0	4	-0
INTERNAL SHELLS								
POINT	T1	S	T2	M2	T1	S	T2	M2
0	32828	-396	379	297	38969	1230	30	50
1	10798	-3916	107	314	13719	-4048	-207	217
2	-3283	-4441	-498	116	-2271	-5091	-1080	80
3	-10869	-3182	-1044	-129	-11044	-3605	-1898	-141
4	-13383	-1096	-1318	-266	-14277	-897	-2280	-239
5	-11617	1047	-1252	-231	-12674	1977	-2086	-130
6	-5387	2533	-896	-73	-5679	3982	-1387	106
7	6395	2536	-428	62	8182	3856	-490	245
8	25367	-0	-179	0	30626	0	4	-0
EXTERNAL NORTH SHELL								
POINT	T1	S	T2	M2	T1	S	T2	M2
0	33107	-401	384	321	38855	1224	26	38
1	10933	-3954	108	321	13640	-4034	-210	212
2	-3244	-4494	-505	108	-2331	-5063	-1079	84
3	-10930	-3232	-1059	-150	-11063	-3568	-1890	-129
4	-13571	-1126	-1340	-296	-14214	-864	-2264	-221
5	-11924	1059	-1276	-262	-12507	1987	-2065	-110
6	-5679	2598	-914	-98	-5475	3950	-1369	123
7	6477	2626	-432	49	8188	3797	-482	254
8	26504	-0	-174	-0	29941	-0	4	-0

L = 46 FT.   TABLE NO.N- 37   W = 23 FT.

SHELL DATA:

```
L  =  46.00 FT.
W  =  23.00 FT.
T  =   2.50 IN.
G  =  50.00 LB/FT*2
QA= 150.00 LB/FT
QV= 150.00 LB/FT

FOR 60DEG.GLAZING:
K  = 11.50 FT.
R  = 20.00 FT.
AL= 20.85 FT.
θ  = 59.73 DEG.

FOR 90DEG.GLAZING:
K  = 11.50 FT.
R  = 32.50 FT.
AL= 26.44 FT.
O  = 46.61 DEG.
```

BEAM DATA:

```
FOR 60DEG.GLAZING:

B1= 0.75 FT.
D1= 3.00 FT.
B2= 0.75 FT.
D2= 2.50 FT.

EXT.NORTH BEAM:

Z'= 1.48 FT.
Y'= 0.94 FT.
IZ= 1.97 FT*4
IY= 1.27 FT*4

FOR 90DEG.GLAZING:

ALL BEAMS:

B = 1.00 FT.
D = 3.00 FT.
```

BEAM RESULTS:

```
FOR 60DEG.GLAZIN

EXT.SOUTH BEAM
F  =     1558 LB.
M1=   107271 LB.
M2=     7112 LB.FT
S1=    -7178 LB.
S2=      420 LB.
INTERNAL BEAMS
F  =      272 LB.
P  =      349 LB/FT
M1=   188321 LB.FT
M2=      590 LB.FT
S1=   -12809 LB.
S2=       19 LB.
EXT.NORTH BEAM
P  =      346 LB/FT
M1=   249292 LB.FT
M2=    47211 LB.FT
S1=   -17026 LB.
S2=    -3224 LB.

FOR 90DEG.GLAZING:

EXT.SOUTH BEAM
F  =    -3992 LB.
M1=   185788 LB.FT
M2=    10960 LB.FT
S1=   -13090 LB.
S2=      876 LB.
INTERNAL BEAMS
F  =   -22720 LB.
P  =      381 LB/FT
M1=   248862 LB.FT
M2=    15497 LB.FT
S1=   -19314 LB.
S2=     1823 LB.
EXT.NORTH BEAM
P  =      386 LB/FT
M1=   246779 LB.FT
S1=   -16854 LB.
```

SHELL RESULTS:

EXTERNAL SOUTH SHELL

	SHELLS WITH 60-DEGREE GLAZING				SHELLS WITH 90-DEGREE GLAZING			
POINT	T1	S	T2	M2	T1	S	T2	M2
0	35101	-106	65	473	30093	273	-61	-61
1	10896	-4047	-231	350	10191	-4120	-426	152
2	-3975	-4540	-909	76	-2476	-4872	-1356	40
3	-11590	-3062	-1497	-190	-9665	-3414	-2200	-159
4	-13794	-733	-1748	-309	-12688	-819	-2587	-230
5	-11579	1588	-1593	-237	-11749	2019	-2363	-90
6	-4819	3121	-1103	-45	-5686	4101	-1582	175
7	7565	2971	-495	100	7545	4047	-563	313
8	27167	0	-168	0	30002	0	11	-0

INTERNAL SHELLS

POINT	T1	S	T2	M2	T1	S	T2	M2
0	35336	-19	345	107	39443	1552	20	-28
1	12241	-4124	53	232	13752	-4244	-240	219
2	-2840	-4852	-661	103	-2145	-5393	-1256	91
3	-11283	-3507	-1318	-100	-10675	-3829	-2211	-151
4	-14309	-1156	-1649	-212	-13849	-974	-2663	-261
5	-12534	1304	-1558	-161	-12487	2085	-2448	-138
6	-5702	3008	-1103	3	-5848	4269	-1638	131
7	7202	2974	-504	122	7923	4189	-582	290
8	27593	0	-174	-0	30883	0	11	0

EXTERNAL NORTH SHELL

POINT	T1	S	T2	M2	T1	S	T2	M2
0	35611	3	347	116	39340	1538	16	-37
1	12358	-4137	55	236	13676	-4238	-243	215
2	-2819	-4876	-663	102	-2210	-5371	-1257	93
3	-11321	-3529	-1324	-106	-10711	-3795	-2206	-142
4	-14388	-1167	-1658	-221	-13812	-939	-2649	-246
5	-12637	1309	-1567	-170	-12348	2100	-2428	-121
6	-5786	3030	-1110	-5	-5658	4244	-1619	147
7	7235	3003	-506	118	7947	4134	-573	299
8	27928	0	-173	0	30267	0	11	0

L = 46 FT. TABLE NO.N- 38 W = 25 FT.

SHELL DATA:

L = 46.00 FT.
W = 25.00 FT.
T = 2.50 IN.
G = 50.00 LB/FT*2
QA= 150.00 LB/FT
QV= 150.00 LB/FT

FOR 60DEG.GLAZING:
K = 12.00 FT.
R = 22.50 FT.
AL= 22.59 FT.
θ = 57.53 DEG.

FOR 90DEG.GLAZING:
K = 12.00 FT.
R = 40.00 FT.
AL= 28.32 FT.
θ = 40.56 DEG.

BEAM DATA:

FOR 60DEG.GLAZING:

B1= 0.75 FT.
D1= 3.00 FT.
B2= 0.75 FT.
D2= 2.50 FT.

EXT.NORTH BEAM:

Z'= 1.48 FT.
Y'= 0.94 FT.
IZ= 1.97 FT*4
IY= 1.27 FT*4

FOR 90DEG.GLAZING:

ALL BEAMS:

B = 1.00 FT.
D = 3.00 FT.

BEAM RESULTS:

FOR 60DEG.GLAZING:

EXT.SOUTH BEAM
F = -2473 LB.
M1= 109331 LB.FT
M2= 7067 LB.FT
S1= -7665 LB.
S2= 525 LB.
INTERNAL BEAMS
F = -3637 LB.
P = 360 LB/FT
M1= 192681 LB.FT
M2= 520 LB.FT
S1= -13511 LB.
S2= 118 LB.
EXT.NORTH BEAM
P = 358 LB/FT
M1= 253179 LB.FT
M2= 48825 LB.FT
S1= -17291 LB.
S2= -3335 LB.

FOR 90DEG.GLAZING:

EXT.SOUTH BEAM
F = -4818 LB.
M1= 194801 LB.FT
M2= 10336 LB.FT
S1= -13791 LB.
S2= 864 LB.
INTERNAL BEAMS
F = -26073 LB.
P = 402 LB/FT
M1= 260904 LB.FT
M2= 14686 LB.FT
S1= -20481 LB.
S2= 1885 LB.
EXT.NORTH BEAM
P = 407 LB/FT
M1= 253688 LB.FT
S1= -17326 LB.

SHELL RESULTS:

	SHELLS WITH 60-DEGREE GLAZING				SHELLS WITH 90-DEGREE GLAZING			
EXTERNAL SOUTH SHELL								
POINT	T1	S	T2	M2	T1	S	T2	M2
0	35162	169	41	433	31028	329	-87	-187
1	10746	-4080	-267	345	10564	-4530	-539	134
2	-3769	-4614	-1020	68	-2522	-5373	-1660	54
3	-10924	-3099	-1671	-214	-10030	-3759	-2677	-146
4	-12951	-727	-1951	-338	-13269	-864	-3145	-221
5	-10998	1644	-1780	-251	-12375	2328	-2876	-67
6	-4789	3244	-1234	-32	-6018	4682	-1928	221
7	7122	3126	-550	129	8038	4617	-685	367
8	26764	-0	-176	0	31947	-0	20	-0
INTERNAL SHELLS								
POINT	T1	S	T2	M2	T1	S	T2	M2
0	35730	248	345	90	40545	1781	-7	-176
1	12137	-4215	34	250	14254	-4617	-323	192
2	-2791	-4991	-769	107	-2093	-5912	-1535	105
3	-10853	-3580	-1507	-126	-10976	-4205	-2678	-135
4	-13663	-1141	-1875	-254	-14400	-1042	-3222	-247
5	-12034	1408	-1767	-192	-13110	2382	-2965	-112
6	-5612	3196	-1248	1	-6199	4848	-1988	180
7	6954	3179	-561	143	8385	4761	-705	344
8	27507	-0	-180	0	32798	-0	20	-0
EXTERNAL NORTH SHELL								
POINT	T1	S	T2	M2	T1	S	T2	M2
0	35896	264	346	96	40469	1766	-10	-186
1	12212	-4222	36	252	14180	-4613	-327	188
2	-2769	-5007	-770	106	-2171	-5890	-1536	107
3	-10865	-3596	-1511	-131	-11028	-4167	-2671	-125
4	-13709	-1152	-1881	-261	-14370	-999	-3206	-231
5	-12109	1409	-1775	-200	-12959	2403	-2941	-92
6	-5687	3212	-1255	-5	-5979	4821	-1964	198
7	6967	3203	-563	139	8422	4695	-693	355
8	27781	-0	-179	0	32090	-0	20	-0

TABLE NO.N- 39

L = 48 FT. W = 22 FT.

SHELL DATA:

L = 48.00 FT.
W = 22.00 FT.
T = 2.50 IN.
G = 50.00 LB/FT*2
QA= 150.00 LB/FT
QV= 150.00 LB/FT

FOR 60DEG.GLAZING:
K = 11.00 FT.
R = 17.50 FT.
AL= 20.15 FT.
θ = 65.96 DEG.

FOR 90DEG.GLAZING:
K = 11.00 FT.
R = 27.50 FT.
AL= 25.50 FT.
θ = 53.13 DEG.

BEAM DATA:

FOR 60DEG.GLAZING:

B1= 0.75 FT.
D1= 3.00 FT.
B2= 0.75 FT.
D2= 2.50 FT.

EXT.NORTH BEAM:

Z'= 1.48 FT.
Y'= 0.94 FT.
IZ= 1.97 FT*4
IY= 1.27 FT*4

FOR 90DEG.GLAZING:

ALL BEAMS:

B = 1.00 FT.
D = 3.00 FT.

BEAM RESULTS:

FOR 60DEG.GLAZING:

EXT.SOUTH BEAM
F = 8306 LB.
M1= 108052 LB.FT
M2= 7509 LB.FT
S1= -6375 LB.
S2= 253 LB.
INTERNAL BEAMS
F = 4439 LB.
P = 338 LB/FT
M1= 194342 LB.FT
M2= 876 LB.FT
S1= -12253 LB.
S2= -72 LB.
EXT.NORTH BEAM
P = 332 LB/FT
M1= 266695 LB.FT
M2= 49352 LB.FT
S1= -17455 LB.
S2= -3230 LB.

FOR 90DEG.GLAZING:

EXT.SOUTH BEAM
F = -3025 LB.
M1= 192528 LB.FT
M2= 12404 LB.FT
S1= -12888 LB.
S2= 898 LB.
INTERNAL BEAMS
F = -20967 LB.
P = 363 LB/FT
M1= 257751 LB.FT
M2= 17424 LB.FT
S1= -18916 LB.
S2= 1810 LB.
EXT.NORTH BEAM
P = 367 LB/FT
M1= 261867 LB.FT
S1= -17139 LB.

SHELL RESULTS:

	SHELLS WITH 60-DEGREE GLAZING				SHELLS WITH 90-DEGREE GLAZING			
EXTERNAL SOUTH SHELL								
POINT	T1	S	T2	M2	T1	S	T2	M2
0	36412	-544	125	652	31698	198	-28	88
1	10782	-4284	-194	436	10500	-4050	-352	202
2	-4940	-4643	-841	93	-2810	-4732	-1178	34
3	-12878	-3083	-1388	-220	-10172	-3290	-1920	-192
4	-14993	-717	-1616	-364	-13111	-793	-2256	-275
5	-12398	1603	-1467	-302	-11998	1902	-2056	-139
6	-5038	3111	-1013	-105	-5759	3859	-1375	125
7	8177	2943	-459	59	7676	3805	-491	275
8	28957	-0	-166	0	30543	-0	-0	-0
INTERNAL SHELLS								
POINT	T1	S	T2	M2	T1	S	T2	M2
0	37049	-291	361	251	41603	1372	55	157
1	12562	-4254	68	298	14137	-4234	-188	288
2	-3423	-4899	-599	115	-2628	-5271	-1101	90
3	-12308	-3515	-1204	-122	-11369	-3694	-1948	-189
4	-15392	-1160	-1507	-253	-14396	-922	-2339	-316
5	-13368	1280	-1422	-210	-12770	1993	-2141	-198
6	-6027	2957	-1008	-46	-5882	4044	-1429	74
7	7660	2919	-467	87	8130	3953	-509	247
8	29209	-0	-174	-0	31506	-0	-0	-0
EXTERNAL NORTH SHELL								
POINT	T1	S	T2	M2	T1	S	T2	M2
0	37415	-280	364	269	41446	1356	51	148
1	12718	-4285	69	305	14049	-4225	-192	284
2	-3402	-4944	-604	111	-2681	-5248	-1102	92
3	-12386	-3555	-1217	-137	-11385	-3664	-1943	-181
4	-15557	-1179	-1524	-274	-14350	-894	-2328	-303
5	-13593	1293	-1441	-232	-12643	2003	-2126	-183
6	-6215	3006	-1021	-63	-5721	4021	-1414	87
7	7739	2981	-470	78	8142	3908	-502	255
8	29976	-0	-171	0	30972	-0	-0	-0

L = 48 FT. TABLE NO.N- 40 W = 24 FT.

SHELL DATA:

```
L =  48.00 FT.
W =  24.00 FT.
T =   2.50 IN.
G =  50.00 LB/FT*2
QA= 150.00 LB/FT
QV= 150.00 LB/FT

FOR 60DEG.GLAZING:
K =  11.50 FT.
R =  20.00 FT.
AL=  21.86 FT.
θ =  62.63 DEG.

FOR 90DEG.GLAZING:
K =  11.50 FT.
R =  32.50 FT.
AL=  27.42 FT.
θ =  48.34 DEG.
```

BEAM DATA:

```
FOR 60DEG.GLAZING:

B1= 0.75 FT.
D1= 3.00 FT.
B2= 0.75 FT.
D2= 2.50 FT.

EXT.NORTH BEAM:

Z'= 1.48 FT.
Y'= 0.94 FT.
IZ= 1.97 FT*4
IY= 1.27 FT*4

FOR 90DEG.GLAZING:

ALL BEAMS:

B = 1.00 FT.
D = 3.00 FT.
```

BEAM RESULTS:

```
FOR 60DEG.GLAZING:

EXT.SOUTH BEAM
 F =      2991 LB.
 M1=    111839 LB.FT
 M2=      7516 LB.FT
 S1=     -7059 LB.
 S2=       390 LB.
INTERNAL BEAMS
 F =      -211 LB.
 P =       350 LB/FT
 M1=    200519 LB.FT
 M2=       765 LB.FT
 S1=    -13118 LB.
 S2=        40 LB.
EXT.NORTH BEAM
 P =       345 LB/FT
 M1=    271190 LB.FT
 M2=     51260 LB.FT
 S1=    -17749 LB.
 S2=     -3355 LB.

FOR 90DEG.GLAZING:

EXT.SOUTH BEAM
 F =     -5605 LB.
 M1=    198683 LB.FT
 M2=     11942 LB.FT
 S1=    -13546 LB.
 S2=       955 LB.
INTERNAL BEAMS
 F =    -26002 LB.
 P =       381 LB/FT
 M1=    267061 LB.FT
 M2=     16854 LB.FT
 S1=    -20021 LB.
 S2=      1941 LB.
EXT.NORTH BEAM
 P =       385 LB/FT
 M1=    268481 LB.FT
 S1=    -17572 LB.
```

SHELL RESULTS:

EXTERNAL SOUTH SHELL

POINT	T1 (60-DEGREE GLAZING)	S	T2	M2	T1 (90-DEGREE GLAZING)	S	T2	M2
0	36837	-196	88	602	32182	367	-53	-9
1	10863	-4288	-238	421	10681	-4269	-422	194
2	-4589	-4711	-954	77	-2686	-5034	-1379	47
3	-12123	-3117	-1561	-246	-10048	-3512	-2244	-189
4	-14097	-703	-1815	-392	-13081	-846	-2641	-280
5	-11785	1673	-1649	-313	-12161	2066	-2417	-132
6	-4992	3248	-1141	-90	-6016	4222	-1624	159
7	7733	3106	-511	88	7731	4196	-582	322
8	28548	0	-174	-0	31543	-0	5	-0

INTERNAL SHELLS

POINT	T1 (60-DEGREE GLAZING)	S	T2	M2	T1 (90-DEGREE GLAZING)	S	T2	M2
0	37680	14	355	223	42309	1702	28	38
1	12610	-4335	46	305	14401	-4426	-235	271
2	-3290	-5045	-705	113	-2461	-5585	-1286	101
3	-11847	-3595	-1388	-150	-11194	-3928	-2266	-185
4	-14749	-1142	-1726	-293	-14298	-981	-2725	-319
5	-12877	1399	-1626	-239	-12869	2152	-2504	-188
6	-5929	3164	-1149	-46	-6114	4398	-1679	110
7	7449	3137	-522	108	8138	4335	-600	295
8	29206	0	-180	-0	32370	-0	5	-0

EXTERNAL NORTH SHELL

POINT	T1 (60-DEGREE GLAZING)	S	T2	M2	T1 (90-DEGREE GLAZING)	S	T2	M2
0	37918	26	357	236	42181	1683	25	29
1	12721	-4352	47	310	14317	-4422	-240	267
2	-3256	-5075	-709	109	-2527	-5565	-1288	102
3	-11875	-3626	-1397	-161	-11230	-3896	-2261	-177
4	-14847	-1162	-1740	-309	-14265	-947	-2713	-305
5	-13040	1403	-1641	-256	-12737	2168	-2486	-171
6	-6089	3199	-1160	-60	-5930	4376	-1661	126
7	7485	3187	-525	101	8166	4283	-591	304
8	29808	0	-177	-0	31774	-0	5	-0

L = 48 FT.

TABLE NO.N- 41

W = 26 FT.

SHELL DATA:

L = 48.00 FT.
W = 26.00 FT.
T = 2.50 IN.
G = 50.00 LB/FT*2
QA= 150.00 LB/FT
QV= 150.00 LB/FT

FOR 60DEG.GLAZING:
K = 12.50 FT.
R = 25.00 FT.
AL= 23.36 FT.
θ = 53.54 DEG.

FOR 90DEG.GLAZING:
K = 12.50 FT.
R = 40.00 FT.
AL= 29.51 FT.
θ = 42.28 DEG

BEAM DATA:

FOR 60DEG.GLAZING:

B1= 0.75 FT.
D1= 3.00 FT.
B2= 0.75 FT.
D2= 2.50 FT.

EXT.NORTH BEAM:

Z'= 1.48 FT.
Y'= 0.94 FT.
IZ= 1.97 FT*4
IY= 1.27 FT*4

FOR 90DEG.GLAZING:

ALL BEAMS:

B = 1.00 FT.
D = 3.00 FT.

BEAM RESULTS:

FOR 60DEG.GLAZING:

EXT.SOUTH BEAM
F = -3540 LB.
M1= 121810 LB.FT
M2= 7591 LB.FT
S1= -8248 LB.
S2= 563 LB.
INTERNAL BEAMS
F = -5079 LB.
P = 363 LB/FT
M1= 214181 LB.FT
M2= 413 LB.FT
S1= -14496 LB.
S2= 142 LB.
EXT.NORTH BEAM
P = 361 LB/FT
M1= 277769 LB.FT
M2= 53686 LB.FT
S1= -18180 LB.
S2= -3514 LB.

FOR 90DEG.GLAZING:

EXT.SOUTH BEAM
F = -7404 LB.
M1= 206285 LB.FT
M2= 11738 LB.FT
S1= -14221 LB.
S2= 1003 LB.
INTERNAL BEAMS
F = -30831 LB.
P = 402 LB/FT
M1= 278324 LB.FT
M2= 16669 LB.FT
S1= -21234 LB.
S2= 2091 LB.
EXT.NORTH BEAM
P = 407 LB/FT
M1= 277270 LB.FT
S1= -18147 LB.

SHELL RESULTS:

EXTERNAL SOUTH SHELL

POINT	T1 (60°)	S (60°)	T2 (60°)	M2 (60°)	T1 (90°)	S (90°)	T2 (90°)	M2 (90°)
	SHELLS WITH 60-DEGREE GLAZING				SHELLS WITH 90-DEGREE GLAZING			
0	38837	232	29	433	33027	485	-77	-133
1	12078	-4446	-309	358	11042	-4646	-522	179
2	-4062	-5063	-1134	73	-2628	-5517	-1668	61
3	-12193	-3402	-1851	-222	-10214	-3865	-2707	-181
4	-14580	-766	-2158	-352	-13453	-930	-3191	-276
5	-12377	1882	-1965	-258	-12671	2313	-2928	-114
6	-5291	3658	-1355	-24	-6392	4750	-1974	206
7	8162	3500	-591	145	8037	4743	-707	379
8	30056	0	-173	0	33234	-0	17	-0

INTERNAL SHELLS

POINT	T1 (60°)	S (60°)	T2 (60°)	M2 (60°)	T1 (90°)	S (90°)	T2 (90°)	M2 (90°)
0	39499	332	344	76	43460	2018	7	-111
1	13559	-4575	6	256	14923	-4772	-301	248
2	-3012	-5448	-870	111	-2317	-6098	-1549	116
3	-12082	-3901	-1674	-132	-11313	-4316	-2720	-173
4	-15291	-1201	-2073	-263	-14644	-1086	-3277	-311
5	-13445	1625	-1947	-194	-13370	2391	-3021	-167
6	-6161	3595	-1367	13	-6499	4924	-2033	159
7	7951	3547	-603	161	8412	4883	-726	354
8	30781	0	-177	0	34011	-0	17	0

EXTERNAL NORTH SHELL

POINT	T1 (60°)	S (60°)	T2 (60°)	M2 (60°)	T1 (90°)	S (90°)	T2 (90°)	M2 (90°)
0	39701	353	345	81	43357	1997	4	-119
1	13643	-4580	8	258	14844	-4772	-306	244
2	-2996	-5463	-870	111	-2391	-6079	-1552	117
3	-12102	-3914	-1677	-135	-11363	-4281	-2716	-165
4	-15334	-1208	-2078	-268	-14622	-1047	-3264	-297
5	-13500	1627	-1953	-199	-13237	2412	-3000	-149
6	-6207	3607	-1371	8	-6300	4902	-2012	-175
7	7966	3563	-604	158	8452	4825	-716	364
8	30954	0	-176	0	33375	-0	17	-0

L = 50 FT. TABLE NO.N- 42 W = 23 FT.

SHELL DATA:

L = 50.00 FT.
W = 23.00 FT.
T = 2.50 IN.
G = 50.00 LB/FT*2
QA= 150.00 LB/FT
QV= 150.00 LB/FT

FOR 60DEG.GLAZING:
K = 11.75 FT.
R = 17.50 FT.
AL= 21.19 FT.
θ = 69.38 DEG.

FOR 90DEG.GLAZING:
K = 11.75 FT.
R = 30.00 FT.
AL= 26.70 FT.
θ = 50.99 DEG.

BEAM DATA:

FOR 60DEG.GLAZING:

B1= 0.75 FT.
D1= 3.25 FT.
B2= 0.75 FT.
D2= 2.75 FT.

EXT.NORTH BEAM:

Z'= 1.60 FT.
Y'= 0.96 FT.
IZ= 2.54 FT*4
IY= 1.60 FT*4

FOR 90DEG.GLAZING:

ALL BEAMS:

B = 1.00 FT.
D = 3.25 FT.

BEAM RESULTS:

FOR 60DEG.GLAZING:

EXT.SOUTH BEAM
F = 8579 LB.
M1= 128088 LB.FT
M2= 7848 LB.FT
S1= -7291 LB.
S2= 256 LB.
INTERNAL BEAMS
F = 5287 LB.
P = 353 LB/FT
M1= 221463 LB.FT
M2= 1110 LB.FT
S1= -13346 LB.
S2= -75 LB.
EXT.NORTH BEAM
P = 346 LB/FT
M1= 312834 LB.FT
M2= 55871 LB.FT
S1= -19656 LB.
S2= -3510 LB.

FOR 90DEG.GLAZING:

EXT.SOUTH BEAM
F = -2158 LB.
M1= 231611 LB.FT
M2= 12277 LB.FT
S1= -14764 LB.
S2= 828 LB.
INTERNAL BEAMS
F = -21523 LB.
P = 382 LB/FT
M1= 308204 LB.FT
M2= 17229 LB.FT
S1= -21551 LB.
S2= 1744 LB.
EXT.NORTH BEAM
P = 387 LB/FT
M1= 304084 LB.FT
S1= -19106 LB.

SHELL RESULTS:

EXTERNAL SOUTH SHELL

	SHELLS WITH 60-DEGREE GLAZING				SHELLS WITH 90-DEGREE GLAZING			
POINT	T1	S	T2	M2	T1	S	T2	M2
0	35438	-539	122	637	31993	136	-60	7
1	10370	-4200	-183	429	10800	-4203	-403	175
2	-4790	-4542	-812	86	-2778	-4926	-1265	33
3	-12295	-3030	-1343	-227	-10504	-3446	-2045	-181
4	-14251	-755	-1565	-366	-13715	-834	-2398	-253
5	-11833	1473	-1425	-291	-12615	2009	-2185	-101
6	-4947	2938	-991	-83	-6022	4076	-1459	178
7	7589	2808	-458	82	8119	4007	-519	323
8	27550	0	-177	0	31832	-0	5	-0

INTERNAL SHELLS

POINT	T1	S	T2	M2	T1	S	T2	M2
0	36156	-332	383	212	41721	1352	15	56
1	12222	-4233	95	292	14491	-4341	-237	251
2	-3320	-4867	-564	119	-2470	-5443	-1180	87
3	-11895	-3515	-1165	-121	-11600	-3852	-2061	-175
4	-14871	-1217	-1469	-252	-14959	-979	-2473	-288
5	-12981	1167	-1394	-203	-13399	2082	-2266	-155
6	-5989	2823	-994	-28	-6183	4248	-1512	130
7	7192	2819	-468	107	8539	4151	-536	297
8	28105	0	-184	0	32789	-0	5	-0

EXTERNAL NORTH SHELL

POINT	T1	S	T2	M2	T1	S	T2	M2
0	36441	-327	387	231	41606	1340	12	45
1	12359	-4262	97	299	14408	-4333	-241	246
2	-3281	-4910	-569	113	-2540	-5420	-1180	90
3	-11944	-3557	-1177	-138	-11635	-3816	-2055	-164
4	-15025	-1242	-1487	-277	-14913	-943	-2459	-271
5	-13229	1176	-1414	-228	-13242	2096	-2247	-135
6	-6224	2874	-1009	-49	-5974	4222	-1494	147
7	7255	2890	-472	96	8561	4094	-528	307
8	29009	0	-180	0	32107	-0	5	-0

L = 50 FT. TABLE NO.N- 43 W = 25 FT.

```
SHELL DATA:

   L =  50.00 FT.
   W =  25.00 FT.
   T =   2.50 IN.
   G =  50.00 LB/FT*2
   QA= 150.00 LB/FT
   QV= 150.00 LB/FT

 FOR 60DEG.GLAZING:
   K =  12.25 FT.
   R =  20.00 FT.
   AL=  22.88 FT.
   θ =  65.54 DEG.

 FOR 90DEG.GLAZING:
   K =  12.25 FT.
   R =  35.00 FT.
   AL=  28.63 FT.
   θ =  46.87 DEG.

BEAM DATA:

 FOR 60DEG.GLAZING:

   B1= 0.75 FT.
   D1= 3.25 FT.
   B2= 0.75 FT.
   D2= 2.75 FT.

   EXT.NORTH BEAM:

   Z'= 1.60 FT.
   Y'= 0.96 FT.
   IZ= 2.54 FT*4
   IY= 1.60 FT*4

 FOR 90DEG.GLAZING:

   ALL BEAMS:

   B = 1.00 FT.
   D = 3.25 FT.
```

```
BEAM RESULTS:

FOR 60DEG.GLAZING:

EXT.SOUTH BEAM
 F =      2899 LB.
 M1=    133411 LB.FT
 M2=      7827 LB.FT
 S1=     -8117 LB.
 S2=       395 LB.
INTERNAL BEAMS
 F =       483 LB.
 P =       366 LB/FT
 M1=    229505 LB.FT
 M2=       969 LB.FT
 S1=    -14346 LB.
 S2=        34 LB.
EXT.NORTH BEAM
 P =       360 LB/FT
 M1=    317972 LB.FT
 M2=     58093 LB.FT
 S1=    -19979 LB.
 S2=     -3650 LB.

FOR 90DEG.GLAZING:

EXT.SOUTH BEAM
 F =     -4281 LB.
 M1=    237582 LB.FT
 M2=     11840 LB.FT
 S1=    -15357 LB.
 S2=       869 LB.
INTERNAL BEAMS
 F =    -25901 LB.
 P =       399 LB/FT
 M1=    317284 LB.FT
 M2=     16682 LB.FT
 S1=    -22570 LB.
 S2=      1850 LB.
EXT.NORTH BEAM
 P =       404 LB/FT
 M1=    310725 LB.FT
 S1=    -19523 LB.
```

SHELL RESULTS:

	SHELLS WITH 60-DEGREE GLAZING				SHELLS WITH 90-DEGREE GLAZING			
EXTERNAL SOUTH SHELL								
POINT	T1	S	T2	M2	T1	S	T2	M2
0	36020	-182	78	573	32396	269	-81	-79
1	10573	-4194	-229	409	10914	-4435	-474	174
2	-4418	-4610	-922	72	-2693	-5230	-1470	47
3	-11631	-3072	-1510	-246	-10385	-3666	-2372	-182
4	-13517	-746	-1758	-385	-13670	-886	-2787	-264
5	-11375	1548	-1603	-294	-12766	2172	-2549	-100
6	-4956	3088	-1116	-60	-6282	4439	-1711	209
7	7266	2983	-509	116	8164	4400	-611	370
8	27455	0	-185	-0	32827	-0	10	0
INTERNAL SHELLS								
POINT	T1	S	T2	M2	T1	S	T2	M2
0	36917	-30	374	172	42304	1627	-7	-50
1	12354	-4310	72	295	14655	-4553	-287	243
2	-3187	-5012	-668	117	-2368	-5760	-1369	100
3	-11515	-3597	-1343	-144	-11445	-4083	-2384	-175
4	-14354	-1200	-1682	-286	-14851	-1034	-2863	-298
5	-12603	1290	-1591	-224	-13482	2243	-2631	-152
6	-5928	3036	-1132	-21	-6410	4605	-1764	163
7	7074	3042	-523	132	8544	4536	-628	345
8	28328	0	-190	-0	33649	-0	10	0
EXTERNAL NORTH SHELL								
POINT	T1	S	T2	M2	T1	S	T2	M2
0	37087	-22	376	186	42210	1611	-10	-60
1	12449	-4325	73	300	14575	-4550	-291	239
2	-3138	-5039	-671	113	-2448	-5740	-1371	102
3	-11520	-3629	-1351	-156	-11498	-4046	-2379	-165
4	-14441	-1224	-1695	-304	-14821	-994	-2849	-281
5	-12781	1289	-1607	-244	-13330	2263	-2611	-132
6	-6122	3072	-1144	-38	-6189	4580	-1743	181
7	7096	3098	-526	124	8583	4475	-618	355
8	29027	0	-187	-0	32940	-0	10	0

L = 50 FT. TABLE NO.N- 44 W = 27 FT.

SHELL DATA:

L = 50.00 FT.
W = 27.00 FT.
T = 2.50 IN.
G = 50.00 LB/FT*2
QA= 150.00 LB/FT
QV= 150.00 LB/FT

FOR 60DEG.GLAZING:
K = 12.75 FT.
R = 25.00 FT.
AL= 24.35 FT.
θ = 55.80 DEG.

FOR 90DEG.GLAZING:
K = 12.75 FT.
R = 40.00 FT.
AL= 30.60 FT.
θ = 43.83 DEG.

BEAM DATA:

FOR 60DEG.GLAZING:

B1= 0.75 FT.
D1= 3.25 FT.
B2= 0.75 FT.
D2= 2.75 FT.

EXT.NORTH BEAM:

Z'= 1.60 FT.
Y'= 0.96 FT.
IZ= 2.54 FT*4
IY= 1.60 FT*4

FOR 90DEG.GLAZING:

ALL BEAMS:

B = 1.00 FT.
D = 3.25 FT.

BEAM RESULTS:

FOR 60DEG.GLAZING:

EXT.SOUTH BEAM
F = -3977 LB.
M1= 147327 LB.FT
M2= 7513 LB.FT
S1= -9587 LB.
S2= 546 LB.
INTERNAL BEAMS
F = -4154 LB.
P = 380 LB/FT
M1= 248238 LB.FT
M2= 295 LB.FT
S1= -16000 LB.
S2= 106 LB.
EXT.NORTH BEAM
P = 379 LB/FT
M1= 324465 LB.FT
M2= 61097 LB.FT
S1= -20387 LB.
S2= -3839 LB.

FOR 90DEG.GLAZING:

EXT.SOUTH BEAM
F = -6352 LB.
M1= 242002 LB.FT
M2= 11527 LB.FT
S1= -15847 LB.
S2= 917 LB.
INTERNAL BEAMS
F = -29968 LB.
P = 414 LB/FT
M1= 324424 LB.FT
M2= 16296 LB.FT
S1= -23435 LB.
S2= 1956 LB.
EXT.NORTH BEAM
P = 419 LB/FT
M1= 316830 LB.FT
S1= -19907 LB.

SHELL RESULTS:

SHELLS WITH 60-DEGREE GLAZING | SHELLS WITH 90-DEGREE GLAZING

EXTERNAL SOUTH SHELL

POINT	T1	S	T2	M2	T1	S	T2	M2
0	38139	250	5	383	32665	399	-96	-156
1	11921	-4351	-313	341	10913	-4651	-541	182
2	-3883	-4974	-1115	71	-2630	-5504	-1673	64
3	-11855	-3365	-1815	-219	-10174	-3866	-2701	-186
4	-14257	-793	-2118	-343	-13472	-950	-3181	-283
5	-12207	1807	-1934	-238	-12805	2297	-2923	-107
6	-5329	3570	-1340	6	-6549	4761	-1973	234
7	7930	3439	-594	173	8078	4773	-709	415
8	29626	0	-186	0	33627	-0	13	-0

INTERNAL SHELLS

POINT	T1	S	T2	M2	T1	S	T2	M2
0	39016	261	354	27	42740	1883	-21	-143
1	13364	-4585	16	251	14657	-4763	-332	246
2	-3025	-5440	-854	118	-2334	-6051	-1560	117
3	-11981	-3902	-1654	-127	-11227	-4288	-2710	-178
4	-15161	-1225	-2052	-260	-14595	-1093	-3259	-316
5	-13371	1582	-1932	-184	-13442	2371	-3006	-158
6	-6188	3547	-1361	35	-6630	4919	-2026	189
7	7826	3513	-607	185	8425	4897	-726	391
8	30531	0	-189	0	34306	-0	13	-0

EXTERNAL NORTH SHELL

POINT	T1	S	T2	M2	T1	S	T2	M2
0	39155	276	354	31	42661	1862	-24	-151
1	13424	-4589	17	253	14581	-4765	-337	242
2	-3010	-5451	-854	118	-2419	-6034	-1563	119
3	-11993	-3912	-1656	-130	-11292	-4253	-2708	-169
4	-15190	-1231	-2056	-264	-14582	-1049	-3246	-300
5	-13413	1582	-1936	-188	-13302	2396	-2985	-138
6	-6228	3556	-1364	31	-6406	4898	-2004	207
7	7835	3526	-608	183	8479	4834	-714	402
8	30678	0	-189	0	33599	-0	13	-0

L = 52 FT. | TABLE NO.N- 45 | W = 24 FT.

SHELL DATA:

```
L =  52.00 FT.
W =  24.00 FT.
T =   2.50 IN.
G =  50.00 LB/FT*2
QA= 150.00 LB/FT
QV= 150.00 LB/FT

FOR 60DEG.GLAZING:
K =  11.75 FT.
R =  20.00 FT.
AL=  21.86 FT.
θ =  62.62 DEG.

FOR 90DEG.GLAZING:
K =  11.75 FT.
R =  30.00 FT.
AL=  27.69 FT.
θ =  52.89 DEG.
```

BEAM DATA:

```
FOR 60DEG.GLAZING:

B1= 0.75 FT.
D1= 3.25 FT.
B2= 0.75 FT.
D2= 2.75 FT.

EXT.NORTH BEAM:

Z'= 1.60 FT.
Y'= 0.96 FT.
IZ= 2.54 FT*4
IY= 1.60 FT*4

FOR 90DEG.GLAZING:

ALL BEAMS:

B = 1.00 FT.
D = 3.25 FT.
```

BEAM RESULTS:

```
FOR 60DEG.GLAZING:

EXT.SOUTH BEAM
 F  =     5734 LB.
 M1 =   148984 LB.FT
 M2 =     8045 LB.FT
 S1 =    -8507 LB.
 S2 =      324 LB.
INTERNAL BEAMS
 F  =     3610 LB.
 P  =      360 LB/FT
 M1 =   252854 LB.FT
 M2 =      584 LB.FT
 S1 =   -14892 LB.
 S2 =      -64 LB.
EXT.NORTH BEAM
 P  =      356 LB/FT
 M1 =   341306 LB.FT
 M2 =    62130 LB.FT
 S1 =   -20620 LB.
 S2 =    -3754 LB.

FOR 90DEG.GLAZING:

EXT.SOUTH BEAM
 F  =    -3622 LB.
 M1 =   245961 LB.FT
 M2 =    13323 LB.FT
 S1 =   -15206 LB.
 S2 =      902 LB.
INTERNAL BEAMS
 F  =   -24538 LB.
 P  =      382 LB/FT
 M1 =   328424 LB.FT
 M2 =    18661 LB.FT
 S1 =   -22238 LB.
 S2 =     1852 LB.
EXT.NORTH BEAM
 P  =      387 LB/FT
 M1 =   328642 LB.FT
 S1 =   -19855 LB.
```

SHELL RESULTS:

EXTERNAL SOUTH SHELL

	SHELLS WITH 60-DEGREE GLAZING				SHELLS WITH 90-DEGREE GLAZING			
POINT	T1	S	T2	M2	T1	S	T2	M2
0	39914	-346	71	588	34000	219	-52	67
1	12269	-4496	-246	419	11229	-4344	-399	220
2	-4900	-4974	-935	91	-3002	-5074	-1287	40
3	-13766	-3335	-1527	-220	-10851	-3532	-2086	-214
4	-16311	-776	-1777	-360	-14043	-857	-2448	-306
5	-13618	1764	-1614	-280	-12970	2049	-2234	-147
6	-5563	3427	-1115	-63	-6334	4184	-1497	160
7	9010	3243	-502	103	8264	4143	-537	330
8	31817	-0	-177	0	33266	-0	-1	-0

INTERNAL SHELLS

POINT	T1	S	T2	M2	T1	S	T2	M2
0	40606	-218	362	169	44471	1482	25	131
1	13977	-4589	51	284	15051	-4516	-232	307
2	-3565	-5333	-677	122	-2817	-5620	-1208	97
3	-13437	-3834	-1342	-117	-12085	-3937	-2112	-212
4	-16944	-1246	-1674	-247	-15336	-981	-2530	-349
5	-14756	1448	-1578	-191	-13711	2143	-2317	-208
6	-6610	3299	-1115	-7	-6423	4363	-1549	107
7	8570	3244	-511	129	8711	4283	-553	302
8	32291	-0	-184	-0	34150	-0	-1	-0

EXTERNAL NORTH SHELL

POINT	T1	S	T2	M2	T1	S	T2	M2
0	40939	-199	365	181	44330	1465	22	121
1	14119	-4608	52	289	14960	-4511	-237	303
2	-3542	-5364	-680	120	-2887	-5598	-1210	99
3	-13492	-3863	-1350	-125	-12121	-3903	-2107	-202
4	-17057	-1260	-1685	-260	-15296	-946	-2518	-334
5	-14905	1456	-1590	-205	-13564	2158	-2300	-189
6	-6732	3331	-1124	-18	-6222	4340	-1532	124
7	8622	3284	-514	123	8738	4229	-545	311
8	32783	-0	-182	-0	33499	-0	-1	0

L = 52 FT. TABLE NO.N- 46 W = 26 FT.

```
SHELL DATA:

  L =  52.00 FT.
  W =  26.00 FT.
  T =   2.50 IN.
  G =  50.00 LB/FT*2
  QA= 150.00 LB/FT
  QV= 150.00 LB/FT

 FOR 60DEG.GLAZING:
  K =  12.75 FT.
  R =  22.50 FT.
  AL=  23.58 FT.
  θ =  60.05 DEG.

 FOR 90DEG.GLAZING:
  K =  12.75 FT.
  R =  37.50 FT.
  AL=  29.73 FT.
  θ =  45.42 DEG.

BEAM DATA:

 FOR 60DEG.GLAZING:

  B1= 0.75 FT.
  D1= 3.25 FT.
  B2= 0.75 FT.
  D2= 2.75 FT.

  EXT.NORTH BEAM:

  Z'= 1.60 FT.
  Y'= 0.96 FT.
  IZ= 2.54 FT*4
  IY= 1.60 FT*4

 FOR 90DEG.GLAZING:

  ALL BEAMS:

  B = 1.00 FT.
  D = 3.25 FT.
```

```
BEAM RESULTS:

FOR 60DEG.GLAZING:

EXT.SOUTH BEAM
 F =     1530 LB.
 M1=   149833 LB.FT
 M2=     8292 LB.FT
 S1=    -8909 LB.
 S2=      440 LB.
INTERNAL BEAMS
 F =     -812 LB.
 P =      368 LB/FT
 M1=   255697 LB.FT
 M2=      780 LB.FT
 S1=   -15502 LB.
 S2=       51 LB.
EXT.NORTH BEAM
 P =      364 LB/FT
 M1=   346430 LB.FT
 M2=    63477 LB.FT
 S1=   -20930 LB.
 S2=    -3835 LB.

FOR 90DEG.GLAZING:

EXT.SOUTH BEAM
 F =    -5488 LB.
 M1=   257515 LB.FT
 M2=    12909 LB.FT
 S1=   -16089 LB.
 S2=      937 LB.
INTERNAL BEAMS
 F =   -29944 LB.
 P =      405 LB/FT
 M1=   344858 LB.FT
 M2=    18201 LB.FT
 S1=   -23764 LB.
 S2=     1992 LB.
EXT.NORTH BEAM
 P =      410 LB/FT
 M1=   339630 LB.FT
 S1=   -20519 LB.
```

SHELL RESULTS:

EXTERNAL SOUTH SHELL

	SHELLS WITH 60-DEGREE GLAZING				SHELLS WITH 90-DEGREE GLAZING			
POINT	T1	S	T2	M2	T1	S	T2	M2
0	39877	-92	54	556	35088	332	-83	-83
1	12093	-4538	-277	413	11789	-4750	-507	192
2	-4640	-5057	-1036	77	-2913	-5606	-1578	52
3	-12982	-3383	-1688	-249	-11181	-3922	-2548	-200
4	-15322	-787	-1963	-391	-14692	-937	-2994	-292
5	-12914	1796	-1785	-293	-13718	2344	-2738	-117
6	-5496	3518	-1234	-47	-6760	4778	-1838	217
7	8438	3368	-550	135	8787	4737	-656	394
8	31011	0	-180	0	35426	-0	13	-0

INTERNAL SHELLS

POINT	T1	S	T2	M2	T1	S	T2	M2
0	40811	49	367	146	45892	1809	-7	-48
1	13890	-4666	40	294	15850	-4877	-307	268
2	-3388	-5472	-769	120	-2573	-6178	-1472	110
3	-12842	-3924	-1508	-145	-12338	-4369	-2563	-192
4	-16149	-1261	-1876	-288	-15969	-1093	-3077	-330
5	-14159	1516	-1767	-218	-14483	2424	-2827	-174
6	-6510	3449	-1247	-3	-6889	4957	-1896	167
7	8192	3417	-562	154	9200	4882	-675	367
8	31839	0	-185	0	36303	-0	13	-0

EXTERNAL NORTH SHELL

POINT	T1	S	T2	M2	T1	S	T2	M2
0	41029	64	369	157	45784	1790	-10	-58
1	13994	-4680	41	298	15762	-4875	-312	263
2	-3353	-5497	-771	117	-2657	-6156	-1474	112
3	-12862	-3951	-1515	-154	-12393	-4331	-2558	-182
4	-16230	-1278	-1887	-301	-15938	-1051	-3062	-312
5	-14298	1518	-1780	-233	-14324	2445	-2806	-153
6	-6648	3477	-1257	-15	-6657	4932	-1874	186
7	8219	3459	-565	148	9241	4818	-664	378
8	32352	0	-183	0	35562	-0	13	-0

L = 52 FT. TABLE NO.N- 47 W = 28 FT.

SHELL DATA:

L = 52.00 FT.
W = 28.00 FT.
T = 2.50 IN.
G = 50.00 LB/FT*2
QA= 150.00 LB/FT
QV= 150.00 LB/FT

FOR 60DEG.GLAZING:
K = 13.25 FT.
R = 27.50 FT.
AL= 25.13 FT.
θ = 52.35 DEG.

FOR 90DEG.GLAZING:
K = 13.25 FT.
R = 45.00 FT.
AL= 31.62 FT.
θ = 40.26 DEG.

BEAM DATA:

FOR 60DEG.GLAZING:

B1= 0.75 FT.
D1= 3.25 FT.
B2= 0.75 FT.
D2= 2.75 FT.

EXT.NORTH BEAM:

Z'= 1.60 FT.
Y'= 0.96 FT.
IZ= 2.54 FT*4
IY= 1.60 FT*4

FOR 90DEG.GLAZING:

ALL BEAMS:

B = 1.00 FT.
D = 3.25 FT.

BEAM RESULTS:

FOR 60DEG.GLAZING:

EXT.SOUTH BEAM
F = -5063 LB.
M1= 162320 LB.FT
M2= 8033 LB.FT
S1= -10213 LB.
S2= 580 LB.
INTERNAL BEAMS
F = -5698 LB.
P = 383 LB/FT
M1= 273386 LB.FT
M2= 188 LB.FT
S1= -17055 LB.
S2= 131 LB.
EXT.NORTH BEAM
P = 383 LB/FT
M1= 353477 LB.FT
M2= 66740 LB.FT
S1= -21355 LB.
S2= -4032 LB.

FOR 90DEG.GLAZING:

EXT.SOUTH BEAM
F = -6398 LB.
M1= 268097 LB.FT
M2= 12267 LB.FT
S1= -16818 LB.
S2= 927 LB.
INTERNAL BEAMS
F = -33676 LB.
P = 425 LB/FT
M1= 359056 LB.FT
M2= 17367 LB.FT
S1= -24990 LB.
S2= 2058 LB.
EXT.NORTH BEAM
P = 430 LB/FT
M1= 348039 LB.FT
S1= -21027 LB.

SHELL RESULTS:

	SHELLS WITH 60-DEGREE GLAZING				SHELLS WITH 90-DEGREE GLAZING			
EXTERNAL SOUTH SHELL								
POINT	T1	S	T2	M2	T1	S	T2	M2
0	41753	306	-6	378	36036	387	-106	-214
1	13224	-4714	-354	353	12162	-5172	-620	178
2	-4178	-5417	-1227	76	-2959	-6119	-1885	69
3	-13108	-3664	-1992	-227	-11537	-4278	-3031	-189
4	-15865	-831	-2322	-357	-15268	-990	-3560	-287
5	-13571	2041	-2116	-245	-14362	2650	-3260	-98
6	-5830	3977	-1459	16	-7129	5369	-2192	264
7	8956	3808	-635	191	9272	5325	-782	451
8	32893	0	-183	-0	37452	0	22	0
INTERNAL SHELLS								
POINT	T1	S	T2	M2	T1	S	T2	M2
0	42779	344	354	10	47013	2035	-30	-201
1	14779	-4944	-11	257	16348	-5266	-390	245
2	-3249	-5895	-953	123	-2528	-6712	-1755	127
3	-13207	-4221	-1820	-132	-12637	-4756	-3036	-178
4	-16782	-1285	-2248	-269	-16518	-1167	-3644	-320
5	-14777	1797	-2109	-185	-15125	2720	-3354	-151
6	-6738	3943	-1478	48	-7274	5547	-2253	216
7	8820	3879	-648	205	9656	5472	-802	425
8	33804	0	-186	-0	38298	0	21	0
EXTERNAL NORTH SHELL								
POINT	T1	S	T2	M2	T1	S	T2	M2
0	42947	364	354	12	46930	2014	-32	-212
1	14847	-4946	-9	258	16263	-5267	-395	241
2	-3239	-5904	-952	123	-2624	-6691	-1758	129
3	-13225	-4229	-1821	-133	-12709	-4714	-3031	-167
4	-16808	-1288	-2250	-271	-16496	-1118	-3628	-301
5	-14800	1799	-2112	-187	-14956	2747	-3328	-128
6	-6751	3948	-1479	46	-7015	5520	-2227	237
7	8829	3885	-648	204	9711	5398	-788	437
8	33858	0	-186	-0	37475	0	22	0

L = 54 FT.　　TABLE NO.N- 48　　W = 25 FT.

SHELL DATA:

```
L =  54.00 FT.
W =  25.00 FT.
T =   2.50 IN.
G =  50.00 LB/FT*2
QA= 150.00 LB/FT
QV= 150.00 LB/FT

FOR 60DEG.GLAZING:
K =  12.50 FT.
R =  20.00 FT.
AL=  22.88 FT.
θ =  65.54 DEG.

FOR 90DEG.GLAZING:
K =  12.50 FT.
R =  32.50 FT.
AL=  28.89 FT.
θ =  50.94 DEG.
```

BEAM DATA:

```
FOR 60DEG.GLAZING:

B1= 0.75 FT.
D1= 3.50 FT.
B2= 0.75 FT.
D2= 3.00 FT.

EXT.NORTH BEAM:

Z'= 1.72 FT.
Y'= 0.99 FT.
IZ= 3.22 FT*4
IY= 1.97 FT*4

FOR 90DEG.GLAZING:

ALL BEAMS:

B = 1.00 FT.
D = 3.50 FT.
```

BEAM RESULTS:

```
FOR 60DEG.GLAZING:

EXT.SOUTH BEAM
 F =     6017 LB.
 M1=   173804 LB.FT
 M2=     8339 LB.FT
 S1=    -9570 LB.
 S2=      322 LB.
INTERNAL BEAMS
 F =     4748 LB.
 P =      374 LB/FT
 M1=   285241 LB.FT
 M2=      806 LB.FT
 S1=   -16083 LB.
 S2=      -75 LB.
EXT.NORTH BEAM
 P =      369 LB/FT
 M1=   394979 LB.FT
 M2=    69454 LB.FT
 S1=   -22979 LB.
 S2=    -4041 LB.

FOR 90DEG.GLAZING:

EXT.SOUTH BEAM
 F =    -2425 LB.
 M1=   291230 LB.FT
 M2=    13161 LB.FT
 S1=   -17181 LB.
 S2=      825 LB.
INTERNAL BEAMS
 F =   -24764 LB.
 P =      400 LB/FT
 M1=   386578 LB.FT
 M2=    18413 LB.FT
 S1=   -25000 LB.
 S2=     1777 LB.
EXT.NORTH BEAM
 P =      406 LB/FT
 M1=   377149 LB.FT
 S1=   -21942 LB.
```

SHELL RESULTS:

EXTERNAL SOUTH SHELL

POINT	T1 (60-DEGREE GLAZING)	S	T2	M2	T1 (90-DEGREE GLAZING)	S	T2	M2
0	38920	-350	65	573	34273	141	-81	-12
1	11872	-4417	-238	412	11521	-4509	-450	195
2	-4749	-4880	-909	84	-2979	-5279	-1377	39
3	-13220	-3288	-1486	-227	-11201	-3695	-2214	-205
4	-15633	-815	-1731	-362	-14677	-901	-2596	-287
5	-13118	1643	-1577	-270	-13622	2159	-2368	-110
6	-5496	3269	-1095	-40	-6618	4410	-1585	213
7	8473	3121	-501	128	8729	4356	-566	381
8	30547	0	-187	-0	34636	0	4	-0

INTERNAL SHELLS

POINT	T1 (60-DEGREE GLAZING)	S	T2	M2	T1 (90-DEGREE GLAZING)	S	T2	M2
0	39697	-276	383	132	44549	1441	-11	32
1	13629	-4579	74	280	15389	-4639	-281	273
2	-3483	-5308	-647	126	-2674	-5804	-1289	95
3	-13068	-3838	-1308	-116	-12339	-4103	-2230	-199
4	-16480	-1301	-1641	-248	-15933	-1040	-2670	-324
5	-14418	1345	-1553	-184	-14378	2237	-2447	-167
6	-6586	3178	-1104	12	-6745	4579	-1636	163
7	8148	3154	-513	151	9145	4493	-582	353
8	31299	0	-193	-0	35521	0	4	0

EXTERNAL NORTH SHELL

POINT	T1 (60-DEGREE GLAZING)	S	T2	M2	T1 (90-DEGREE GLAZING)	S	T2	M2
0	39956	-263	385	145	44445	1425	-14	21
1	13750	-4597	76	285	15302	-4635	-286	268
2	-3450	-5339	-650	122	-2759	-5782	-1291	98
3	-13103	-3868	-1316	-127	-12393	-4065	-2225	-188
4	-16588	-1320	-1653	-264	-15896	-1000	-2656	-306
5	-14587	1350	-1567	-201	-14209	2256	-2427	-145
6	-6745	3212	-1114	-2	-6504	4553	-1616	183
7	8191	3201	-516	143	9183	4430	-572	365
8	31905	0	-191	-0	34750	0	4	0

L = 54 FT. TABLE NO.N- 49 W = 27 FT.

SHELL DATA:

```
  L =  54.00 FT.
  W =  27.00 FT.
  T =   2.50 IN.
  G =  50.00 LB/FT*2
  QA= 150.00 LB/FT
  QV= 150.00 LB/FT

 FOR 60DEG.GLAZING:
  K =  13.00 FT.
  R =  22.50 FT.
  AL=  24.59 FT.
  θ =  62.63 DEG.

 FOR 90DEG.GLAZING:
  K =  13.00 FT.
  R =  40.00 FT.
  AL=  30.72 FT.
  θ =  44.00 DEG.
```

BEAM DATA:

```
 FOR 60DEG.GLAZING:

  B1= 0.75 FT.
  D1= 3.50 FT.
  B2= 0.75 FT.
  D2= 3.00 FT.

  EXT.NORTH BEAM:

  Z'= 1.72 FT.
  Y'= 0.99 FT.
  IZ= 3.22 FT*4
  IY= 1.97 FT*4

 FOR 90DEG.GLAZING:

  ALL BEAMS:

  B = 1.00 FT.
  D = 3.50 FT.
```

BEAM RESULTS:

```
FOR 60DEG.GLAZING:

EXT.SOUTH BEAM
 F =       870 LB.
 M1=    178217 LB.FT
 M2=      8278 LB.FT
 S1=    -10277 LB.
 S2=       436 LB.
INTERNAL BEAMS
 F =       -13 LB.
 P =       385 LB/FT
 M1=    293030 LB.FT
 M2=       699 LB.FT
 S1=    -17024 LB.
 S2=        27 LB.
EXT.NORTH BEAM
 P =       382 LB/FT
 M1=    400602 LB.FT
 M2=     71832 LB.FT
 S1=    -23306 LB.
 S2=     -4179 LB.

FOR 90DEG.GLAZING:

EXT.SOUTH BEAM
 F =     -2942 LB.
 M1=    306185 LB.FT
 M2=     12252 LB.FT
 S1=    -18105 LB.
 S2=       790 LB.
INTERNAL BEAMS
 F =    -28371 LB.
 P =       422 LB/FT
 M1=    406022 LB.FT
 M2=     17230 LB.FT
 S1=    -26500 LB.
 S2=      1817 LB.
EXT.NORTH BEAM
 P =       428 LB/FT
 M1=    387004 LB.FT
 S1=    -22515 LB.
```

SHELL RESULTS:

EXTERNAL SOUTH SHELL

	SHELLS WITH 60-DEGREE GLAZING				SHELLS WITH 90-DEGREE GLAZING			
POINT	T1	S	T2	M2	T1	S	T2	M2
0	39216	-51	32	515	35444	171	-112	-167
1	11907	-4441	-280	399	12070	-4963	-572	167
2	-4452	-4964	-1018	74	-3033	-5837	-1686	53
3	-12565	-3340	-1655	-249	-11773	-4081	-2693	-188
4	-14879	-813	-1927	-386	-15576	-942	-3154	-269
5	-12645	1715	-1758	-277	-14522	2520	-2878	-78
6	-5516	3421	-1223	-19	-7018	5068	-1926	270
7	8120	3301	-554	163	9473	4985	-684	442
8	30416	-0	-195	-0	37166	-0	15	0

INTERNAL SHELLS

POINT	T1	S	T2	M2	T1	S	T2	M2
0	40311	1	378	100	45896	1651	-45	-149
1	13666	-4671	52	291	16132	-5047	-375	234
2	-3387	-5460	-753	127	-2570	-6380	-1573	109
3	-12671	-3923	-1490	-142	-12824	-4527	-2697	-176
4	-15916	-1286	-1859	-288	-16816	-1113	-3227	-301
5	-14001	1466	-1755	-211	-15311	2581	-2962	-129
6	-6521	3393	-1244	16	-7193	5238	-1982	224
7	7993	3380	-568	177	9860	5129	-703	416
8	31453	-0	-199	0	38068	-0	15	-0

EXTERNAL NORTH SHELL

POINT	T1	S	T2	M2	T1	S	T2	M2
0	40459	11	380	110	45821	1635	-48	-162
1	13744	-4680	53	295	16046	-5045	-379	228
2	-3350	-5480	-754	125	-2671	-6357	-1575	113
3	-12674	-3945	-1495	-150	-12896	-4483	-2692	-164
4	-15976	-1303	-1868	-300	-16785	-1064	-3211	-280
5	-14123	1465	-1766	-225	-15121	2607	-2936	-104
6	-6655	3416	-1252	4	-6913	5207	-1956	247
7	8007	3417	-571	170	9911	5053	-690	430
8	31930	-0	-197	-0	37174	-0	15	-0

L = 54 FT. TABLE NO.N- 50 W = 29 FT.

SHELL DATA:

L = 54.00 FT.
W = 29.00 FT.
T = 2.50 IN.
G = 50.00 LB/FT*2
QA= 150.00 LB/FT
QV= 150.00 LB/FT

FOR 60DEG.GLAZING:
K = 14.00 FT.
R = 30.00 FT.
AL= 25.92 FT.
θ = 49.50 DEG.

FOR 90DEG.GLAZING:
K = 14.00 FT.
R = 42.50 FT.
AL= 33.03 FT.
θ = 44.53 DEG.

BEAM DATA:

FOR 60DEG.GLAZING:

B1= 0.75 FT.
D1= 3.50 FT.
B2= 0.75 FT.
D2= 3.00 FT.

EXT.NORTH BEAM:

Z'= 1.72 FT.
Y'= 0.99 FT.
IZ= 3.22 FT*4
IY= 1.97 FT*4

FOR 90DEG.GLAZING:

ALL BEAMS:

B = 1.00 FT.
D = 3.50 FT.

BEAM RESULTS:

FOR 60DEG.GLAZING:

EXT.SOUTH BEAM
F = -6120 LB.
M1= 200105 LB.FT
M2= 7970 LB.FT
S1= -12163 LB.
S2= 580 LB.
INTERNAL BEAMS
F = -4244 LB.
P = 400 LB/FT
M1= 322060 LB.FT
M2= -136 LB.FT
S1= -19149 LB.
S2= 76 LB.
EXT.NORTH BEAM
P = 401 LB/FT
M1= 409923 LB.FT
M2= 75476 LB.FT
S1= -23848 LB.
S2= -4391 LB.

FOR 90DEG.GLAZING:

EXT.SOUTH BEAM
F = -7571 LB.
M1= 300306 LB.FT
M2= 12894 LB.FT
S1= -18235 LB.
S2= 963 LB.
INTERNAL BEAMS
F = -34538 LB.
P = 432 LB/FT
M1= 402808 LB.FT
M2= 18183 LB.FT
S1= -26942 LB.
S2= 2053 LB.
EXT.NORTH BEAM
P = 436 LB/FT
M1= 393262 LB.FT
S1= -22879 LB.

SHELL RESULTS:

	SHELLS WITH 60-DEGREE GLAZING				SHELLS WITH 90-DEGREE GLAZING			
EXTERNAL SOUTH SHELL								
POINT	T1	S	T2	M2	T1	S	T2	M2
0	42394	356	-50	216	34798	440	-112	-171
1	14037	-4781	-392	284	11550	-4922	-576	213
2	-3855	-5593	-1278	75	-2750	-5825	-1767	74
3	-13498	-3845	-2069	-183	-10622	-4115	-2851	-216
4	-16747	-904	-2417	-288	-14148	-1065	-3364	-329
5	-14439	2122	-2206	-164	-13670	2367	-3103	-127
6	-6134	4164	-1520	93	-7238	5029	-2106	268
7	9542	3971	-659	245	8444	5102	-760	481
8	34234	-0	-186	0	36226	-0	15	0
INTERNAL SHELLS								
POINT	T1	S	T2	M2	T1	S	T2	M2
0	43394	247	359	-146	45541	2009	-41	-162
1	15381	-5131	-25	195	15467	-5043	-363	280
2	-3173	-6145	-999	124	-2506	-6390	-1654	131
3	-13792	-4433	-1900	-90	-11759	-4536	-2863	-208
4	-17777	-1361	-2348	-204	-15293	-1194	-3443	-365
5	-15684	1888	-2203	-108	-14265	2452	-3184	-181
6	-7033	4141	-1540	123	-7264	5185	-2155	221
7	9437	4049	-672	257	8794	5216	-775	456
8	35184	-0	-188	-0	36809	-0	15	0
EXTERNAL NORTH SHELL								
POINT	T1	S	T2	M2	T1	S	T2	M2
0	43547	269	359	-149	45464	1985	-43	-169
1	15435	-5128	-23	195	15391	-5049	-369	276
2	-3177	-6146	-996	127	-2595	-6376	-1659	132
3	-13811	-4431	-1897	-86	-11833	-4502	-2862	-200
4	-17777	-1357	-2345	-199	-15290	-1149	-3432	-350
5	-15649	1890	-2199	-103	-14129	2481	-3164	-161
6	-6980	4133	-1537	127	-7032	5167	-2133	240
7	9440	4034	-671	260	8859	5153	-763	467
8	34999	-0	-189	-0	36082	-0	15	0

L = 26 FT. TABLE NO.N- 51 W = 26 FT.

SHELL DATA:

L = 50.00 FT.
W = 26.00 FT.
T = 2.50 IN.
G = 50.00 LB/FT*2
QA= 150.00 LB/FT
QV= 150.00 LB/FT

FOR 60DEG.GLAZING:
K = 13.00 FT.
R = 22.20 FT.
AL= 23.58 FT.
Ø = 60.05 DEG.

FOR 90DEG.GLAZING:
K = 13.00 FT.
R = 32.00 FT.
AL= 29.98 FT.
Ø = 49.07 DEG.

BEAM DATA:

FOR 60DEG.GLAZING:

B1= 0.75 FT.
D1= 3.20 FT.
B2= 0.75 FT.
D2= 3.00 FT.

EXT.NORTH BEAM:

Z'= 1.72 FT.
Y'= 0.99 FT.
IZ= 3.22 FT*4
IY= 1.97 FT*4

FOR 90DEG.GLAZING:

ALL BEAMS:

B = 1.00 FT.
D = 3.20 FT.

BEAM RESULTS:

FOR 60DEG.GLAZING:

EXT.SOUTH BEAM
F = 4278 LB.
M1= 194349 LB.FT
M2= 8809 LB.FT
S1= -10526 LB.
S2= 374 LB.
INTERNAL BEAMS
F = 3272 LB.
P = 377 LB/FT
M1= 316738 LB.FT
M2= 596 LB.FT
S1= -17419 LB.
S2= -52 LB.
EXT.NORTH BEAM
P = 375 LB/FT
M1= 428309 LB.FT
M2= 75798 LB.FT
S1= -24028 LB.
S2= -4252 LB.

FOR 90DEG.GLAZING:

EXT.SOUTH BEAM
F = -3396 LB.
M1= 314741 LB.FT
M2= 14197 LB.FT
S1= -17981 LB.
S2= 880 LB.
INTERNAL BEAMS
F = -28621 LB.
P = 407 LB/FT
M1= 418820 LB.FT
M2= 19881 LB.FT
S1= -26294 LB.
S2= 1904 LB.
EXT.NORTH BEAM
P = 412 LB/FT
M1= 409948 LB.FT
S1= -22998 LB.

SHELL RESULTS:

SHELLS WITH 60-DEGREE GLAZING | SHELLS WITH 90-DEGREE GLAZING

EXTERNAL SOUTH SHELL

POINT	T1	S	T2	M2	T1	S	T2	M2
0	42912	-240	37	534	36984	191	-85	-19
1	13495	-4738	-286	405	12417	-4827	-486	212
2	-4927	-5310	-1018	88	-3209	-5657	-1486	44
3	-14588	-3591	-1652	-224	-12043	-3953	-2389	-222
4	-17498	-856	-1922	-357	-15765	-948	-2800	-315
5	-14721	1885	-1748	-258	-14625	2339	-2554	-125
6	-6061	3690	-1207	-17	-7102	4756	-1710	224
7	9694	3499	-540	152	9403	4695	-609	406
8	34236	0	-183	0	37332	0	8	-0

INTERNAL SHELLS

POINT	T1	S	T2	M2	T1	S	T2	M2
0	43732	-184	375	85	48134	1606	-13	30
1	15257	-4920	44	268	16609	-4964	-305	296
2	-3658	-5757	-741	128	-2877	-6222	-1393	104
3	-14426	-4159	-1462	-111	-13271	-4390	-2407	-216
4	-18340	-1362	-1823	-239	-17116	-1097	-2880	-354
5	-16035	1565	-1718	-167	-15436	2424	-2639	-186
6	-7189	3582	-1212	39	-7235	4939	-1764	170
7	9312	3522	-550	177	9853	4842	-627	377
8	34926	0	-189	-0	38281	0	8	-0

EXTERNAL NORTH SHELL

POINT	T1	S	T2	M2	T1	S	T2	M2
0	44036	-162	376	95	48017	1588	-17	18
1	15389	-4933	46	273	16514	-4959	-310	290
2	-3632	-5782	-742	126	-2968	-6198	-1395	107
3	-14468	-4183	-1467	-118	-13328	-4350	-2402	-204
4	-18431	-1375	-1832	-249	-17077	-1054	-2866	-334
5	-16155	1571	-1727	-178	-15256	2444	-2618	-168
6	-7288	3606	-1219	30	-6980	4911	-1743	191
7	9350	3553	-552	172	9893	4776	-617	389
8	35314	0	-187	0	37465	0	8	-0

L = 56 FT. TABLE NO.N- 52 W = 28 FT.

SHELL DATA:

```
L =  56.00 FT.
W =  28.00 FT.
T =   2.50 IN.
G =  50.00 LB/FT*2
QA= 150.00 LB/FT
QV= 150.00 LB/FT
```

FOR 60DEG.GLAZING:

```
K =  13.50 FT.
R =  25.00 FT.
AL=  25.32 FT.
θ =  58.04 DEG.
```

FOR 90DEG.GLAZING:

```
K =  13.50 FT.
R =  40.00 FT.
AL=  31.93 FT.
θ =  45.73 DEG.
```

BEAM DATA:

FOR 60DEG.GLAZING:

```
B1= 0.75 FT.
D1= 3.50 FT.
B2= 0.75 FT.
D2= 3.00 FT.
```

EXT.NORTH BEAM:

```
Z'= 1.72 FT.
Y'= 0.99 FT.
IZ= 3.22 FT*4
IY= 1.97 FT*4
```

FOR 90DEG.GLAZING:

ALL BEAMS:

```
B = 1.00 FT.
D = 3.50 FT.
```

BEAM RESULTS:

FOR 60DEG.GLAZING:

```
EXT.SOUTH BEAM
 F =     -389 LB.
 M1=   197199 LB.FT
 M2=     8738 LB.FT
 S1=   -11080 LB.
 S2=      473 LB.
INTERNAL BEAMS
 F =    -1376 LB.
 P =      388 LB/FT
 M1=   323052 LB.FT
 M2=      523 LB.FT
 S1=   -18233 LB.
 S2=       45 LB.
EXT.NORTH BEAM
 P =      386 LB/FT
 M1=   433826 LB.FT
 M2=    78050 LB.FT
 S1=   -24338 LB.
 S2=    -4379 LB.
```

FOR 90DEG.GLAZING:

```
EXT.SOUTH BEAM
 F =    -5868 LB.
 M1=   321274 LB.FT
 M2=    13782 LB.FT
 S1=   -18591 LB.
 S2=      929 LB.
INTERNAL BEAMS
 F =   -33553 LB.
 P =      423 LB/FT
 M1=   429029 LB.FT
 M2=    19367 LB.FT
 S1=   -27352 LB.
 S2=     2016 LB.
EXT.NORTH BEAM
 P =      428 LB/FT
 M1=   417967 LB.FT
 S1=   -23448 LB.
```

SHELL RESULTS:

	SHELLS WITH 60-DEGREE GLAZING				SHELLS WITH 90-DEGREE GLAZING			
EXTERNAL SOUTH SHELL								
POINT	T1	S	T2	M2	T1	S	T2	M2
0	42955	22	13	494	37356	329	-103	-109
1	13355	-4789	-324	403	12500	-5058	-554	215
2	-4694	-5406	-1130	79	-3115	-5958	-1689	61
3	-13889	-3645	-1829	-253	-11865	-4174	-2716	-224
4	-16636	-853	-2127	-392	-15646	-1012	-3191	-329
5	-14143	1956	-1937	-276	-14746	2488	-2924	-128
6	-6048	3841	-1339	-4	-7398	5113	-1967	255
7	9258	3678	-594	184	9389	5095	-704	457
8	33908	0	-191	0	38320	-0	12	0
INTERNAL SHELLS								
POINT	T1	S	T2	M2	T1	S	T2	M2
0	44176	77	374	72	48698	1882	-32	-78
1	15174	-5028	22	290	16730	-5180	-353	293
2	-3598	-5918	-852	131	-2785	-6538	-1582	121
3	-13983	-4247	-1653	-144	-13072	-4621	-2731	-217
4	-17682	-1346	-2051	-290	-16938	-1160	-3273	-368
5	-15531	1689	-1929	-205	-15484	2572	-3010	-187
6	-7098	3800	-1358	35	-7490	5289	-2022	203
7	9092	3752	-607	200	9801	5232	-722	428
8	34931	0	-195	0	39126	-0	12	0
EXTERNAL NORTH SHELL								
POINT	T1	S	T2	M2	T1	S	T2	M2
0	44361	93	376	79	48601	1860	-35	-89
1	15259	-5036	24	293	16639	-5182	-358	288
2	-3572	-5935	-853	129	-2884	-6518	-1586	123
3	-13998	-4265	-1657	-149	-13145	-4581	-2728	-206
4	-17736	-1358	-2058	-299	-16916	-1113	-3259	-349
5	-15620	1690	-1937	-215	-15313	2598	-2987	-164
6	-7185	3818	-1364	27	-7228	5264	-1999	224
7	9109	3778	-609	196	9857	5162	-710	441
8	35251	0	-193	0	38296	-0	12	0

L = 56 FT. TABLE NO.N- 53 W = 30 FT.

SHELL DATA:

L = 56,00 FT.
W = 30,00 FT.
T = 2,50 IN.
G = 50,00 LB/FT*2
QA= 150,00 LB/FT
QV= 150,00 LB/FT

FOR 60DEG,GLAZING:
K = 14,00 FT.
R = 27,50 FT.
AL= 27,08 FT.
Ø = 56,42 DEG.

FOR 90DEG,GLAZING:
K = 14,00 FT.
R = 42,00 FT.
AL= 33,90 FT.
Ø = 43,17 DEG.

BEAM DATA:

FOR 60DEG,GLAZING:

B1= 0,75 FT.
D1= 3,50 FT.
B2= 0,75 FT.
D2= 3,00 FT.

EXT,NORTH BEAM:

Z'= 1,72 FT.
Y'= 0,99 FT.
IZ= 3,22 FT*4
IY= 1,97 FT*4

FOR 90DEG,GLAZING:

ALL BEAMS:

B = 1,00 FT.
D = 3,50 FT.

BEAM RESULTS:

FOR 60DEG,GLAZING:

EXT,SOUTH BEAM
F = -5083 LB.
M1= 199665 LB,FT
M2= 8695 LB,FT
S1= -11615 LB.
S2= 574 LB.
INTERNAL BEAMS
F = -6398 LB.
P = 400 LB/FT
M1= 329065 LB,FT
M2= 474 LB,FT
S1= -19067 LB.
S2= 150 LB.
EXT,NORTH BEAM
P = 398 LB/FT
M1= 439761 LB,FT
M2= 80542 LB,FT
S1= -24671 LB.
S2= -4518 LB.

FOR 90DEG,GLAZING:

EXT,SOUTH BEAM
F = -8346 LB.
M1= 326359 LB,FT
M2= 13472 LB,FT
S1= -19121 LB.
S2= 982 LB.
INTERNAL BEAMS
F = -38253 LB.
P = 438 LB/FT
M1= 437428 LB,FT
M2= 18991 LB,FT
S1= -28286 LB.
S2= 2129 LB.
EXT,NORTH BEAM
P = 443 LB/FT
M1= 425474 LB,FT
S1= -23869 LB.

SHELL RESULTS:

	SHELLS WITH 60-DEGREE GLAZING				SHELLS WITH 90-DEGREE GLAZING			
EXTERNAL SOUTH SHELL								
POINT	T1	S	T2	M2	T1	S	T2	M2
0	42945	285	-8	447	37617	468	-117	-192
1	13159	-4827	-357	402	12482	-5271	-619	226
2	-4485	-5483	-1236	75	-3043	-6229	-1891	81
3	-13176	-3690	-1999	-275	-11604	-4376	-3045	-229
4	-15748	-863	-2327	-422	-15388	-1085	-3586	-350
5	-13570	1995	-2124	-289	-14766	2599	-3301	-137
6	-6083	3959	-1473	14	-7703	5430	-2235	281
7	8763	3841	-650	219	9256	5476	-805	507
8	33564	-0	-199	-0	39147	-0	15	-0
INTERNAL SHELLS								
POINT	T1	S	T2	M2	T1	S	T2	M2
0	44564	359	375	45	49141	2146	-45	-178
1	15047	-5112	6	311	16710	-5391	-397	299
2	-3531	-6049	-956	138	-2754	-6828	-1773	141
3	-13470	-4317	-1837	-169	-12811	-4826	-3058	-220
4	-16930	-1341	-2272	-334	-16624	-1226	-3670	-389
5	-14975	1774	-2137	-238	-15421	2688	-3388	-196
6	-7043	3971	-1504	37	-7744	5599	-2288	231
7	8751	3952	-666	227	9636	5600	-822	479
8	34770	-0	-201	-0	39803	-0	15	-0
EXTERNAL NORTH SHELL								
POINT	T1	S	T2	M2	T1	S	T2	M2
0	44651	368	376	51	49059	2120	-48	-187
1	15095	-5115	7	313	16625	-5397	-403	295
2	-3504	-6060	-957	137	-2855	-6812	-1778	143
3	-13465	-4331	-1839	-174	-12895	-4787	-3056	-211
4	-16959	-1353	-2278	-342	-16618	-1176	-3657	-372
5	-15044	1772	-2144	-246	-15264	2719	-3365	-173
6	-7124	3984	-1510	30	-7481	5577	-2264	252
7	8752	3975	-668	223	9707	5530	-809	492
8	35053	-0	-200	-0	38979	-0	15	-0

L = 58 FT. TABLE NO.N- 54 W = 27 FT.

SHELL DATA:

L = 58.00 FT.
W = 27.00 FT.
T = 2.50 IN.
G = 50.00 LB/FT*2
QA= 150.00 LB/FT
QV= 150.00 LB/FT

FOR 60DEG.GLAZING:
K = 13.25 FT.
-R20 25.00 FT.

K = 13.25 FT.
R = 35.00 FT.
AL= 31.09 FT.
θ = 50.89 DEG.

BEAM DATA:

FOR 60DEG.GLAZING:

B1= 0.75 FT.
D1= 3.75 FT.
B2= 0.75 FT.
D2= 3.25 FT.

EXT.NORTH BEAM:
Z'= 1.84 FT.
Y'= 1.02 FT.
IZ= 4.01 FT*4
IY= 2.40 FT*4

FOR 90DEG.GLAZING:

ALL BEAMS:

B = 1.00 FT.
D = 3.75 FT.

BEAM RESULTS:

FOR 60DEG.GLAZING:

EXT.SOUTH BEAM
F = 546 LB.
M1= 243294 LB.FT
M2= 8428 LB.FT
S1= -13113 LB.
S2= 419 LB.
INTERNAL BEAMS
F = 3446 LB.
M2= -128 LB.FT
S2= -91 LB.
EXT.NORTH BEAM
P = 400 LB/FT
M1= 493490 LB.FT
M2= 86869 LB.FT
S1= -26730 LB.
S2= -4705 LB.

5 -324 296

FOR 90DEG.GLAZING:

EXT.SOUTH BEAM
F = -2730 LB.
M1= 360076 LB.FT
M2= 14041 LB.FT
S2= 823 LB.
INTERNAL BEAMS
F = -28196 LB.
P = 418 LB/FT
M1= 476848 LB.FT
M2= 19583 LB.FT
S1= -28681 LB.
S2= 1810 LB.
EXT.NORTH BEAM
P = 424 LB/FT
M1= 460915 LB.FT
S1= -24966 LB.

SHELL RESULTS:

EXTERNAL SOUTH SHELL

	SHELLS WITH 60-DEGREE GLAZING				SHELLS WITH 90-DEGREE GLAZING			
POINT	T1	S	T2	M2	T1	S	T2	M2
0	44199	-30	-33	315	36535	148	-101	-35
1	14743	-4732	-336	310	12239	-4811	-496	217
2	-4382	-5455	-1077	85	-3166	-5629	-1486	46
3	-14948	-3757	-1736	-165	-11873	-3947	-2381	-230
6	-6432	3910	-1274	76	-7249	4743	-1713	-20130 L
8	36020	0	-191	0	37499	-0	3	-0

INTERNAL SHELLS

POINT	T1	S	T2	M2	T1	S	T2	M2
0	45048	-187	368	-111	47348	1527	-36	4
2	-3553	-6015	-795	129	-2867	-6163	-1397	105
3	-15144	-4370	-1552	-58	-13056	-4356	-2397	-223
4	-19530	-1419	-1934	-157	-16892	-1109	-2865	-362
5	-17106	1692	-1822	-74	-15372	2385	-2629	-179
6	-7529	3825	-1283	125	-7341	4909	-1761	200
7	10095	3732	-580	235	S1=	-19772 LB.		

EXTERNAL NORTH SHELL

POINT	T1	S	T2	M2	T1	S	T2	M2
0	45311	-153	369	-111	47254	1509	-39	-7
1	16264	-5092	27	191	16188	-4935	-329	291
2	-3539	-6020	-792	132	-2966	-6142	-1400	107
3	-15166	-4374	-1550	-54	-13128	-4317	-2394	-212
4	-19547	-1420	-1933	-153	-16866	-1063	-2852	-342
5	-17095	1693	-1820	-70	-15193	2409	-2608	-155
6	-7498	3822	-1281	127	-7071	4883	-1740	222
7	10101	3724	-580	236	9794	4771	-618	427
8	36684	0	-196	0	37459	-0	3	-0

L = 58 FT. TABLE NO.N- 55 W = 29 FT.

SHELL DATA:

L = 58.00 FT.
W = 29.00 FT.
T = 2.50 IN.
G = 50.00 LB/FT*2
QA= 150.00 LB/FT
QV= 150.00 LB/FT

FOR 60DEG.GLAZING:
K = 14.75 FT.
R = 25.00 FT.
AL= 26.31 FT.
θ = 60.31 DEG.

FOR 90DEG.GLAZING:
K = 14.75 FT.
R = 42.50 FT.
AL= 33.39 FT.
θ = 45.01 DEG.

BEAM DATA:

FOR 60DEG.GLAZING:

B1= 0.75 FT.
D1= 3.75 FT.
B2= 0.75 FT.
D2= 3.25 FT.

EXT.NORTH BEAM:

Z'= 1.84 FT.
Y'= 1.02 FT.
IZ= 4.01 FT*4
IY= 2.40 FT*4

FOR 90DEG.GLAZING:

ALL BEAMS:

B = 1.00 FT.
D = 3.75 FT.

BEAM RESULTS:

FOR 60DEG.GLAZING:

EXT.SOUTH BEAM
F = 480 LB.
M1= 225066 LB.FT
M2= 9279 LB.FT
S1= -12134 LB.
S2= 468 LB.
INTERNAL BEAMS
F = 252 LB.
P = 398 LB/FT
M1= 358145 LB.FT
M2= 964 LB.FT
S1= -19350 LB.
S2= 33 LB.
EXT.NORTH BEAM
P = 395 LB/FT
M1= 496595 LB.FT
M2= 85656 LB.FT
S1= -26898 LB.
S2= -4640 LB.

FOR 90DEG.GLAZING:

EXT.SOUTH BEAM
F = -4716 LB.
M1= 369766 LB.FT
M2= 14198 LB.FT
S1= -20500 LB.
S2= 888 LB.
INTERNAL BEAMS
F = -33811 LB.
P = 439 LB/FT
M1= 492639 LB.FT
M2= 19934 LB.FT
S1= -30109 LB.
S2= 1985 LB.
EXT.NORTH BEAM
P = 444 LB/FT
M1= 475152 LB.FT
S1= -25737 LB.

SHELL RESULTS:

	SHELLS WITH 60-DEGREE GLAZING				SHELLS WITH 90-DEGREE GLAZING			
EXTERNAL SOUTH SHELL								
POINT	T1	S	T2	M2	T1	S	T2	M2
0	41899	-26	12	487	37339	255	-123	-157
1	12940	-4723	-312	396	12600	-5196	-594	208
2	-4561	-5319	-1094	71	-3048	-6127	-1766	62
3	-13386	-3612	-1773	-259	-11939	-4327	-2831	-228
4	-16027	-914	-2065	-390	-15924	-1094	-3329	-330
5	-13711	1808	-1885	-258	-15175	2522	-3056	-109
6	-6013	3655	-1309	27	-7727	5265	-2060	301
7	8735	3533	-588	215	9562	5272	-736	510
8	32612	0	-195	0	39299	-0	20	0
INTERNAL SHELLS								
POINT	T1	S	T2	M2	T1	S	T2	M2
0	43037	-14	399	30	48556	1831	-56	-142
1	14809	-5007	52	280	16821	-5299	-389	280
2	-3461	-5883	-807	132	-2682	-6696	-1654	122
3	-13570	-4258	-1597	-140	-13102	-4770	-2840	-218
4	-17218	-1429	-1996	-284	-17180	-1246	-3405	-366
5	-15228	1545	-1886	-189	-15891	2599	-3138	-166
6	-7116	3635	-1334	63	-7818	5432	-2112	251
7	8627	3628	-603	229	9951	5402	-753	483
8	33815	0	-198	0	40053	-0	19	0
EXTERNAL NORTH SHELL								
POINT	T1	S	T2	M2	T1	S	T2	M2
0	43173	-3	400	39	48477	1810	-58	-152
1	14886	-5015	54	284	16736	-5303	-395	275
2	-3419	-5901	-808	130	-2784	-6678	-1658	123
3	-13567	-4281	-1602	-149	-13182	-4731	-2837	-208
4	-17274	-1448	-2004	-296	-17166	-1197	-3392	-348
5	-15351	1543	-1896	-203	-15725	2627	-3116	-143
6	-7255	3658	-1343	51	-7552	5409	-2089	272
7	8636	3666	-606	222	10015	5333	-741	496
8	34304	0	-196	0	39220	-0	20	0

L = 58 FT. TABLE NO.N- 56 W = 31 FT.

SHELL DATA:

L = 58.00 FT.
W = 31.00 FT.
T = 2.50 IN.
G = 50.00 LB/FT*2
QA= 150.00 LB/FT
QV= 150.00 LB/FT

FOR 60DEG.GLAZING:
K = 14.75 FT.
R = 27.50 FT.
AL= 28.06 FT.
θ = 58.46 DEG.

FOR 90DEG.GLAZING:
K = 14.75 FT.
R = 45.00 FT.
AL= 35.22 FT.
θ = 44.85 DEG.

BEAM DATA:

FOR 60DEG.GLAZING:

B1= 0.75 FT.
D1= 3.75 FT.
B2= 0.75 FT.
D2= 3.25 FT.

EXT.NORTH BEAM:

Z'= 1.84 FT.
Y'= 1.02 FT.
IZ= 4.01 FT*4
IY= 2.40 FT*4

FOR 90DEG.GLAZING:

ALL BEAMS:

B = 1.00 FT.
D = 3.75 FT.

BEAM RESULTS:

FOR 60DEG.GLAZING:

EXT.SOUTH BEAM
F = -4961 LB.
M1= 231239 LB.FT
M2= 8906 LB.FT
S1= -12948 LB.
S2= 562 LB.
INTERNAL BEAMS
F = -5049 LB.
P = 414 LB/FT
M1= 369713 LB.FT
M2= 629 LB.FT
S1= -20517 LB.
S2= 125 LB.
EXT.NORTH BEAM
P = 411 LB/FT
M1= 502889 LB.FT
M2= 89289 LB.FT
S1= -27239 LB.
S2= -4836 LB.

FOR 90DEG.GLAZING:

EXT.SOUTH BEAM
F = -8116 LB.
M1= 370595 LB.FT
M2= 13769 LB.FT
S1= -20891 LB.
S2= 958 LB.
INTERNAL BEAMS
F = -38385 LB.
P = 449 LB/FT
M1= 495555 LB.FT
M2= 19345 LB.FT
S1= -30732 LB.
S2= 2078 LB.
EXT.NORTH BEAM
P = 454 LB/FT
M1= 479257 LB.FT
S1= -25959 LB.

SHELL RESULTS:

EXTERNAL SOUTH SHELL

	SHELLS WITH 60-DEGREE GLAZING				SHELLS WITH 90-DEGREE GLAZING			
POINT	T1	S	T2	M2	T1	S	T2	M2
0	42153	269	-22	415	37080	440	-132	-205
1	12931	-4751	-355	390	12269	-5229	-622	237
2	-4322	-5404	-1209	72	-2937	-6177	-1875	86
3	-12801	-3664	-1954	-274	-11278	-4373	-3017	-240
4	-15368	-911	-2278	-413	-15088	-1152	-3561	-368
5	-13361	1890	-2085	-268	-14731	2500	-3291	-140
6	-6129	3838	-1453	48	-7941	5367	-2239	308
7	8453	3754	-649	252	9032	5475	-810	548
8	32912	0	-208	0	39255	0	14	0

INTERNAL SHELLS

POINT	T1	S	T2	M2	T1	S	T2	M2
0	43817	273	391	-6	48365	2079	-65	-199
1	14804	-5109	25	303	16357	-5348	-408	307
2	-3473	-6032	-926	144	-2707	-6753	-1763	145
3	-13253	-4328	-1798	-164	-12471	-4795	-3030	-232
4	-16693	-1396	-2233	-328	-16256	-1274	-3639	-406
5	-14847	1684	-2107	-223	-15305	2590	-3369	-196
6	-7091	3874	-1489	64	-7933	5521	-2286	259
7	8506	3882	-666	256	9386	5582	-824	522
8	34235	0	-209	0	39781	0	14	0

EXTERNAL NORTH SHELL

POINT	T1	S	T2	M2	T1	S	T2	M2
0	43860	279	391	-1	48297	2053	-67	-206
1	14839	-5111	26	305	16279	-5357	-414	303
2	-3440	-6041	-926	142	-2807	-6740	-1769	145
3	-13236	-4342	-1800	-169	-12561	-4759	-3030	-224
4	-16716	-1409	-2239	-336	-16262	-1225	-3628	-389
5	-14922	1680	-2114	-232	-15160	2623	-3348	-174
6	-7189	3887	-1495	56	-7675	5503	-2262	280
7	8500	3908	-669	252	9466	5514	-812	535
8	34562	0	-208	0	38973	0	14	0

L = 60 FT. TABLE NO.N- 57 W = 28 FT.

SHELL DATA:

L = 60.00 FT.
W = 28.00 FT.
T = 2.50 IN.
G = 50.00 LB/FT*2
QA= 150.00 LB/FT
QV= 150.00 LB/FT

FOR 60DEG.GLAZING:
K = 14.00 FT.
R = 22.50 FT.
AL= 25.61 FT.
θ = 65.21 DEG.

FOR 90DEG.GLAZING:
K = 14.00 FT.
R = 37.50 FT.
AL= 32.29 FT.
θ = 49.34 DEG.

BEAM DATA:

FOR 60DEG.GLAZING:

B1= 0.75 FT.
D1= 4.00 FT.
B2= 0.75 FT.
D2= 3.50 FT.

EXT.NORTH BEAM:

Z'= 1.96 FT.
Y'= 1.05 FT.
IZ= 4.93 FT*4
IY= 2.88 FT*4

FOR 90DEG.GLAZING:

ALL BEAMS:

B = 1.00 FT.
D = 4.00 FT.

BEAM RESULTS:

FOR 60DEG.GLAZING:

EXT.SOUTH BEAM
F = 4451 LB.
M1= 258460 LB.FT
M2= 9035 LB.FT
S1= -13110 LB.
S2= 356 LB.
INTERNAL BEAMS
F = 6106 LB.
P = 406 LB/FT
M1= 398543 LB.FT
M2= 718 LB.FT
S1= -20202 LB.
S2= -99 LB.
EXT.NORTH BEAM
P = 403 LB/FT
M1= 557008 LB.FT
M2= 93492 LB.FT
S1= -29165 LB.
S2= -4895 LB.

FOR 90DEG.GLAZING:

EXT.SOUTH BEAM
F = -831 LB.
M1= 416375 LB.FT
M2= 13866 LB.FT
S1= -21880 LB.
S2= 737 LB.
INTERNAL BEAMS
F = -27621 LB.
P = 434 LB/FT
M1= 548376 LB.FT
M2= 19317 LB.FT
S1= -31595 LB.
S2= 1721 LB.
EXT.NORTH BEAM
P = 440 LB/FT
M1= 520129 LB.FT
S1= -27234 LB.

SHELL RESULTS:

	SHELLS WITH 60-DEGREE GLAZING				SHELLS WITH 90-DEGREE GLAZING			
EXTERNAL SOUTH SHELL								
POINT	T1	S	T2	M2	T1	S	T2	M2
0	41254	-233	14	498	36819	44	-125	-104
1	12859	-4601	-284	393	12492	-4998	-547	199
2	-4650	-5153	-986	76	-3185	-5847	-1580	46
3	-13679	-3516	-1596	-239	-12248	-4117	-2516	-226
4	-16427	-916	-1860	-363	-16277	-1021	-2947	-311
5	-13993	1704	-1699	-239	-15325	2418	-2694	-90
6	-6032	3468	-1185	30	-7565	4980	-1807	302
7	8888	3338	-546	204	9803	4937	-644	494
8	32542	0	-206	0	38996	0	8	-0
INTERNAL SHELLS								
POINT	T1	S	T2	M2	T1	S	T2	M2
0	42338	-320	409	22	47440	1446	-65	-81
1	14629	-4953	76	271	16554	-5091	-374	270
2	-3654	-5751	-707	138	-2794	-6365	-1484	104
3	-13960	-4174	-1426	-116	-13355	-4528	-2524	-216
4	-17705	-1436	-1791	-251	-17515	-1167	-3015	-346
5	-15585	1435	-1697	-164	-16064	2486	-2767	-145
6	-7189	3439	-1207	70	-7685	5139	-1854	253
7	8751	3424	-561	220	10194	5065	-660	467
8	33735	0	-210	0	39809	0	8	0
EXTERNAL NORTH SHELL								
POINT	T1	S	T2	M2	T1	S	T2	M2
0	42502	-309	410	31	47369	1429	-68	-93
1	14713	-4963	77	274	16466	-5092	-379	265
2	-3619	-5770	-708	136	-2902	-6345	-1487	107
3	-13969	-4195	-1431	-124	-13439	-4486	-2521	-203
4	-17770	-1452	-1799	-263	-17492	-1117	-3001	-324
5	-15705	1436	-1706	-177	-15874	2514	-2745	-119
6	-7315	3461	-1214	59	-7393	5113	-1831	277
7	8769	3459	-563	214	10256	4992	-648	481
8	34188	0	-208	0	38890	0	8	0

L = 60 FT. TABLE NO.N- 58 W = 30 FT.

```
SHELL DATA:                BEAM RESULTS:

  L =  60.00 FT.           FOR 60DEG.GLAZING:
  W =  30.00 FT.
  T =   2.50 IN.           EXT.SOUTH BEAM
  G =  50.00 LB/FT*2        F =    -2025 LB.
  QA= 150.00 LB/FT          M1=   278539 LB.FT
  QV= 150.00 LB/FT          M2=     8516 LB.FT
                            S1=   -14756 LB.
 FOR 60DEG.GLAZING:         S2=      464 LB.
  K =  14.50 FT.           INTERNAL BEAMS
  R =  27.50 FT.            F =     1637 LB.
  AL=  27.06 FT.            P =      419 LB/FT
  θ =  56.39 DEG.           M1=   426334 LB.FT
                            M2=        7 LB.FT
 FOR 90DEG.GLAZING:         S1=   -22129 LB.
  K =  14.50 FT.            S2=      -43 LB.
  R =  42.50 FT.           EXT.NORTH BEAM
  AL=  34.24 FT.            P =      419 LB/FT
  θ =  46.16 DEG.           M1=   565535 LB.FT
                            M2=    97343 LB.FT
BEAM DATA:                  S1=   -29611 LB.
                            S2=    -5097 LB.
 FOR 60DEG.GLAZING:

  B1= 0.75 FT.             FOR 90DEG.GLAZING:
  D1= 4.00 FT.
  B2= 0.75 FT.              EXT.SOUTH BEAM
  D2= 3.50 FT.               F =    -2771 LB.
                             M1=   423988 LB.FT
  EXT.NORTH BEAM:            M2=    13446 LB.FT
                             S1=   -22483 LB.
  Z'= 1.96 FT.               S2=      768 LB.
  Y'= 1.05 FT.             INTERNAL BEAMS
  IZ= 4.93 FT*4              F =   -31807 LB.
  IY= 2.88 FT*4              P =      449 LB/FT
                             M1=   559957 LB.FT
 FOR 90DEG.GLAZING:          M2=    18785 LB.FT
                             S1=   -32641 LB.
  ALL BEAMS:                 S2=     1806 LB.
                            EXT.NORTH BEAM
  B = 1.00 FT.               P =      455 LB/FT
  D = 4.00 FT.               M1=   528816 LB.FT
                             S1=   -27689 LB.
```

SHELL RESULTS:

	SHELLS WITH 60-DEGREE GLAZING				SHELLS WITH 90-DEGREE GLAZING			
EXTERNAL SOUTH SHELL								
POINT	T1	S	T2	M2	T1	S	T2	M2
0	43143	106	-54	287	37182	145	-141	-192
1	14127	-4793	-372	318	12564	-5242	-619	207
2	-4177	-5532	-1178	74	-3118	-6154	-1785	65
3	-14008	-3813	-1895	-209	-12108	-4343	-2845	-230
4	-17311	-953	-2210	-317	-16207	-1088	-3340	-331
5	-14955	1989	-2020	-177	-15504	2568	-3065	-97
6	-6437	3984	-1401	105	-7900	5348	-2067	332
7	9700	3820	-625	269	9825	5350	-740	546
8	35092	-0	-205	-0	40139	-0	13	-0
INTERNAL SHELLS								
POINT	T1	S	T2	M2	T1	S	T2	M2
0	44527	-86	390	-130	47960	1665	-82	-185
1	15654	-5263	14	225	16650	-5325	-427	273
2	-3565	-6202	-896	138	-2739	-6688	-1678	123
3	-14566	-4486	-1734	-101	-13207	-4763	-2850	-219
4	-18677	-1447	-2155	-225	-17401	-1233	-3409	-365
5	-16479	1761	-2029	-121	-16181	2638	-3141	-151
6	-7463	3990	-1428	132	-7983	5502	-2114	284
7	9674	3925	-641	278	10188	5469	-755	520
8	36336	-0	-207	-0	40835	-0	12	-0
EXTERNAL NORTH SHELL								
POINT	T1	S	T2	M2	T1	S	T2	M2
0	44647	-71	390	-130	47903	1644	-84	-195
1	15701	-5263	15	225	16566	-5331	-432	268
2	-3561	-6205	-895	139	-2852	-6672	-1683	125
3	-14580	-4488	-1734	-100	-13303	-4722	-2849	-208
4	-18688	-1447	-2155	-224	-17394	-1180	-3396	-345
5	-16477	1762	-2029	-120	-16002	2671	-3117	-126
6	-7450	3990	-1428	133	-7689	5479	-2090	308
7	9679	3923	-640	279	10265	5394	-743	534
8	36296	-0	-207	-0	39913	-0	12	0

L = 60 FT. TABLE NO.N- 59 W = 32 FT.

SHELL DATA:

L = 60.00 FT.
W = 32.00 FT.
T = 2.50 IN.
G = 50.00 LB/FT*2
QA= 150.00 LB/FT
QV= 150.00 LB/FT

FOR 60DEG.GLAZING:
K = 15.50 FT.
R = 30.00 FT.
AL= 28.81 FT.
A = 55.03 DEG.

FOR 90DEG.GLAZING:
K = 15.50 FT.
R = 47.50 FT.
AL= 36.44 FT.
A = 43.96 DEG.

BEAM DATA:

FOR 60DEG.GLAZING:

B1= 0.75 FT.
D1= 4.00 FT.
B2= 0.75 FT.
D2= 3.50 FT.

EXT.NORTH BEAM:

Z'= 1.96 FT.
Y'= 1.05 FT.
IZ= 4.93 FT*4
IY= 2.88 FT*4

FOR 90DEG.GLAZING:

ALL BEAMS:

B = 1.00 FT.
D = 4.00 FT.

BEAM RESULTS:

FOR 60DEG.GLAZING:

EXT.SOUTH BEAM
F = -5408 LB.
M1= 277793 LB.FT
M2= 8742 LB.FT
S1= -15028 LB.
S2= 545 LB.
INTERNAL BEAMS
F = -2609 LB.
P = 426 LB/FT
M1= 428124 LB.FT
M2= 234 LB.FT
S1= -22670 LB.
S2= 54 LB.
EXT.NORTH BEAM
P = 426 LB/FT
M1= 572062 LB.FT
M2= 98967 LB.FT
S1= -29953 LB.
S2= -5182 LB.

FOR 90DEG.GLAZING:

EXT.SOUTH BEAM
F = -5830 LB.
M1= 426647 LB.FT
M2= 13648 LB.FT
S1= -22943 LB.
S2= 860 LB.
INTERNAL BEAMS
F = -37165 LB.
P = 463 LB/FT
M1= 567053 LB.FT
M2= 19129 LB.FT
S1= -33575 LB.
S2= 1966 LB.
EXT.NORTH BEAM
P = 469 LB/FT
M1= 538964 LB.FT
S1= -28220 LB.

SHELL RESULTS:

	SHELLS WITH 60-DEGREE GLAZING				SHELLS WITH 90-DEGREE GLAZING			
EXTERNAL SOUTH SHELL								
POINT	T1	S	T2	M2	T1	S	T2	M2
0	42916	283	-63	269	37293	305	-152	-262
1	13820	-4870	-399	330	12471	-5419	-675	228
2	-4041	-5636	-1277	72	-2969	-6391	-1971	88
3	-13358	-3883	-2057	-237	-11623	-4542	-3154	-241
4	-16481	-985	-2402	-355	-15696	-1205	-3719	-365
5	-14420	2011	-2201	-197	-15398	2606	-3438	-119
6	-6478	4086	-1529	119	-8285	5603	-2337	354
7	9207	3968	-676	304	9477	5708	-844	600
8	34643	0	-207	0	40770	-0	19	-0
INTERNAL SHELLS								
POINT	T1	S	T2	M2	T1	S	T2	M2
0	44545	137	397	-142	48343	1946	-91	-269
1	15453	-5347	3	250	16552	-5512	-462	292
2	-3453	-6332	-989	145	-2671	-6950	-1854	147
3	-13996	-4571	-1903	-130	-12757	-4964	-3160	-230
4	-17899	-1469	-2362	-274	-16843	-1336	-3792	-400
5	-15928	1814	-2225	-155	-15983	2686	-3513	-173
6	-7429	4130	-1566	134	-8300	5751	-2383	307
7	9282	4099	-694	307	9815	5814	-858	575
8	35990	0	-209	0	41302	-0	18	-0
EXTERNAL NORTH SHELL								
POINT	T1	S	T2	M2	T1	S	T2	M2
0	44599	143	397	-141	48292	1920	-93	-277
1	15476	-5348	3	250	16476	-5523	-468	288
2	-3447	-6335	-989	145	-2778	-6939	-1860	148
3	-14000	-4574	-1904	-131	-12857	-4927	-3161	-222
4	-17907	-1471	-2362	-275	-16853	-1284	-3781	-383
5	-15939	1814	-2226	-156	-15830	2722	-3491	-150
6	-7439	4132	-1567	133	-8024	5733	-2358	329
7	9284	4102	-694	307	9904	5742	-844	588
8	36025	0	-209	0	40442	-0	19	-0

L = 62 FT. TABLE NO.N- 60 W = 29 FT.

SHELL DATA:

```
  L =  62.00 FT.
  W =  29.00 FT.
  T =   2.50 IN.
  G =  50.00 LB/FT*2
  QA= 150.00 LB/FT
  QV= 200.00 LB/FT

FOR 60DEG.GLAZING:
  K =  14.50 FT.
  R =  20.00 FT.
  AL=  27.15 FT.
  Θ =  77.79 DEG.

FOR 90DEG.GLAZING:
  K =  14.50 FT.
  R =  35.00 FT.
  AL=  33.71 FT.
  Θ =  55.19 DEG.
```

BEAM DATA:

```
FOR 60DEG.GLAZING:

  B1= 0.75 FT.
  D1= 4.00 FT.
  B2= 0.75 FT.
  D2= 3.50 FT.

  EXT.NORTH BEAM:

  Z'= 1.96 FT.
  Y'= 1.05 FT.
  IZ= 4.93 FT*4
  IY= 2.88 FT*4

FOR 90DEG.GLAZING:

  ALL BEAMS:

  B = 1.00 FT.
  D = 4.00 FT.
```

BEAM RESULTS:

```
FOR 60DEG.GLAZING:

EXT.SOUTH BEAM
 F =      8635 LB.
 M1=    253633 LB.FT
 M2=     11159 LB.FT
 S1=    -12071 LB.
 S2=       363 LB.
INTERNAL BEAMS
 F =      4336 LB.
 P =       409 LB/FT
 M1=    401024 LB.FT
 M2=      2204 LB.FT
 S1=    -19851 LB.
 S2=         6 LB.
EXT.NORTH BEAM
 P =       401 LB/FT
 M1=    595846 LB.FT
 M2=     99311 LB.FT
 S1=    -30192 LB.
 S2=     -5032 LB.

FOR 90DEG.GLAZING:

EXT.SOUTH BEAM
 F =     -9414 LB.
 M1=    438572 LB.FT
 M2=     17041 LB.FT
 S1=    -23168 LB.
 S2=      1089 LB.
INTERNAL BEAMS
 F =    -37231 LB.
 P =       430 LB/FT
 M1=    578576 LB.FT
 M2=     23379 LB.FT
 S1=    -33078 LB.
 S2=      2112 LB.
EXT.NORTH BEAM
 P =       435 LB/FT
 M1=    555143 LB.FT
 S1=    -28130 LB.
```

SHELL RESULTS:

EXTERNAL SOUTH SHELL

	SHELLS WITH 60-DEGREE GLAZING				SHELLS WITH 90-DEGREE GLAZING			
POINT	T1	S	T2	M2	T1	S	T2	M2
0	42278	-438	141	850	39098	477	-94	31
1	11141	-4806	-192	571	12768	-4850	-453	289
2	-5964	-5077	-909	79	-3088	-5723	-1456	62
3	-13209	-3319	-1497	-364	-11526	-4058	-2371	-283
4	-14594	-866	-1737	-544	-15190	-1135	-2805	-411
5	-12246	1492	-1586	-406	-14676	2138	-2591	-186
6	-5737	3111	-1117	-79	-7900	4690	-1762	252
7	7393	3087	-530	163	8794	4804	-642	493
8	30436	0	-223	0	38567	-0	-4	-0

INTERNAL SHELLS

POINT	T1	S	T2	M2	T1	S	T2	M2
0	43299	-220	493	241	50568	1887	-29	77
1	13838	-4954	170	416	16760	-5021	-284	377
2	-4132	-5640	-610	164	-3048	-6269	-1379	123
3	-13179	-4045	-1314	-205	-12863	-4430	-2397	-283
4	-16021	-1461	-1671	-408	-16404	-1218	-2879	-459
5	-14159	1200	-1597	-324	-15213	2246	-2660	-251
6	-7099	3115	-1151	-42	-7824	4843	-1800	197
7	7345	3220	-550	175	9188	4901	-652	464
8	32053	0	-227	0	39064	-0	-4	-0

EXTERNAL NORTH SHELL

POINT	T1	S	T2	M2	T1	S	T2	M2
0	43365	-213	496	259	50485	1861	-31	67
1	13935	-4961	173	423	16668	-5028	-290	371
2	-4012	-5667	-610	158	-3164	-6255	-1384	124
3	-13102	-4089	-1320	-224	-12964	-4391	-2397	-273
4	-16093	-1507	-1685	-437	-16404	-1166	-2869	-439
5	-14441	1184	-1617	-358	-15036	2280	-2640	-225
6	-7478	3159	-1169	-70	-7520	4823	-1778	222
7	7318	3309	-555	160	9275	4829	-641	479
8	33319	0	-222	0	38118	-0	-4	-0

L = 62 FT. TABLE NO.N- 61 W = 31 FT.

SHELL DATA:

L = 62.00 FT.
W = 31.00 FT.
T = 2.50 IN.
G = 50.00 LB/FT*2
QA= 150.00 LB/FT
QV= 200.00 LB/FT

FOR 60DEG.GLAZING:
K = 15.00 FT.
R = 25.00 FT.
AL= 28.35 FT.
θ = 64.96 DEG.

FOR 90DEG.GLAZING:
K = 15.00 FT.
R = 40.00 FT.
AL= 35.60 FT.
θ = 51.00 DEG.

BEAM DATA:

FOR 60DEG.GLAZING:

B1= 0.75 FT.
D1= 4.00 FT.
B2= 0.75 FT.
D2= 3.50 FT.

EXT.NORTH BEAM:

Z'= 1.96 FT.
Y'= 1.05 FT.
IZ= 4.93 FT*4
IY= 2.88 FT*4

FOR 90DEG.GLAZING:

ALL BEAMS:

B = 1.00 FT.
D = 4.00 FT.

BEAM RESULTS:

FOR 60DEG.GLAZING:

EXT.SOUTH BEAM
F = -3071 LB.
M1= 285738 LB.FT
M2= 10377 LB.FT
S1= -14736 LB.
S2= 556 LB.
INTERNAL BEAMS
F = -1916 LB.
P = 417 LB/FT
M1= 438428 LB.FT
M2= 1149 LB.FT
S1= -22384 LB.
S2= 79 LB.
EXT.NORTH BEAM
P = 415 LB/FT
M1= 603917 LB.FT
M2= 102825 LB.FT
S1= -30601 LB.
S2= -5210 LB.

FOR 90DEG.GLAZING:

EXT.SOUTH BEAM
F = -12112 LB.
M1= 449402 LB.FT
M2= 16331 LB.FT
S1= -23991 LB.
S2= 1124 LB.
INTERNAL BEAMS
F = -42577 LB.
P = 446 LB/FT
M1= 594212 LB.FT
M2= 22484 LB.FT
S1= -34414 LB.
S2= 2205 LB.
EXT.NORTH BEAM
P = 452 LB/FT
M1= 565132 LB.FT
S1= -28636 LB.

SHELL RESULTS:

SHELLS WITH 60-DEGREE GLAZING / SHELLS WITH 90-DEGREE GLAZING

EXTERNAL SOUTH SHELL

POINT	T1	S	T2	M2	T1	S	T2	M2
0	45221	156	22	594	39582	614	-118	-87
1	13472	-4891	-299	475	12960	-5084	-524	286
2	-4955	-5482	-1108	80	-2997	-6035	-1656	83
3	-13689	-3691	-1807	-318	-11470	-4294	-2693	-279
4	-16085	-942	-2106	-480	-15266	-1208	-3192	-420
5	-13824	1811	-1928	-331	-14988	2293	-2959	-183
6	-6342	3713	-1347	4	-8279	5073	-2021	288
7	8623	3641	-613	232	8925	5236	-738	549
8	33885	-0	-216	-0	40001	0	1	-0

INTERNAL SHELLS

POINT	T1	S	T2	M2	T1	S	T2	M2
0	46175	97	452	67	51231	2157	-55	-63
1	15454	-5250	97	355	17032	-5237	-336	366
2	-3834	-6138	-806	161	-2917	-6597	-1566	143
3	-14041	-4417	-1633	-182	-12777	-4679	-2715	-276
4	-17543	-1495	-2050	-369	-16446	-1294	-3267	-465
5	-15566	1559	-1944	-267	-15494	2399	-3029	-246
6	-7523	3735	-1383	31	-8192	5222	-2060	236
7	8630	3775	-631	240	9293	5327	-748	522
8	35397	0	-219	0	40425	0	1	0

EXTERNAL NORTH SHELL

POINT	T1	S	T2	M2	T1	S	T2	M2
0	46262	106	453	73	51167	2129	-57	-71
1	15507	-5253	98	357	16944	-5249	-343	361
2	-3799	-6148	-806	160	-3040	-6585	-1573	144
3	-14029	-4432	-1635	-187	-12892	-4639	-2716	-267
4	-17572	-1509	-2055	-377	-16460	-1238	-3256	-446
5	-15648	1555	-1951	-277	-15320	2439	-3007	-219
6	-7627	3748	-1389	23	-7875	5203	-2035	262
7	8627	3802	-633	236	9397	5250	-735	537
8	35748	0	-217	0	39437	0	1	-0

L = 62 FT. TABLE NO.N- 62 W = 33 FT.

SHELL DATA:

L = 62.00 FT.
W = 33.00 FT.
T = 2.50 IN.
G = 50.00 LB/FT*2
QA= 150.00 LB/FT
QV= 200.00 LB/FT

FOR 60DEG.GLAZING:
K = 15.50 FT.
R = 27.50 FT.
AL= 30.07 FT.
θ = 62.66 DEG.

FOR 90DEG.GLAZING:
K = 15.50 FT.
R = 40.00 FT.
AL= 37.86 FT.
θ = 54.22 DEG.

BEAM DATA:

FOR 60DEG.GLAZING:

B1= 0.75 FT.
D1= 4.00 FT.
B2= 0.75 FT.
D2= 3.50 FT.

EXT.NORTH BEAM:

Z'= 1.96 FT.
Y'= 1.05 FT.
IZ= 4.93 FT*4
IY= 2.88 FT*4

FOR 90DEG.GLAZING:

ALL BEAMS:

B = 1.00 FT.
D = 4.00 FT.

BEAM RESULTS:

FOR 60DEG.GLAZING:

EXT.SOUTH BEAM
F = -9050 LB.
M1= 291485 LB.FT
M2= 10284 LB.FT
S1= -15558 LB.
S2= 669 LB.
INTERNAL BEAMS
F = -7856 LB.
P = 430 LB/FT
M1= 449012 LB.FT
M2= 1039 LB.FT
S1= -23525 LB.
S2= 189 LB.
EXT.NORTH BEAM
P = 429 LB/FT
M1= 611902 LB.FT
M2= 106290 LB.FT
S1= -31006 LB.
S2= -5386 LB.

FOR 90DEG.GLAZING:

EXT.SOUTH BEAM
F = -19086 LB.
M1= 439615 LB.FT
M2= 17143 LB.FT
S1= -24202 LB.
S2= 1342 LB.
INTERNAL BEAMS
F = -50160 LB.
P = 455 LB/FT
M1= 587163 LB.FT
M2= 23644 LB.FT
S1= -34825 LB.
S2= 2456 LB.
EXT.NORTH BEAM
P = 460 LB/FT
M1= 571085 LB.FT
S1= -28937 LB.

SHELL RESULTS:

EXTERNAL SOUTH SHELL

	SHELLS WITH 60-DEGREE GLAZING				SHELLS WITH 90-DEGREE GLAZING			
POINT	T1	S	T2	M2	T1	S	T2	M2
0	45469	459	-11	514	38708	967	-123	-109
1	13463	-4906	-334	462	12021	-4844	-488	329
2	-4682	-5553	-1208	80	-2808	-5749	-1627	112
3	-13040	-3743	-1966	-329	-9877	-4118	-2669	-296
4	-15360	-965	-2295	-495	-13099	-1312	-3192	-474
5	-13444	1845	-2107	-330	-13600	1958	-3005	-235
6	-6484	3841	-1478	34	-8515	4771	-2097	278
7	8247	3823	-669	276	7403	5173	-786	585
8	33977	-0	-225	-0	38488	-0	-9	-0

INTERNAL SHELLS

POINT	T1	S	T2	M2	T1	S	T2	M2
0	46804	398	447	10	50649	2542	-57	-84
1	15453	-5320	81	366	15808	-5062	-299	412
2	-3755	-6262	-904	171	-3065	-6319	-1551	169
3	-13598	-4485	-1804	-199	-11292	-4444	-2707	-302
4	-16905	-1493	-2257	-401	-14130	-1327	-3268	-524
5	-15130	1638	-2140	-287	-13840	2095	-3063	-296
6	-7532	3905	-1523	45	-8240	4894	-2121	232
7	8371	3980	-689	276	7717	5217	-789	562
8	35529	-0	-225	-0	38511	-0	-9	-0

EXTERNAL NORTH SHELL

POINT	T1	S	T2	M2	T1	S	T2	M2
0	46814	401	448	14	50590	2512	-58	-88
1	15473	-5320	81	367	15741	-5077	-305	409
2	-3727	-6266	-904	170	-3162	-6315	-1559	167
3	-13577	-4494	-1805	-202	-11396	-4414	-2711	-297
4	-16914	-1505	-2260	-407	-14166	-1278	-3262	-510
5	-15184	1632	-2145	-294	-13721	2135	-3046	-275
6	-7611	3913	-1527	39	-7976	4886	-2099	254
7	8360	4000	-691	272	7831	5153	-777	576
8	35785	-0	-224	-0	37677	-0	-9	-0

L = 64 FT. TABLE NO.N- 63 W = 30 FT.

SHELL DATA:

```
   L =  64.00 FT.
   W =  30.00 FT.
   T =   2.50 IN.
   G =  50.00 LB/FT*2
   QA= 150.00 LB/FT
   QV= 200.00 LB/FT

 FOR 60DEG.GLAZING:
   K =  14.50 FT.
   R =  22.50 FT.
   AL=  27.70 FT.
   θ =  70.54 DEG.

 FOR 90DEG.GLAZING:
   K =  14.50 FT.
   R =  32.50 FT.
   AL=  34.99 FT.
   θ =  61.68 DEG.
```

BEAM DATA:

```
 FOR 60DEG.GLAZING:

   B1= 0.75 FT.
   D1= 4.00 FT.
   B2= 0.75 FT.
   D2= 3.50 FT.

   EXT.NORTH BEAM:

   Z'= 1.96 FT.
   Y'= 1.05 FT.
   IZ= 4.93 FT*4
   IY= 2.88 FT*4

 FOR 90DEG.GLAZING:

   ALL BEAMS:

   B = 1.00 FT.
   D = 4.00 FT.
```

BEAM RESULTS:

```
FOR 60DEG.GLAZING:

EXT.SOUTH BEAM
 F =      4218 LB.
 M1=    288552 LB.FT
 M2=     11093 LB.FT
 S1=    -13785 LB.
 S2=       432 LB.
INTERNAL BEAMS
 F =      2267 LB.
 P =       410 LB/FT
 M1=    448515 LB.FT
 M2=      1574 LB.FT
 S1=    -21763 LB.
 S2=        15 LB.
EXT.NORTH BEAM
 P =       405 LB/FT
 M1=    637007 LB.FT
 M2=    107033 LB.FT
 S1=    -31269 LB.
 S2=     -5254 LB.

FOR 90DEG.GLAZING:

EXT.SOUTH BEAM
 F =    -14120 LB.
 M1=    448546 LB.FT
 M2=     19373 LB.FT
 S1=    -23394 LB.
 S2=      1282 LB.
INTERNAL BEAMS
 F =    -42504 LB.
 P =       429 LB/FT
 M1=    596471 LB.FT
 M2=     26541 LB.FT
 S1=    -33439 LB.
 S2=      2326 LB.
EXT.NORTH BEAM
 P =       434 LB/FT
 M1=    590547 LB.FT
 S1=    -28988 LB.
```

SHELL RESULTS:

EXTERNAL SOUTH SHELL

	SHELLS WITH 60-DEGREE GLAZING				SHELLS WITH 90-DEGREE GLAZING			
POINT	T1	S	T2	M2	T1	S	T2	M2
0	46637	-207	81	795	40361	693	-74	142
1	13249	-5072	-258	563	12411	-4721	-392	361
2	-5893	-5521	-1033	88	-3290	-5515	-1353	73
3	-14680	-3655	-1687	-359	-10841	-3891	-2223	-330
4	-16799	-905	-1958	-545	-13992	-1169	-2641	-488
5	-14187	1793	-1785	-403	-13881	1892	-2459	-256
6	-6400	3629	-1246	-64	-8127	4393	-1694	217
7	8901	3544	-574	184	7832	4649	-631	498
8	34835	0	-218	0	38116	-0	-22	-0

INTERNAL SHELLS

POINT	T1	S	T2	M2	T1	S	T2	M2
0	47752	-111	468	211	52680	2086	-3	215
1	15687	-5318	120	415	16264	-4978	-232	464
2	-4330	-6129	-730	168	-3647	-6082	-1298	132
3	-14787	-4388	-1502	-207	-12405	-4210	-2267	-342
4	-18242	-1496	-1889	-412	-15148	-1179	-2718	-549
5	-16068	1498	-1792	-320	-14184	2038	-2519	-329
6	-7751	3621	-1278	-24	-7843	4533	-1721	161
7	8819	3667	-592	199	8217	4709	-636	469
8	36386	-0	-222	-0	38259	-0	-21	-0

EXTERNAL NORTH SHELL

POINT	T1	S	T2	M2	T1	S	T2	M2
0	47885	-101	470	224	52575	2058	-5	210
1	15777	-5327	122	420	16181	-4987	-238	460
2	-4262	-6150	-731	164	-3741	-6072	-1304	132
3	-14762	-4418	-1507	-220	-12493	-4180	-2269	-335
4	-18307	-1523	-1899	-431	-15162	-1136	-2712	-534
5	-16249	1492	-1806	-342	-14051	2069	-2504	-308
6	-7973	3652	-1289	-42	-7590	4520	-1703	182
7	8818	3721	-596	189	8305	4651	-627	482
8	37149	-0	-219	0	37466	0	-21	0

L = 64 FT. TABLE NO.N- 64 W = 32 FT.

SHELL DATA:

L = 64.00 FT.
W = 32.00 FT.
T = 2.50 IN.
G = 50.00 LB/FT*2
QA= 150.00 LB/FT
QV= 200.00 LB/FT

FOR 60DEG.GLAZING:
K = 15.50 FT.
R = 22.50 FT.
AL= 29.86 FT.
Θ = 76.04 DEG.

FOR 90DEG.GLAZING:
K = 15.50 FT.
R = 37.50 FT.
AL= 37.04 FT.
Θ = 56.60 DEG.

BEAM DATA:

FOR 60DEG.GLAZING:

B1= 0.75 FT.
D1= 4.00 FT.
B2= 0.75 FT.
D2= 3.50 FT.

EXT.NORTH BEAM:

Z'= 1.96 FT.
Y'= 1.05 FT.
IZ= 4.93 FT*4
IY= 2.88 FT*4

FOR 90DEG.GLAZING:

ALL BEAMS:

B = 1.00 FT.
D = 4.00 FT.

BEAM RESULTS:

FOR 60DEG.GLAZING:

EXT.SOUTH BEAM
F = 1818 LB.
M1= 272549 LB.FT
M2= 12037 LB.FT
S1= -13208 LB.
S2= 523 LB.
INTERNAL BEAMS
F = -4379 LB.
P = 427 LB/FT
M1= 435732 LB.FT
M2= 2557 LB.FT
S1= -21791 LB.
S2= 185 LB.
EXT.NORTH BEAM
P = 419 LB/FT
M1= 647432 LB.FT
M2= 110611 LB.FT
S1= -31781 LB.
S2= -5430 LB.

FOR 90DEG.GLAZING:

EXT.SOUTH BEAM
F = -18218 LB.
M1= 459147 LB.FT
M2= 18978 LB.FT
S1= -24318 LB.
S2= 1367 LB.
INTERNAL BEAMS
F = -49931 LB.
P = 446 LB/FT
M1= 613604 LB.FT
M2= 26146 LB.FT
S1= -35011 LB.
S2= 2493 LB.
EXT.NORTH BEAM
P = 451 LB/FT
M1= 603990 LB.FT
S1= -29648 LB.

SHELL RESULTS:

SHELLS WITH 60-DEGREE GLAZING / SHELLS WITH 90-DEGREE GLAZING

EXTERNAL SOUTH SHELL

POINT	T1	S	T2	M2	T1	S	T2	M2
0	44868	-89	109	864	40853	894	-103	10
1	11291	-4953	-237	599	12664	-4919	-453	353
2	-6119	-5219	-1028	69	-3087	-5809	-1537	95
3	-12784	-3367	-1671	-420	-10644	-4136	-2525	-323
4	-13836	-871	-1930	-612	-13957	-1284	-3010	-496
5	-11874	1527	-1767	-443	-14182	1986	-2820	-253
6	-6093	3252	-1253	-61	-8604	4730	-1956	257
7	6935	3319	-593	214	7847	5071	-729	561
8	31590	0	-238	-0	39554	0	-11	0

INTERNAL SHELLS

POINT	T1	S	T2	M2	T1	S	T2	M2
0	46505	215	489	236	53508	2451	-33	59
1	14345	-5133	159	465	16659	-5161	-272	449
2	-4311	-5870	-712	178	-3[illegible]3	-6399	-1469	154
3	-13010	-4162	-1493	-254	-12185	-4472	-2567	-333
4	-15515	-1477	-1884	-488	-15090	-1298	-3090	-554
5	-13882	1277	-1801	-385	-14463	2132	-2881	-323
6	-7379	3318	-1302	-47	-8314	4867	-1982	204
7	7069	3498	-615	214	8207	5125	-734	534
8	33419	-0	-239	-0	39634	0	-11	0

EXTERNAL NORTH SHELL

POINT	T1	S	T2	M2	T1	S	T2	M2
0	46423	221	491	250	53423	2421	-35	55
1	14408	-5126	163	471	16582	-5174	-278	445
2	-4151	-5884	-709	173	-3490	-6392	-1476	153
3	-12864	-4206	-1495	-271	-12283	-4442	-2570	-327
4	-15532	-1535	-1897	-518	-15116	-1251	-3084	-539
5	-14163	1245	-1822	-421	-14335	2168	-2865	-301
6	-7819	3356	-1323	-79	-8052	4856	-1963	226
7	6985	3596	-623	196	8310	5063	-723	547
8	34827	0	-234	-0	38811	0	-11	0

L = 64 FT. TABLE NO.N- 65 W = 34 FT.

SHELL DATA:

L = 64.00 FT.
W = 34.00 FT.
T = 2.50 IN.
G = 50.00 LB/FT*2
QA= 150.00 LB/FT
QV= 200.00 LB/FT

FOR 60DEG.GLAZING:
K = 16.00 FT.
R = 27.50 FT.
AL= 31.09 FT.
θ = 64.78 DEG.

FOR 90DEG.GLAZING:
K = 16.00 FT.
R = 42.50 FT.
AL= 38.92 FT.
θ = 52.47 DEG.

BEAM DATA:

FOR 60DEG.GLAZING:

B1= 0.75 FT.
D1= 4.00 FT.
B2= 0.75 FT.
D2= 3.50 FT.

EXT.NORTH BEAM:

Z'= 1.96 FT.
Y'= 1.05 FT.
IZ= 4.93 FT*4
IY= 2.88 FT*4

FOR 90DEG.GLAZING:

ALL BEAMS:

B = 1.00 FT.
D = 4.00 FT.

BEAM RESULTS:

FOR 60DEG.GLAZING:

EXT.SOUTH BEAM
F = -9184 LB.
M1= 301669 LB.FT
M2= 11234 LB.FT
S1= -15583 LB.
S2= 695 LB.
INTERNAL BEAMS
F = -10588 LB.
P = 434 LB/FT
M1= 470118 LB.FT
M2= 1533 LB.FT
S1= -24092 LB.
S2= 256 LB.
EXT.NORTH BEAM
P = 431 LB/FT
M1= 655172 LB.FT
M2= 113860 LB.FT
S1= -32161 LB.
S2= -5589 LB.

FOR 90DEG.GLAZING:

EXT.SOUTH BEAM
F = -21218 LB.
M1= 470760 LB.FT
M2= 18187 LB.FT
S1= -25184 LB.
S2= 1494 LB.
INTERNAL BEAMS
F = -55712 LB.
P = 462 LB/FT
M1= 630170 LB.FT
M2= 25137 LB.FT
S1= -36393 LB.
S2= 2588 LB.
EXT.NORTH BEAM
P = 467 LB/FT
M1= 614293 LB.FT
S1= -30154 LB.

SHELL RESULTS:

EXTERNAL SOUTH SHELL

	SHELLS WITH 60-DEGREE GLAZING				SHELLS WITH 90-DEGREE GLAZING			
POINT	T1	S	T2	M2	T1	S	T2	M2
0	47447	451	9	625	41356	1042	-127	-121
1	13428	-5079	-329	523	12852	-5145	-522	351
2	-5209	-5654	-1225	79	-3021	-6107	-1735	121
3	-13291	-3762	-1988	-378	-10598	-4362	-2843	-315
4	-15299	-964	-2313	-563	-14023	-1366	-3396	-506
5	-13403	1834	-2124	-385	-14501	2114	-3193	-254
6	-6684	3848	-1494	10	-9024	5093	-2225	291
7	8092	3873	-679	280	7956	5505	-832	617
8	34811	-0	-230	0	41072	0	-5	0

INTERNAL SHELLS

POINT	T1	S	T2	M2	T1	S	T2	M2
0	49087	520	458	76	54186	2735	-60	-93
1	15831	-5438	94	419	16936	-5372	-321	440
2	-4029	-6375	-909	181	-3280	-6715	-1654	181
3	-13812	-4542	-1820	-235	-12109	-4712	-2883	-322
4	-16943	-1524	-2277	-462	-15133	-1384	-3477	-560
5	-15214	1619	-2162	-342	-14767	2260	-3256	-320
6	-7797	3924	-1543	18	-8734	5227	-2251	241
7	8248	4047	-701	278	8295	5554	-836	592
8	36491	-0	-230	0	41109	0	-5	0

EXTERNAL NORTH SHELL

POINT	T1	S	T2	M2	T1	S	T2	M2
0	49054	522	458	83	54120	2703	-61	-97
1	15855	-5436	95	421	16863	-5388	-328	436
2	-3967	-6381	-908	179	-3383	-6710	-1663	179
3	-13756	-4561	-1821	-242	-12219	-4681	-2887	-316
4	-16951	-1548	-2282	-474	-15170	-1333	-3471	-545
5	-15325	1606	-2171	-356	-14639	2301	-3238	-297
6	-7969	3939	-1553	5	-8454	5217	-2229	264
7	8215	4087	-704	270	8415	5486	-823	607
8	37040	-0	-229	-0	40226	0	-5	0

L = 66 FT. TABLE NO.N- 66 W = 30 FT.

SHELL DATA:

```
L  =  66.00 FT.
W  =  30.00 FT.
T  =   2.50 IN.
G  =  50.00 LB/FT*2
QA= 150.00 LB/FT
QV= 200.00 LB/FT

FOR 60DEG.GLAZING:
K  =  15.00 FT.
R  =  20.00 FT.
AL=  28.28 FT.
θ  =  81.01 DEG.

FOR 90DEG.GLAZING:
K  =  15.00 FT.
R  =  32.50 FT.
AL=  35.24 FT.
θ  =  62.13 DEG.
```

BEAM DATA:

```
FOR 60DEG.GLAZING:

B1= 0.75 FT.
D1= 4.50 FT.
B2= 0.75 FT.
D2= 4.00 FT.

EXT.NORTH BEAM:

Z'= 2.19 FT.
Y'= 1.11 FT.
IZ= 7.20 FT*4
IY= 4.03 FT*4

FOR 90DEG.GLAZING:

ALL BEAMS:

B = 1.00 FT.
D = 4.50 FT.
```

BEAM RESULTS:

```
FOR 60DEG.GLAZING:

EXT.SOUTH BEAM
F  =     9523 LB.
M1=   336095 LB.FT
M2=    11381 LB.FT
S1=   -15078 LB.
S2=      333 LB.
INTERNAL BEAMS
F  =     9506 LB.
P  =      435 LB/FT
M1=   500948 LB.FT
M2=     1991 LB.FT
S1=   -22798 LB.
S2=      -99 LB.
EXT.NORTH BEAM
P  =      429 LB/FT
M1=   752149 LB.FT
M2=   120459 LB.FT
S1=   -35802 LB.
S2=    -5734 LB.

FOR 90DEG.GLAZING:

EXT.SOUTH BEAM
F  =    -8527 LB.
M1=   541832 LB.FT
M2=    17933 LB.FT
S1=   -26695 LB.
S2=     1042 LB.
INTERNAL BEAMS
F  =   -35441 LB.
P  =      444 LB/FT
M1=   710104 LB.FT
M2=    24364 LB.FT
S1=   -37585 LB.
S2=     1985 LB.
EXT.NORTH BEAM
P  =      449 LB/FT
M1=   678696 LB.FT
S1=   -32306 LB.
```

SHELL RESULTS:

EXTERNAL SOUTH SHELL

	SHELLS WITH 60-DEGREE GLAZING				SHELLS WITH 90-DEGREE GLAZING			
POINT	T1	S	T2	M2	T1	S	T2	M2
0	41994	-453	109	775	38033	406	-106	66
1	11409	-4735	-195	536	12237	-4655	-420	308
2	-5670	-5053	-874	77	-2957	-5470	-1335	60
3	-13152	-3365	-1440	-344	-10871	-3923	-2176	-305
4	-14804	-951	-1678	-507	-14470	-1207	-2585	-434
5	-12576	1404	-1540	-354	-14385	1895	-2404	-187
6	-5956	3038	-1093	-20	-8176	4402	-1649	286
7	7413	3031	-533	208	8250	4608	-610	543
8	30463	-0	-245	-0	38075	0	-20	-0

INTERNAL SHELLS

POINT	T1	S	T2	M2	T1	S	T2	M2
0	42800	-452	520	126	48960	1687	-49	108
1	13867	-5056	193	385	15887	-4834	-274	393
2	-4151	-5733	-573	176	-3070	-5966	-1274	117
3	-13457	-4148	-1266	-177	-12189	-4234	-2203	-307
4	-16503	-1554	-1624	-371	-15551	-1250	-2648	-482
5	-14639	1134	-1560	-278	-14761	2006	-2458	-249
6	-7336	3070	-1132	10	-8006	4529	-1676	235
7	7457	3183	-553	216	8597	4674	-616	517
8	32242	-0	-248	0	38334	0	-20	-0

EXTERNAL NORTH SHELL

POINT	T1	S	T2	M2	T1	S	T2	M2
0	42835	-447	522	140	48897	1662	-51	101
1	13942	-5059	195	391	15808	-4845	-279	388
2	-4047	-5752	-572	171	-3180	-5957	-1280	116
3	-13384	-4183	-1270	-192	-12296	-4201	-2204	-299
4	-16555	-1592	-1635	-395	-15569	-1202	-2641	-465
5	-14871	1119	-1576	-306	-14607	2041	-2441	-225
6	-7656	3105	-1146	-14	-7716	4514	-1657	259
7	7428	3255	-557	203	8696	4609	-606	532
8	33301	0	-244	-0	37434	-0	-20	-0

L = 66 FT. TABLE NO.N- 67 W = 32 FT.

SHELL DATA:

L = 66.00 FT.
W = 32.00 FT.
T = 2.50 IN.
G = 50.00 LB/FT*2
QA= 150.00 LB/FT
QV= 200.00 LB/FT

FOR 60DEG.GLAZING:
K = 16.00 FT.
R = 22.50 FT.
AL= 29.86 FT.
θ = 76.03 DEG.

FOR 90DEG.GLAZING:
K = 16.00 FT.
R = 37.50 FT.
AL= 37.30 FT.
θ = 56.98 DEG.

BEAM DATA:

FOR 60DEG.GLAZING:

B1= 0.75 FT.
D1= 4.50 FT.
B2= 0.75 FT.
D2= 4.00 FT.

EXT.NORTH BEAM:

Z'= 2.19 FT.
Y'= 1.11 FT.
IZ= 7.20 FT*4
IY= 4.03 FT*4

FOR 90DEG.GLAZING:

ALL BEAMS:

B = 1.00 FT.
D = 4.50 FT.

BEAM RESULTS:

FOR 60DEG.GLAZING:

EXT.SOUTH BEAM
F = 2332 LB.
M1= 347039 LB.FT
M2= 11448 LB.FT
S1= -16284 LB.
S2= 471 LB.
INTERNAL BEAMS
F = 3696 LB.
P = 444 LB/FT
M1= 515083 LB.FT
M2= 1950 LB.FT
S1= -24097 LB.
S2= 7 LB.
EXT.NORTH BEAM
P = 438 LB/FT
M1= 761341 LB.FT
M2= 123171 LB.FT
S1= -36240 LB.
S2= -5863 LB.

FOR 90DEG.GLAZING:

EXT.SOUTH BEAM
F = -11719 LB.
M1= 553550 LB.FT
M2= 17498 LB.FT
S1= -27596 LB.
S2= 1100 LB.
INTERNAL BEAMS
F = -41585 LB.
P = 460 LB/FT
M1= 728637 LB.FT
M2= 23893 LB.FT
S1= -39127 LB.
S2= 2112 LB.
EXT.NORTH BEAM
P = 465 LB/FT
M1= 692511 LB.FT
S1= -32963 LB.

SHELL RESULTS:

SHELLS WITH 60-DEGREE GLAZING / SHELLS WITH 90-DEGREE GLAZING

EXTERNAL SOUTH SHELL

POINT	T1	S	T2	M2	T1	S	T2	M2
0	42672	-111	57	681	38488	558	-134	-56
1	11830	-4722	-241	503	12468	-4870	-487	304
2	-5189	-5136	-970	67	-2787	-5775	-1523	81
3	-12598	-3447	-1586	-350	-10725	-4174	-2479	-302
4	-14378	-989	-1849	-506	-14497	-1316	-2955	-447
5	-12479	1449	-1702	-334	-14732	2007	-2764	-188
6	-6157	3187	-1210	26	-8655	4760	-1907	324
7	7265	3217	-580	261	8323	5039	-705	606
8	30891	0	-246	-0	39616	-0	-9	0

INTERNAL SHELLS

POINT	T1	S	T2	M2	T1	S	T2	M2
0	43585	-176	508	55	49689	1979	-77	-35
1	14145	-5122	173	376	16236	-5035	-321	383
2	-3931	-5877	-660	176	-2858	-6292	-1449	138
3	-13124	-4252	-1418	-189	-12027	-4501	-2504	-302
4	-16159	-1576	-1809	-387	-15559	-1364	-3020	-493
5	-14484	1215	-1735	-278	-15088	2118	-2819	-247
6	-7413	3254	-1255	41	-8478	4884	-1934	276
7	7410	3390	-601	261	8649	5100	-711	581
8	32705	0	-247	0	39817	-0	-9	-0

EXTERNAL NORTH SHELL

POINT	T1	S	T2	M2	T1	S	T2	M2
0	43557	-172	510	65	49639	1952	-79	-40
1	14189	-5120	175	380	16162	-5050	-328	378
2	-3837	-5888	-659	172	-2970	-6285	-1456	138
3	-13044	-4279	-1421	-200	-12141	-4468	-2507	-295
4	-16179	-1610	-1817	-405	-15588	-1313	-3014	-476
5	-14660	1198	-1748	-300	-14943	2157	-2802	-223
6	-7678	3279	-1268	21	-8185	4872	-1913	300
7	7368	3449	-605	251	8760	5033	-700	596
8	33562	-0	-243	0	38905	-0	-9	-0

L = 66 FT. | TABLE NO.N- 68 | W = 34 FT.

SHELL DATA:

L = 66.00 FT.
W = 34.00 FT.
T = 2.50 IN.
G = 50.00 LB/FT*2
QA= 150.00 LB/FT
QV= 200.00 LB/FT

FOR 60DEG.GLAZING:
K = 16.50 FT.
R = 27.50 FT.
AL= 31.08 FT.
θ = 64.75 DEG.

FOR 90DEG.GLAZING:
K = 16.50 FT.
R = 45.00 FT.
AL= 39.00 FT.
θ = 49.66 DEG.

BEAM DATA:

FOR 60DEG.GLAZING:

B1= 0.75 FT.
D1= 4.50 FT.
B2= 0.75 FT.
D2= 4.00 FT.

EXT.NORTH BEAM:

Z'= 2.19 FT.
Y'= 1.11 FT.
IZ= 7.20 FT*4
IY= 4.03 FT*4

FOR 90DEG.GLAZING:

ALL BEAMS:

B = 1.00 FT.
D = 4.50 FT.

BEAM RESULTS:

FOR 60DEG.GLAZING:

EXT.SOUTH BEAM
F = -6597 LB.
M1= 378542 LB.FT
M2= 10449 LB.FT
S1= -18633 LB.
S2= 591 LB.
INTERNAL BEAMS
F = -1223 LB.
P = 450 LB/FT
M1= 554425 LB.FT
M2= 977 LB.FT
S1= -26499 LB.
S2= 55 LB.
EXT.NORTH BEAM
P = 449 LB/FT
M1= 768726 LB.FT
M2= 126137 LB.FT
S1= -36591 LB.
S2= -6004 LB.

FOR 90DEG.GLAZING:

EXT.SOUTH BEAM
F = -11077 LB.
M1= 577828 LB.FT
M2= 15986 LB.FT
S1= -28684 LB.
S2= 1016 LB.
INTERNAL BEAMS
F = -44228 LB.
P = 478 LB/FT
M1= 758787 LB.FT
M2= 21926 LB.FT
S1= -40846 LB.
S2= 2085 LB.
EXT.NORTH BEAM
P = 484 LB/FT
M1= 705062 LB.FT
S1= -33561 LB.

SHELL RESULTS:

EXTERNAL SOUTH SHELL

	SHELLS WITH 60-DEGREE GLAZING				SHELLS WITH 90-DEGREE GLAZING			
POINT	T1	S	T2	M2	T1	S	T2	M2
0	44778	314	-39	438	39596	527	-164	-224
1	13682	-4867	-340	423	13094	-5357	-614	286
2	-4409	-5550	-1167	74	-2874	-6367	-1840	103
3	-13128	-3813	-1895	-309	-11489	-4591	-2971	-287
4	-15817	-1063	-2218	-454	-15673	-1371	-3532	-441
5	-13988	1763	-2044	-271	-15874	2384	-3295	-166
6	-6743	3780	-1439	102	-9134	5465	-2266	378
7	8446	3767	-661	329	9260	5718	-830	673
8	34255	0	-236	0	42632	0	5	0

INTERNAL SHELLS

POINT	T1	S	T2	M2	T1	S	T2	M2
0	46038	58	476	-97	50882	2105	-110	-229
1	15480	-5441	100	330	17079	-5476	-424	355
2	-3760	-6366	-859	177	-2756	-6905	-1744	163
3	-13976	-4607	-1740	-172	-12712	-4962	-2986	-280
4	-17590	-1602	-2190	-360	-16765	-1454	-3599	-480
5	-15801	1569	-2083	-232	-16316	2481	-3357	-222
6	-7813	3863	-1487	108	-9029	5596	-2299	331
7	8633	3937	-681	326	9582	5793	-839	648
8	35930	-0	-236	-0	42935	0	5	0

EXTERNAL NORTH SHELL

POINT	T1	S	T2	M2	T1	S	T2	M2
0	46030	59	476	-95	50849	2076	-112	-235
1	15489	-5441	100	331	17002	-5494	-431	350
2	-3741	-6368	-859	176	-2882	-6898	-1752	162
3	-13959	-4613	-1741	-175	-12842	-4924	-2990	-272
4	-17594	-1609	-2192	-364	-16799	-1395	-3590	-461
5	-15837	1566	-2086	-237	-16153	2527	-3336	-195
6	-7868	3868	-1490	103	-8701	5581	-2273	358
7	8624	3950	-682	324	9707	5714	-825	664
8	36105	-0	-236	-0	41914	0	5	-0

L = 66 FT. TABLE NO.N- 69 W = 36 FT.

SHELL DATA:

L = 66.00 FT.
W = 36.00 FT.
T = 2.50 IN.
G = 50.00 LB/FT*2
QA= 150.00 LB/FT
QV= 200.00 LB/FT

FOR 60DEG.GLAZING:
K = 17.00 FT.
R = 30.00 FT.
AL= 32.80 FT.
θ = 62.65 DEG.

FOR 90DEG.GLAZING:
K = 17.00 FT.
R = 45.00 FT.
AL= 41.24 FT.
θ = 52.51 DEG.

BEAM DATA:

FOR 60DEG.GLAZING:

B1= 0.75 FT.
D1= 4.50 FT.
B2= 0.75 FT.
D2= 4.00 FT.

EXT.NORTH BEAM:

Z'= 2.19 FT.
Y'= 1.11 FT.
IZ= 7.20 FT*4
IY= 4.03 FT*4

ALL BEAMS:

B = 1.00 FT.
D = 4.50 FT.

BEAM RESULTS:

FOR 60DEG.GLAZING:

EXT.SOUTH BEAM
F = -11924 LB.
M1= 384313 LB.FT
M2= 10314 LB.FT
S1= -19416 LB.
S2= 683 LB.
INTERNAL BEAMS
F = -6823 LB.
P = 463 LB/FT
M1= 566037 LB.FT
M2= 872 LB.FT
S1= -27654 LB.
S2= 152 LB.
EXT.NORTH BEAM
P = 462 LB/FT
M1= 777472 LB.FT
M2= 129889 LB.FT
S1= -37008 LB.
S2= -6183 LB.

FOR 90DEG.GLAZING:

EXT.SOUTH BEAM
F = -18513 LB.
M1= 567141 LB.FT
M2= 16740 LB.FT
S1= -28972 LB.
S2= 1229 LB.
INTERNAL BEAMS
F = -52237 LB.
P = 486 LB/FT
M1= 751237 LB.FT
S1= -41345 LB.
S2= 2326 LB.
EXT.NORTH BEAM
P = 491 LB/FT
M1= 711422 LB.FT
S1= -33864 LB.

SHELL RESULTS:

SHELLS WITH 60-DEGREE GLAZING — SHELLS WITH 90-DEGREE GLAZING

EXTERNAL SOUTH SHELL

POINT	T1	S	T2	M2	T1	S	T2	M2
0	44934	568	-67	359	38801	881	-171	-259
1	13638	-4901	-374	416	12233	-5110	-578	329
2	-4188	-5635	-1265	79	-2675	-6080	-1807	138
3	-12569	-3880	-2053	-318	-9931	-4424	-2944	-300
4	-15214	-1097	-2408	-471	-13569	-1494	-3531	-494
5	-13733	1797	-2228	-273	-14580	2034	-3347	-222
6	-6951	3923	-1575	130	-9459	5168	-2351	364
7	8150	3968	-718	374	7777	5679	-884	709
8	34611	-0	-244	0	41382	0	-5	0

INTERNAL SHELLS

POINT	T1	S	T2	M2	T1	S	T2	M2
0	46597	325	472	-155	50337	2486	-116	-264
1	15449	-5518	81	345	15969	-5289	-387	401
2	-3715	-6487	-958	190	-2853	-6627	-1725	194
3	-13576	-4674	-1911	-188	-11253	-4743	-2972	-301
4	-17007	-1603	-2396	-393	-14537	-1513	-3599	-538
5	-15430	1645	-2280	-253	-14797	2159	-3399	-277
6	-7866	4035	-1628	122	-9194	5278	-2372	323
7	8410	4148	-739	364	8056	5714	-886	688
8	36204	0	-243	0	41354	0	-5	-0

EXTERNAL NORTH SHELL

POINT	T1	S	T2	M2	T1	S NG:		M2= 22
0	46540	319	472	-154	50307	2457	-116	-266
1	15435	-5517	81	345	15911	-5308	-394	398
2	-3704	-6486	-958	190	-2954	-6626	-1734	192
3	-13558	-4676	-1911	-189	-11370	-4714	-2978	-297
4	-17002	-1608	-2397	-395	-14590	-1461	-3595	-524
5	-15448	1641	-2281	-255	-14682	2205	-3382	-254
6	-7899	4037	-1630	120	-8911	5273	-2348	347
7	8401	4156	-740	363	8192	5646	-873	703
8	36307	0	-242	0	40451	0	-5	0

L = 68 FT. TABLE NO.N- 70 W = 30 FT.

SHELL DATA:

L = 68.00 FT.
W = 30.00 FT.
T = 2.50 IN.
G = 50.00 LB/FT*2
QA= 150.00 LB/FT
QV= 200.00 LB/FT

FOR 60DEG.GLAZING:
K = 15.00 FT.
R = 22.50 FT.
AL= 27.70 FT.
θ = 70.53 DEG.

FOR 90DEG.GLAZING:
K = 15.00 FT.
R = 32.50 FT.
AL= 35.24 FT.
θ = 62.13 DEG.

BEAM DATA:

FOR 60DEG.GLAZING:

B1= 0.75 FT.
D1= 4.50 FT.
B2= 0.75 FT.
D2= 4.00 FT.

EXT.NORTH BEAM:

Z'= 2.19 FT.
Y'= 1.11 FT.
IZ= 7.20 FT*4
IY= 4.03 FT*4

FOR 90DEG.GLAZING:

ALL BEAMS:

B = 1.00 FT.
D = 4.50 FT.

BEAM RESULTS:

FOR 60DEG.GLAZING:

EXT.SOUTH BEAM
F = 6949 LB.
M1= 386623 LB.FT
M2= 10974 LB.FT
S1= -17203 LB.
S2= 350 LB.
INTERNAL BEAMS
F = 11185 LB.
P = 427 LB/FT
M1= 563322 LB.FT
M2= 1066 LB.FT
S1= -24832 LB.
S2= -166 LB.
EXT.NORTH BEAM
P = 425 LB/FT
M1= 796390 LB.FT
M2= 126696 LB.FT
S1= -36793 LB.
S2= -5853 LB.

FOR 90DEG.GLAZING:

EXT.SOUTH BEAM
F = -7818 LB.
M1= 573973 LB.FT
M2= 18996 LB.FT
S1= -27320 LB.
S2= 1043 LB.
INTERNAL BEAMS
F = -36241 LB.
P = 442 LB/FT
M1= 751796 LB.FT
M2= 25799 LB.FT
S1= -38488 LB.
S2= 2009 LB.
EXT.NORTH BEAM
P = 448 LB/FT
M1= 719800 LB.FT
S1= -33255 LB.

SHELL RESULTS:

EXTERNAL SOUTH SHELL

	SHELLS WITH 60-DEGREE GLAZING				SHELLS WITH 90-DEGREE GLAZING			
POINT	T1	S	T2	M2	T1	S	T2	M2
0	46852	-321	37	652	40345	361	-99	110
1	14313	-5027	-268	488	12970	-4851	-425	327
2	-5476	-5582	-985	91	-3273	-5676	-1350	56
3	-15434	-3799	-1605	-294	-11773	-4044	-2195	-322
4	-18322	-1019	-1872	-446	-15533	-1200	-2600	-454
5	-15562	1765	-1712	-299	-15214	2010	-2408	-201
6	-6780	3641	-1197	20	-8429	4559	-1645	281
7	9739	3522	-559	229	8886	4724	-605	544
8	36106	0	-224	0	40001	-0	-20	-0

INTERNAL SHELLS

POINT	T1	S	T2	M2	T1	S	T2	M2
0	47581	-517	487	42	51896	1674	-41	166
1	16280	-5473	134	334	16851	-5039	-278	420
2	-4317	-6296	-678	171	-3377	-6193	-1289	117
3	-15825	-4577	-1421	-139	-13176	-4372	-2225	-326
4	-19947	-1626	-1801	-306	-16707	-1249	-2667	-507
5	-17560	1460	-1715	-208	-15650	2126	-2466	-268
6	-8206	3622	-1225	67	-8269	4697	-1674	226
7	9617	3634	-576	247	9272	4801	-612	515
8	37656	0	-229	0	40344	-0	-20	-0

EXTERNAL NORTH SHELL

POINT	T1	S	T2	M2	T1	S	T2	M2
0	47765	-502	488	50	51819	1649	-43	157
1	16369	-5479	135	338	16763	-5049	-284	415
2	-4284	-6312	-678	169	-3495	-6182	-1295	117
3	-15835	-4594	-1424	-145	-13286	-4336	-2226	-316
4	-20002	-1639	-1807	-316	-16718	-1199	-2659	-487
5	-17660	1460	-1721	-219	-15477	2160	-2448	-242
6	-8310	3639	-1230	58	-7958	4680	-1655	251
7	9631	3661	-578	243	9370	4732	-602	530
8	38030	-0	-228	0	39379	-0	-20	-0

L = 68 FT. TABLE NO.N- 71 W = 32 FT.

SHELL DATA:

L = 68.00 FT.
W = 32.00 FT.
T = 2.50 IN.
G = 50.00 LB/FT*2
QA= 150.00 LB/FT
QV= 200.00 LB/FT

FOR 60DEG.GLAZING:
K = 16.00 FT.
R = 22.50 FT.
θ = 76.03 DEG.

FOR 90DEG.GLAZING:
K = 16.00 FT.
R = 37.50 FT.
AL= 37.30 FT.
θ = 56.98 DEG.

BEAM DATA:

FOR 60DEG.GLAZING:

B1= 0.75 FT.
D1= 4.50 FT.
B2= 0.75 FT.
D2= 4.00 FT.

EXT.NORTH BEAM:

Z'= 2.19 FT.
Y'= 1.11 FT.
IZ= 7.20 FT*4
IY= 4.03 FT*4

FOR 90DEG.GLAZING:

ALL BEAMS:

B = 1.00 FT.
D = 4.50 FT.

BEAM RESULTS:

FOR 60DEG.GLAZING:

EXT.SOUTH BEAM
F = 5135 LB.
M1= 364880 LB.FT
M2= 11977 LB.FT
S1= -16369 LB.
S2= 429 LB.
INTERNAL BEAMS
F = 5125 LB.
M1= 543766 LB.FT
M2= 2063 LB.FT
S1= -24561 LB.
S2= -15 LB.
EXT.NORTH BEAM
P = 435 LB/FT
M1= 806504 LB.FT
M2= 129780 LB.FT
S1= -37260 LB.
S2= -5996 LB.

FOR 90DEG.GLAZING:

EXT.SOUTH BEAM
F = -11386 LB.
M1= 586529 LB.FT
M2= 18560 LB.FT
S1= -28272 LB.
S2= 1108 LB.
INTERNAL BEAMS
F = -43031 LB.
P = 459 LB/FT
M1= 771758 LB.FT
M2= 25336 LB.FT
S1= -40117 LB.
S2= 2149 LB.
EXT.NORTH BEAM
P = 465 LB/FT
M1= 734753 LB.FT
S1= -33946 LB.

SHELL RESULTS:

EXTERNAL SOUTH SHELL

	SHELLS WITH 60-DEGREE GLAZING				SHELLS WITH 90-DEGREE GLAZING			
POINT	T1	S	T2	M2	T1	S	T2	M2
0	45019	-237	74	762	40834	526	-128	-16
1	12429	-4957	-243	546	13239	-5068	-492	321
2	-5700	-5358	-985	73	-3071	-5990	-1538	76
3	-13658	-3575	22	-994 -1873	-540	-15569	-1310	-29
5	-13335	1550	-1718	-367	-15578	2130	-2766	-199
6	-6419	3336	-1215	3	-8915	4926	-1900	322
7	7903	3338	-579	250	8969	5161	-699	608
8	32842	0	-243	0	41563	-0	-9	0

INTERNAL SHELLS

POINT	T1	S	T2	M2	T1	S	T2	M2
0	46089	-237	513	108	52685	1988	-71	17
1	14978	-5317	169	403	17255	-5240	-326	407
2	-4232	-6088	-670	180	-3123	-6529	-1464	137
3	-14078	-4396	-1433	-202	-12996	-4648	-2526	-320
4	-17324	-1610	-1823	-408	-16722	-1365	-3039	-515
5	-15448	1288	-1745	-298	-15992	2245	-2826	-264
6	-7792	3388	-1259	26	-8747	5062	-1930	269
7	8000	3508	-600	255	9331	5233	-707	581
8	34723	0	-245	0	41841	-0	-9	-0

EXTERNAL NORTH SHELL

POINT	T1	S	T2	M2	T1	S	T2	M2
0	46093	-232	514	120	52624	1960	-73	10
1	15038	-5317	171	408	17171	-5253	-332	402
2	-4131	-6103	-670	176	-3245	-6520	-1471	137
3	-14000	-4428	-1436	-215	-13116	-4612	-2528	-312
4	-17359	-1647	-1833	-430	-16746	-1312	-3031	-497
5	-15652	1271	-1760	-324	-15828	2284	-2807	-237
6	-8086	3417	-1272	4	-8432	5047	-1908	295
7	7963	3573	-604	243	9443	5161	-695	596
8	35682	0	-241	0	40863	-0	-9	-0

L = 68 FT. TABLE NO.N- 72 W = 34 FT.

SHELL DATA:

L = 68.00 FT.
W = 34.00 FT.
T = 2.50 IN.
G = 50.00 LB/FT*2
QA= 150.00 LB/FT
QV= 200.00 LB/FT

FOR 60DEG.GLAZING:

FOR 90DEG.GLAZING:
K = 16.50 FT.
R = 40.00 FT.
AL= 39.36 FT.
θ = 56.38 DEG.

BEAM DATA:

FOR 60DEG.GLAZING:

B1= 0.75 FT.
D1= 4.50 FT.
B2= 0.75 FT.
D2= 4.00 FT.

EXT.NORTH BEAM:

Z'= 2.19 FT.
Y'= 1.11 FT.
IZ= 7.20 FT*4
IY= 4.03 FT*4

FOR 90DEG.GLAZING:

ALL BEAMS:

B = 1.00 FT.
D = 4.50 FT.

BEAM RESULTS:

FOR 60DEG.GLAZING:

EXT.SOUTH BEAM
F = -2856 LB.
M1= 377897 LB.FT
M2= 11762 LB.FT
S1= -17709 LB.
S2= -1894 -353
M2= 1836 LB.FT
S1= -26105 LB.
S2= 94 LB.
EXT.NORTH BEAM
P = 449 LB/FT
M1= 816087 LB.FT
M2= 133935 LB.FT
S1= -37703 LB.
S2= -6188 LB.

FOR 90DEG.GLAZING:

EXT.SOUTH BEAM
F = -16570 LB.
M1= 587796 LB.FT
M2= 18583 LB.FT
S1= -28870 LB.
S2= 1230 LB.
INTERNAL BEAMS
F = -49891 LB.
P = 472 LB/FT
M1= 777594 LB.FT
M2= 25421 LB.FT
S1= -41101 LB.
S2= 2313 LB.
EXT.NORTH BEAM
P = 477 LB/FT
M1= 744187 LB.FT
S1= -34381 LB.

SHELL RESULTS:

SHELLS WITH 60-DEGREE GLAZING / SHELLS WITH 90-DEGREE GLAZING

EXTERNAL SOUTH SHELL

POINT	T1	S	T2	M2	T1	S	T2	M2
0	45700	132	20	653	40679	766	-145	-93
-15100	2041	-2974	-225					
6	-6665	3511	-1343	43	-9282	4982	-2068	335
7	7757	3558	-633	300	8301	5365	-772	657
8	33524	-0	-250	-0	41709	0	-12	0

INTERNAL SHELLS

POINT	T1	S	T2	M2	T1	S	T2	M2
0	47079	71	497	34	52751	2305	-87	-71
1	15223	-5388	144	403	16825	-5260	-336	425
2	-4102	-6228	-769	188	-3107	-6560	-1549	163
3	-13782	-4481	-1600	-214	-12213	-4668	-2678	-328
4	-16950	-1607	-2024	-431	-15623	-1432	-3232	-549
5	-15253	1394	-1935	-308	-15379	2170	-3028	-288
6	-7865	3600	-1394	49	-9024	5105	-2092	287
7	7963	3744	-654	296	8626	5413	-776	632
8	35373	-0	-250	-0	41774	0	-11	-0

EXTERNAL NORTH SHELL

POINT	T1	S	T2	M2	T1	S	T2	M2
0	47020	72	498	42	52701	2275	-88	-75
1	15250	-5384	146	406	16753	-5277	-342	421
2	-4020	-6235	-768	185	-3217	-6556	-1557	162
3	-13705	-4504	-1601	-223	-12332	-4637	-2682	-322
4	-16957	-1637	-2030	-447	-15663	-1380	-3226	-533
5	-15396	1377	-1946	-326	-15243	2212	-3011	-264
6	-8091	3619	-1405	32	-8726	5095	-2070	312
7	7919	3794	-658	287	8751	5345	-764	647
8	36094	-0	-247	-0	40841	0	-11	0

L = 68 FT. TABLE NO.N- 73 W = 36 FT.

SHELL DATA:

```
L =  68.00 FT.
W =  36.00 FT.
T =   2.50 IN.
G =  50.00 LB/FT*2
QA= 150.00 LB/FT
QV= 200.00 LB/FT

FOR 60DEG.GLAZING:
K =  17.00 FT.
R =  27.50 FT.
AL=  33.17 FT.
θ =  69.10 DEG.

FOR 90DEG.GLAZING:
K =  17.00 FT.
R =  47.50 FT.
AL=  41.08 FT.
θ =  49.55 DEG.
```

BEAM DATA:

```
FOR 60DEG.GLAZING:

B1= 0.75 FT.
D1= 4.50 FT.
B2= 0.75 FT.
D2= 4.00 FT.

EXT.NORTH BEAM:

Z'= 2.19 FT.
Y'= 1.11 FT.
IZ= 7.20 FT*4
IY= 4.03 FT*4

FOR 90DEG.GLAZING:

ALL BEAMS:

B = 1.00 FT.
D = 4.50 FT.
```

BEAM RESULTS:

```
FOR 60DEG.GLAZING:

EXT.SOUTH BEAM
 F =    -9791 LB.
 M1=   388000 LB.FT
 M2=    11573 LB.FT
 S1=   -18817 LB.
 S2=      679 LB.
INTERNAL BEAMS
 F =    -7877 LB.
 P =      466 LB/FT
 M1=   578043 LB.FT
 M2=     1649 LB.FT
 S1=   -27501 LB.
 S2=      198 LB.
EXT.NORTH BEAM
 P =      463 LB/FT
 M1=   825694 LB.FT
 M2=   138104 LB.FT
 S1=   -38147 LB.
 S2=    -6380 LB.

FOR 90DEG.GLAZING:

EXT.SOUTH BEAM
 F =   -16162 LB.
 M1=   611567 LB.FT
 M2=    17095 LB.FT
 S1=   -29927 LB.
 S2=     1155 LB.
INTERNAL BEAMS
 F =   -52902 LB.
 P =      489 LB/FT
 M1=   807347 LB.FT
 M2=    23479 LB.FT
 S1=   -42789 LB.
 S2=     2296 LB.
EXT.NORTH BEAM
 P =      494 LB/FT
 M1=   757110 LB.FT
 S1=   -34978 LB.
```

SHELL RESULTS:

EXTERNAL SOUTH SHELL

	SHELLS WITH 60-DEGREE GLAZING				SHELLS WITH 90-DEGREE GLAZING			
POINT	T1	S	T2	M2	T1	S	T2	M2
0	46166	452	-23	549	41751	747	-172	-261
1	13021	-4940	-329	497	13516	-5547	-634	327
2	-4892	-5512	-1183	74	-2989	-6597	-1939	129
3	-12539	-3721	-1917	-375	-11514	-4760	-3140	-311
4	-14571	-1062	-2240	-547	-15672	-1471	-3744	-496
5	-13115	1646	-2073	-344	-16208	2401	-3513	-208
6	-6930	3666	-1476	79	-9773	5674	-2435	388
7	7542	3772	-690	349	9186	6043	-901	725
8	34139	-0	-257	0	44654	-0	3	0

INTERNAL SHELLS

POINT	T1	S	T2	M2	T1	S	T2	M2
0	47904	364	487	-42	53909	2444	-117	-266
1	15335	-5462	123	408	17657	-5698	-433	402
2	-4030	-6357	-866	199	-2996	-7171	-1843	191
3	-13446	-4555	-1766	-224	-12863	-5130	-3163	-307
4	-16481	-1609	-2224	-456	-16778	-1530	-3816	-541
5	-14987	1476	-2126	-319	-16572	2518	-3574	-268
6	-7954	3783	-1532	70	-9584	5804	-2465	340
7	7824	3965	-711	338	9509	6104	-908	700
8	35882	-0	-255	0	44818	-0	3	0

EXTERNAL NORTH SHELL

POINT	T1	S	T2	M2	T1	S	T2	M2
0	47797	360	487	-37	53875	2412	-118	-270
1	15335	-5456	124	410	17583	-5717	-441	397
2	-3963	-6358	-865	198	-3120	-7167	-1853	189
3	-13373	-4570	-1766	-230	-12998	-5094	-3167	-301
4	-16471	-1634	-2228	-466	-16824	-1470	-3809	-523
5	-15089	1460	-2134	-333	-16420	2567	-3553	-241
6	-8130	3796	-1540	57	-9253	5793	-2438	367
7	7778	4004	-715	331	9649	6025	-893	717
8	36435	0	-253	-0	43780	-0	3	-0

L = 70 FT. TABLE NO.N- 74 W = 30 FT.

SHELL DATA:

```
  L =  70.00 FT.
  W =  30.00 FT.
  T =   2.50 IN.
  G =  50.00 LB/FT*2
  QA= 150.00 LB/FT
  QV= 200.00 LB/FT

 FOR 60DEG.GLAZING:
  K =  15.00 FT.
  R =  20.00 FT.
  AL=  28.28 FT.
  θ =  81.01 DEG.

 FOR 90DEG.GLAZING:
  K =  15.00 FT.
  R =  30.00 FT.
  AL=  35.59 FT.
  θ =  67.98 DEG.
```

BEAM DATA:

```
 FOR 60DEG.GLAZING:

  B1= 0.75 FT.
  D1= 4.50 FT.
  B2= 0.75 FT.
  D2= 4.00 FT.

  EXT.NORTH BEAM:

  Z'= 2.19 FT.
  Y'= 1.11 FT.
  IZ= 7.20 FT*4
  IY= 4.03 FT*4

 FOR 90DEG.GLAZING:

  ALL BEAMS:

  B = 1.00 FT.
  D = 4.50 FT.
```

BEAM RESULTS:

```
 FOR 60DEG.GLAZING:

 EXT.SOUTH BEAM
  F =     16809 LB.
  M1=    369738 LB.FT
  M2=     12355 LB.FT
  S1=    -15069 LB.
  S2=       226 LB.
 INTERNAL BEAMS
  F =     13623 LB.
  P =       429 LB/FT
  M1=    556492 LB.FT
  M2=      2203 LB.FT
  S1=    -23566 LB.
  S2=      -159 LB.
 EXT.NORTH BEAM
  P =       421 LB/FT
  M1=    841998 LB.FT
  M2=    133146 LB.FT
  S1=    -37789 LB.
  S2=     -5976 LB.

 FOR 90DEG.GLAZING:

 EXT.SOUTH BEAM
  F =     -9087 LB.
  M1=    586960 LB.FT
  M2=     21598 LB.FT
  S1=    -27250 LB.
  S2=      1154 LB.
 INTERNAL BEAMS
  F =    -37383 LB.
  P =       438 LB/FT
  M1=    772483 LB.FT
  M2=     29285 LB.FT
  S1=    -38430 LB.
  S2=      2129 LB.
 EXT.NORTH BEAM
  P =       442 LB/FT
  M1=    759302 LB.FT
  S1=    -34077 LB.
```

SHELL RESULTS:

EXTERNAL SOUTH SHELL

	SHELLS WITH 60-DEGREE GLAZING				SHELLS WITH 90-DEGREE GLAZING			
POINT	T1	S	T2	M2	T1	S	T2	M2
0	46546	-754	146	941	41811	408	-69	246
1	12518	-5224	-195	628	12834	-4811	-370	385
2	-6801	-5507	-902	95	-3608	-5555	-1249	55
3	-15420	-3632	-1485	-383	-11650	-3928	-2043	-362
4	-17260	-970	-1725	-574	-15048	-1206	-2422	-510
5	-14429	1602	-1574	-422	-14849	1850	-2250	-252
6	-6532	3344	-1108	-70	-8555	4324	-1545	248
7	8784	3284	-532	184	8393	4555	-576	532
8	34626	-0	-239	0	39864	-0	-33	-0

INTERNAL SHELLS

POINT	T1	S	T2	M2	T1	S	T2	M2
0	47731	-611	529	240	54088	1678	-4	333
1	15463	-5455	185	446	16650	-5069	-231	496
2	-4840	-6152	-593	188	-4015	-6092	-1204	116
3	-15467	-4431	-1296	-201	-13258	-4226	-2085	-376
4	-18936	-1621	-1654	-414	-16225	-1209	-2493	-574
5	-16650	1276	-1582	-321	-15149	1994	-2303	-328
6	-8128	3334	-1141	-21	-8254	4460	-1569	189
7	8688	3417	-552	202	8793	4613	-580	502
8	36423	-0	-244	-0	40013	-0	-32	-0

EXTERNAL NORTH SHELL

POINT	T1	S	T2	M2	T1	S	T2	M2
0	47854	-605	533	261	53985	1651	-6	326
1	15579	-5467	188	454	16565	-5078	-237	491
2	-4726	-6183	-594	181	-4114	-6083	-1209	116
3	-15412	-4477	-1302	-222	-13351	-4196	-2087	-368
4	-19037	-1665	-1668	-446	-16238	-1167	-2487	-558
5	-16958	1265	-1601	-357	-15005	2024	-2290	-305
6	-8515	3382	-1157	-51	-7985	4446	-1553	211
7	8683	3505	-556	186	8884	4556	-572	515
8	37744	-0	-239	0	39174	-0	-32	-0

L = 70 FT. TABLE NO.N- 75 W = 32 FT.

SHELL DATA:

L = 70.00 FT.
W = 32.00 FT.
T = 2.50 IN.
G = 50.00 LB/FT*2
QA= 150.00 LB/FT
QV= 200.00 LB/FT

FOR 60DEG.GLAZING:
K = 16.00 FT.
R = 25.00 FT.
AL= 29.37 FT.
θ = 67.32 DEG.

FOR 90DEG.GLAZING:
K = 16.00 FT.
R = 35.00 FT.
AL= 37.55 FT.
θ = 61.47 DEG.

BEAM DATA:

FOR 60DEG.GLAZING:

B1= 0.75 FT.
D1= 4.50 FT.
B2= 0.75 FT.
D2= 4.00 FT.

EXT.NORTH BEAM:

Z'= 2.19 FT.
Y'= 1.11 FT.
IZ= 7.20 FT*4
IY= 4.03 FT*4

FOR 90DEG.GLAZING:

ALL BEAMS:

B = 1.00 FT.
D = 4.50 FT.

BEAM RESULTS:

FOR 60DEG.GLAZING:

EXT.SOUTH BEAM
F = 3707 LB.
M1= 412423 LB.FT
M2= 11695 LB.FT
S1= -18160 LB.
S2= 429 LB.
INTERNAL BEAMS
F = 7015 LB.
P = 434 LB/FT
M1= 604239 LB.FT
M2= 1247 LB.FT
S1= -26381 LB.
S2= -82 LB.
EXT.NORTH BEAM
P = 431 LB/FT
M1= 852603 LB.FT
M2= 136349 LB.FT
S1= -38265 LB.
S2= -6119 LB.

FOR 90DEG.GLAZING:

EXT.SOUTH BEAM
F = -13479 LB.
M1= 604279 LB.FT
M2= 20794 LB.FT
S1= -28471 LB.
S2= 1221 LB.
INTERNAL BEAMS
F = -45594 LB.
P = 455 LB/FT
M1= 798400 LB.FT
M2= 28347 LB.FT
S1= -40424 LB.
S2= 2276 LB.
EXT.NORTH BEAM
P = 460 LB/FT
M1= 775673 LB.FT
S1= -34812 LB.

SHELL RESULTS:

EXTERNAL SOUTH SHELL

POINT	T1 (60-DEGREE GLAZING)	S	T2	M2	T1 (90-DEGREE GLAZING)	S	T2	M2
0	49715	-166	20	659	42453	605	-108	96
1	15116	-5299	-308	510	13326	-5021	-441	368
2	-5646	-5911	-1096	87	-3312	-5890	-1438	74
3	-15942	-4012	-1779	-330	-11592	-4212	-2348	-353
4	-18929	-1055	-2072	-492	-15306	-1320	-2793	-514
5	-16184	1916	-1893	-325	-15412	1993	-2608	-242
6	-7181	3938	-1321	34	-9086	4728	-1799	294
7	10148	3825	-606	266	8637	5018	-669	599
8	38274	-0	-225	-0	41765	0	-20	-0

INTERNAL SHELLS

POINT	T1 (60-DEGREE GLAZING)	S	T2	M2	T1 (90-DEGREE GLAZING)	S	T2	M2
0	50780	-315	488	40	6	-46	154	
1	17300	-5752	114	362	17359	-5248	-283	468
2	-4432	-6663	-777	176	-3591	-6448	-1378	136
3	-16402	-4828	-1592	-171	-13146	-4535	-2386	-362
4	-20665	-1679	-2007	-356	-16477	-1338	-2865	-573
5	-18263	1620	-1904	-241	-15730	2131	-2664	-314
6	-8616	3945	-1354	74	-8811	4863	-1824	238
7	10104	3961	-624	281	9016	5076	-674	570
8	39997	-0	-228	0	41902	0	-19	0

EXTERNAL NORTH SHELL

POINT	T1 (60-DEGREE GLAZING)	S	T2	M2	T1 (90-DEGREE GLAZING)	S	T2	M2
0	50922	-302	489	47	54967	2017	-48	148
1	17374	-5757	115	365	17276	-5261	-289	463
2	-4397	-6677	-777	174	-3701	-6440	-1385	135
3	-16400	-4844	-1595	-177	-13256	-4503	-2388	-355
4	-20709	-1692	-2012	-365	-16502	-1290	-2859	-556
5	-18358	1618	-1911	-251	-15581	2168	-2648	-290
6	-8724	3960	-1360	65	-8518	4850	-1805	262
7	10109	3988	-626	276	9124	5011	-664	585
8	40372	-0	-227	-0	40989	0	-20	-0

L = 70 FT. TABLE NO.N- 76 W = 34 FT.

SHELL DATA:

L = 70.00 FT.
W = 34.00 FT.
T = 2.50 IN.
G = 50.00 LB/FT*2
QA= 150.00 LB/FT
QV= 200.00 LB/FT

FOR 60DEG.GLAZING:
K = 16.50 FT.
R = 25.00 FT.
AL= 31.49 FT.
θ = 72.17 DEG.

K = 16.50 FT.
R = 40.00 FT.
AL= 39.36 FT.
θ = 56.38 DEG.

BEAM DATA:

FOR 60DEG.GLAZING:

B1= 0.75 FT.
D1= 4.50 FT.
B2= 0.75 FT.
D2= 4.00 FT.

EXT.NORTH BEAM:

Z'= 2.19 FT.
Y'= 1.11 FT.
IZ= 7.20 FT*4
IY= 4.03 FT*4

FOR 90DEG.GLAZING:

ALL BEAMS:

B = 1.00 FT.
D = 4.50 FT.

BEAM RESULTS:

FOR 60DEG.GLAZING:

EXT.SOUTH BEAM
F = -381 LB.
M1= 396996 LB.FT
M2= 12307 LB.FT
S1= -17836 LB.
S2= 526 LB.
INTERNAL BEAMS
F = -386 LB.
P = 451 LB/FT
M1= 592567 LB.FT
M2= 1941 LB.FT
S2= 74 LB.
EXT.NORTH BEAM
P = 446 LB/FT
M1= 863046 LB.FT
M2= 140918 LB.FT
S1= -38733 LB.
S2= -6324 LB.

FOR 90DEG.GLAZING:

EXT.SOUTH BEAM
F = -16479 LB.
M1= 621545 LB.FT
M2= 19678 LB.FT
S1= -29550 LB.
S2= 1241 LB.
INTERNAL BEAMS
F = -51740 LB.
P = 471 LB/FT
M1= 822025 LB.FT
M2= 26912 LB.FT
S1= -42106 LB.
S2= 2353 LB.
EXT.NORTH BEAM
P = 476 LB/FT
M1= 788233 LB.FT
S1= -35376 LB.

SHELL RESULTS:

SHELLS WITH 60-DEGREE GLAZING / SHELLS WITH 90-DEGREE GLAZING

EXTERNAL SOUTH SHELL

POINT	T1	S	T2	M2	T1	S	T2	M2
0	48165	17	35	735	43070	740	-139	-50
1	13468	-5170	-293	559	13687	-5266	-516	359
2	-5737	-5657	-1103	75	-3190	-6229	-1638	99
3	-14155	-3778	-1788	-396	-11675	-4475	-2668	-344
4	-16225	-1025	-2080	-577	-15619	-1400	-3179	-519
5	-14109	1714	-1910	-386	-15945	2168	-2977	-238
6	-6929	3666	-1349	0DEG.GLAZING:		S1=	-26606 LB.	
8	35527	0	-247	0	43663	0	-12	0

INTERNAL SHELLS

POINT	T1	S	T2	M2	T1	S	T2	M2
0	49706	17	501	89	55822	2322	-81	-16
1	16101	-5586	140	431	17863	-5466	-340	450
2	-4401	-6446	-780	191	-3368	-6803	-1565	162
3	-14754	-4629	-1616	-228	-13179	-4819	-2701	-348
4	-18140	-1640	-2040	-455	-16780	-1431	-3252	-574
5	-16236	1473	-1946	-330	-16276	2301	-3035	-306
6	-8247	3742	-1398	33	-9289	5287	-2087	281
7	8578	3868	-654	289	9304	5547	-770	632
8	37468	-0	-248	-0	43793	0	-11	0

EXTERNAL NORTH SHELL

POINT	T1	S	T2	M2	T1	S	T2	M2
0	49672	20	502	99	55761	2291	-82	-21
1	16141	-5584	142	435	17782	-5482	-347	446
2	-4312	-6456	-779	188	-3489	-6796	-1573	160
3	-14677	-4655	-1618	-239	-13304	-4784	-2705	-341
4	-18157	-1672	-2047	-472	-16817	-1377	-3246	-556
5	-16403	1457	-1958	-351	-16122	2344	-3017	-279
6	-8499	3765	-1409	15	-8970	5275	-2064	307
7	8537	3923	-658	279	9430	5475	-758	648
8	38277	0	-245	-0	42797	0	-11	0

L = 70 FT. TABLE NO.N- 77 W = 36 FT.

SHELL DATA:

L = 70.00 FT.
W = 36.00 FT.
T = 2.50 IN.
G = 50.00 LB/FT*2
QA= 150.00 LB/FT
QV= 200.00 LB/FT

FOR 60DEG.GLAZING:
K = 17.00 FT.
R = 27.50 FT.
AL= 33.17 FT.
θ = 69.10 DEG.

FOR 90DEG.GLAZING:
K = 17.00 FT.
R = 45.00 FT.
AL= 41.24 FT.
θ = 52.51 DEG.

BEAM DATA:

FOR 60DEG.GLAZING:

B1= 0.75 FT.
D1= 4.50 FT.
B2= 0.75 FT.
D2= 4.00 FT.

EXT.NORTH BEAM:

Z'= 2.19 FT.
Y'= 1.11 FT.
IZ= 7.20 FT*4
IY= 4.03 FT*4

FOR 90DEG.GLAZING:

ALL BEAMS:

B = 1.00 FT.
D = 4.50 FT.

BEAM RESULTS:

FOR 60DEG.GLAZING:

EXT.SOUTH BEAM
F = -7942 LB.
M1= 407994 LB.FT
M2= 12134 LB.FT
S1= -19009 LB.
S2= 651 LB.
INTERNAL BEAMS
F = -7251 LB.
P = 464 LB/FT
M1= 609560 LB.FT
M2= 1751 LB.FT
S1= -28064 LB.
S2= 185 LB.
EXT.NORTH BEAM
P = 460 LB/FT
M1= 873356 LB.FT
M2= 145411 LB.FT
S1= -39196 LB.
S2= -6526 LB.

FOR 90DEG.GLAZING:

EXT.SOUTH BEAM
F = -19152 LB.
M1= 635307 LB.FT
M2= 18860 LB.FT
S1= -30439 LB.
S2= 1266 LB.
INTERNAL BEAMS
F = -57188 LB.
P = 486 LB/FT
M1= 841458 LB.FT
M2= 25872 LB.FT
S1= -43529 LB.
S2= 2432 LB.
EXT.NORTH BEAM
P = 491 LB/FT
M1= 800165 LB.FT
S1= -35911 LB.

SHELL RESULTS:

SHELLS WITH 60-DEGREE GLAZING | SHELLS WITH 90-DEGREE GLAZING

EXTERNAL SOUTH SHELL

POINT	T1	S	T2	M2	T1	S	T2	M2
0	48689	356	-9	629	43569	860	-162	-182
1	13700	-5168	-333	537	13863	-5507	-587	360
2	-5355	-5734	-1201	77	-3142	-6538	-1838	127
3	-13558	-3848	-1942	-399	-11646	-4709	-2988	-337
4	-15678	-1061	-2263	-582	-15708	-1484	-3565	-533
5	-13943	1755	-2087	-376	-16297	2304	-3350	-241
6	-7183	3822	-1480	58	-9991	5533	-2329	368
7	8170	3895	-688	340	9076	5939	-867	719
8	36087	0	-255	-0	45295	0	-5	0

INTERNAL SHELLS

POINT	T1	S	T2	M2	T1	S	T2	M2
0	50587	325	490	12	56481	2567	-105	-168
1	16243	-5658	119	435	18119	-5694	-393	445
2	-4304	-6578	-877	202	-3282	-7130	-1753	190
3	-14394	-4706	-1782	-240	-13128	-5067	-3019	-339
4	-17646	-1642	-2240	-480	-16849	-1519	-3640	-584
5	-15950	1560	-2137	-342	-16612	2438	-3410	-306
6	-8326	3931	-1535	55	-9734	5664	-2356	317
7	8430	4093	-710	331	9418	5992	-872	693
8	37955	0	-254	0	45381	0	-5	0

EXTERNAL NORTH SHELL

POINT	T1	S	T2	M2	T1	S	T2	M2
0	50495	324	490	18	56435	2534	-106	-172
1	16253	-5653	121	437	18042	-5713	-400	441
2	-4230	-6581	-875	200	-3405	-7126	-1762	188
3	-14317	-4725	-1783	-247	-13262	-5032	-3024	-333
4	-17642	-1669	-2245	-492	-16896	-1460	-3634	-567
5	-16069	1543	-2146	-358	-16462	2486	-3391	-279
6	-8525	3946	-1545	41	-9403	5654	-2331	345
7	8384	4136	-714	323	9559	5915	-858	710
8	38581	-0	-252	-0	44343	0	-5	-0

L = 70 FT. TABLE NO.N- 78 W = 38 FT.

SHELL DATA:

L = 70.00 FT.
W = 38.00 FT.
T = 2.50 IN.
G = 50.00 LB/FT*2
QA= 150.00 LB/FT
QV= 200.00 LB/FT

FOR 60DEG.GLAZING:
K = 18.00 FT.
R = 30.00 FT.
AL= 34.85 FT.
θ = 66.56 DEG.

FOR 90DEG.GLAZING:
K = 18.00 FT.
R = 47.50 FT.
AL= 43.56 FT.
θ = 52.54 DEG.

BEAM DATA:

FOR 60DEG.GLAZING:

B1= 0.75 FT.
D1= 4.50 FT.
B2= 0.75 FT.
D2= 4.00 FT.

EXT.NORTH BEAM:

Z'= 2.19 FT.
Y'= 1.11 FT.
IZ= 7.20 FT*4
IY= 4.03 FT*4

FOR 90DEG.GLAZING:

ALL BEAMS:

B = 1.00 FT.
D = 4.50 FT.

BEAM RESULTS:

FOR 60DEG.GLAZING:

EXT.SOUTH BEAM
F = -13871 LB.
M1= 413583 LB.FT
M2= 12296 LB.FT
S1= -19793 LB.
S2= 761 LB.
INTERNAL BEAMS
F = -13308 LB.
P = 474 LB/FT
M1= 619403 LB.FT
M2= 1828 LB.FT
S1= -29120 LB.
S2= 292 LB.
EXT.NORTH BEAM
P = 471 LB/FT
M1= 884205 LB.FT
M2= 148755 LB.FT
S1= -39683 LB.
S2= -6676 LB.

FOR 90DEG.GLAZING:

EXT.SOUTH BEAM
F = -25344 LB.
M1= 632867 LB.FT
M2= 19502 LB.FT
S1= -30955 LB.
S2= 1435 LB.
INTERNAL BEAMS
F = -65249 LB.
P = 498 LB/FT
M1= 844771 LB.FT
M2= 26860 LB.FT
S1= -44493 LB.
S2= 2658 LB.
EXT.NORTH BEAM
P = 503 LB/FT
M1= 812690 LB.FT
S1= -36473 LB.

SHELL RESULTS:

EXTERNAL SOUTH SHELL

POINT	T1 (60°)	S (60°)	T2 (60°)	M2 (60°)	T1 (90°)	S (90°)	T2 (90°)	M2 (90°)
0	48950	623	-37	546	43310	1137	-176	-261
1	13695	-5196	-363	523	13306	-5464	-595	387
2	-5093	-5810	-1289	79	-3034	-6484	-1904	160
3	-12945	-3916	-2084	-405	-10567	-4700	-3108	-342
4	-15033	-1117	-2434	-590	-14272	-1621	-3733	-570
5	-13683	1747	-2255	-368	-15556	2094	-3551	-277
6	-7425	3919	-1606	97	-10455	5471	-2510	376
7	7804	4066	-741	393	8063	6105	-950	772
8	36300	0	-257	0	45201	0	-5	0

(60° = SHELLS WITH 60-DEGREE GLAZING; 90° = SHELLS WITH 90-DEGREE GLAZING)

INTERNAL SHELLS

POINT	T1 (60°)	S (60°)	T2 (60°)	M2 (60°)	T1 (90°)	S (90°)	T2 (90°)	M2 (90°)
0	51061	597	489	-62	56587	2928	-116	-255
1	16226	-5723	110	438	17442	-5693	-389	473
2	-4192	-6691	-959	213	-3375	-7095	-1824	221
3	-13918	-4782	-1930	-249	-12117	-5029	-3149	-349
4	-17015	-1676	-2423	-501	-15323	-1613	-3810	-625
5	-15583	1584	-2314	-348	-15712	2248	-3604	-339
6	-8431	4050	-1665	82	-10091	5588	-2527	331
7	8118	4269	-764	378	8368	6129	-951	750
8	38069	0	-255	0	45051	0	-5	0

EXTERNAL NORTH SHELL

POINT	T1 (60°)	S (60°)	T2 (60°)	M2 (60°)	T1 (90°)	S (90°)	T2 (90°)	M2 (90°)
0	50926	592	489	-57	56548	2897	-117	-256
1	16220	-5715	111	440	17382	-5713	-396	469
2	-4121	-6690	-957	211	-3477	-7095	-1834	218
3	-13835	-4797	-1930	-255	-12239	-5001	-3155	-345
4	-16998	-1703	-2427	-512	-15382	-1560	-3806	-611
5	-15685	1566	-2323	-362	-15599	2296	-3587	-316
6	-8619	4062	-1675	69	-9798	5584	-2504	356
7	8063	4310	-768	371	8516	6059	-937	766
8	38650	-0	-253	-0	44109	0	-5	0

L = 72 FT. TABLE NO.N- 79 W = 30 FT.

SHELL DATA:

L = 72.00 FT.
W = 30.00 FT.
T = 2.50 IN.
G = 50.00 LB/FT*2
QA= 150.00 LB/FT
QV= 200.00 LB/FT

FOR 60DEG.GLAZING:
K = 15.25 FT.
R = 20.00 FT.
AL= 28.28 FT.
θ = 81.02 DEG.

FOR 90DEG.GLAZING:
K = 15.25 FT.
R = 32.50 FT.
AL= 35.37 FT.
θ = 62.36 DEG.

BEAM DATA:

FOR 60DEG.GLAZING:

B1= 0.75 FT.
D1= 4.75 FT.
B2= 0.75 FT.
D2= 4.00 FT.

EXT.NORTH BEAM:

Z'= 2.25 FT.
Y'= 1.16 FT.
IZ= 7.93 FT*4
IY= 4.59 FT*4

FOR 90DEG.GLAZING:

ALL BEAMS:

B = 1.00 FT.
D = 4.75 FT.

BEAM RESULTS:

FOR 60DEG.GLAZING:

EXT.SOUTH BEAM
F = 21503 LB.
M1= 385289 LB.FT
M2= 13058 LB.FT
S1= -14918 LB.
S2= 168 LB.
INTERNAL BEAMS
F = 18760 LB.
P = 437 LB/FT
M1= 615845 LB.FT
M2= 1926 LB.FT
S1= -24894 LB.
S2= -252 LB.
EXT.NORTH BEAM
P = 426 LB/FT
M1= 912594 LB.FT
M2= 142586 LB.FT
S1= -39819 LB.
S2= -6221 LB.

FOR 90DEG.GLAZING:

EXT.SOUTH BEAM
F = -3496 LB.
M1= 680050 LB.FT
M2= 19860 LB.FT
S1= -30024 LB.
S2= 926 LB.
INTERNAL BEAMS
F = -33428 LB.
P = 447 LB/FT
M1= 884034 LB.FT
M2= 26872 LB.FT
S1= -42024 LB.
S2= 1880 LB.
EXT.NORTH BEAM
P = 453 LB/FT
M1= 835753 LB.FT
S1= -36467 LB.

SHELL RESULTS:

	SHELLS WITH 60-DEGREE GLAZING				SHELLS WITH 90-DEGREE GLAZING			
EXTERNAL SOUTH SHELL								
POINT	T1	S	T2	M2	T1	S	T2	M2
0	48882	-938	170	1024	42749	153	-105	137
1	12936	-5487	-192	676	13963	-5117	-444	328
2	-7583	-5725	-911	113	-3641	-5957	-1363	45
3	-16712	-3734	-1498	-383	-13190	-4230	-2200	-334
4	-18451	-947	-1731	-580	-17392	-1206	-2593	-455
5	-15069	1699	-1570	-424	-16723	2175	-2387	-185
6	-6428	3435	-1097	-66	-8845	4787	-1618	308
7	9474	3310	-525	188	10059	4877	-590	566
8	35331	-0	-238	-0	42969	-0	-20	-0
INTERNAL SHELLS								
POINT	T1	S	T2	M2	T1	S	T2	M2
0	47490	-819	541	160	54582	1459	-50	199
1	15693	-5538	193	408	18083	-5286	-304	426
2	-4841	-6241	-570	189	-3625	-6473	-1301	109
3	-15963	-4529	-1262	-168	-14606	-4577	-2226	-335
4	-19760	-1691	-1618	-363	-18668	-1277	-2659	-508
5	-17378	1251	-1551	-268	-17279	2283	-2447	-256
6	-8362	3331	-1120	22	-8752	4936	-1651	249
7	9043	3400	-546	225	10485	4970	-599	535
8	36964	0	-248	0	43492	-0	-19	-0
EXTERNAL NORTH SHELL								
POINT	T1	S	T2	M2	T1	S	T2	M2
0	47769	-805	546	190	54505	1435	-53	187
1	15887	-5560	197	419	17983	-5293	-310	419
2	-4695	-6289	-572	180	-3759	-6458	-1307	111
3	-15920	-4594	-1272	-198	-14724	-4536	-2225	-323
4	-19927	-1747	-1638	-408	-18667	-1223	-2649	-485
5	-17809	1242	-1576	-318	-17070	2318	-2428	-225
6	-8876	3399	-1141	-19	-8399	4913	-1630	278
7	9059	3518	-552	203	10583	4894	-588	552
8	38748	0	-241	0	42393	-0	-19	-0

L = 72 FT

TABLE NO.N- 80

W = 32 FT.

SHELL DATA:

L = 72.00 FT.
W = 32.00 FT.
T = 2.50 IN.
G = 50.00 LB/FT*2
QA= 150.00 LB/FT
QV= 200.00 LB/FT

FOR 60DEG.GLAZING:
K = 16.25 FT.
R = 20.00 FT.
AL= 30.62 FT.
θ = 87.71 DEG.

FOR 90DEG.GLAZING:
K = 16.25 FT.
R = 37.50 FT.
AL= 37.42 FT.
θ = 57.18 DEG.

BEAM DATA:

FOR 60DEG.GLAZING:

B1= 0.75 FT.
D1= 4.75 FT.
B2= 0.75 FT.
D2= 4.00 FT.

EXT.NORTH BEAM:

Z'= 2.25 FT.
Y'= 1.16 FT.
IZ= 7.93 FT*4
IY= 4.59 FT*4

FOR 90DEG.GLAZING:

ALL BEAMS:

B = 1.00 FT.
D = 4.75 FT.

BEAM RESULTS:

FOR 60DEG.GLAZING:

EXT.SOUTH BEAM
F = 22250 LB.
M1= 356159 LB.FT
M2= 14209 LB.FT
S1= -13585 LB.
S2= 207 LB.
INTERNAL BEAMS
F = 12299 LB.
P = 457 LB/FT
M1= 593965 LB.FT
M2= 3082 LB.FT
S1= -24612 LB.
S2= -96 LB.
EXT.NORTH BEAM
P = 442 LB/FT
M1= 927090 LB.FT
M2= 147867 LB.FT
S1= -40452 LB.
S2= -6452 LB.

FOR 90DEG.GLAZING:

EXT.SOUTH BEAM
F = -7138 LB.
M1= 694678 LB.FT
M2= 19410 LB.FT
S1= -31041 LB.
S2= 989 LB.
INTERNAL BEAMS
F = -40526 LB.
P = 465 LB/FT
M1= 907165 LB.FT
M2= 26391 LB.FT
S1= -43770 LB.
S2= 2018 LB.
EXT.NORTH BEAM
P = 471 LB/FT
M1= 852711 LB.FT
S1= -37207 LB.

SHELL RESULTS:

EXTERNAL SOUTH SHELL

	SHELLS WITH 60-DEGREE GLAZING				SHELLS WITH 90-DEGREE GLAZING			
POINT	T1	S	T2	M2	T1	S	T2	M2
0	46539	-971	229	1114	43281	311	-134	10
1	10402	-5448	-164	713	14277	-5347	-514	320
2	-8046	-5435	-909	76	-3412	-6289	-1554	63
3	-14641	-3420	-1486	-465	-13046	-4501	-2506	-333
4	-15074	-883	-1703	-661	-17474	-1312	-2964	-468
5	-12368	1447	-1547	-468	-17132	2315	-2742	-181
6	-5980	3048	-1098	-61	-9338	5179	-1868	353
7	7298	3062	-544	221	10188	5326	-681	633
8	31526	0	-265	0	44618	-0	-8	-0

INTERNAL SHELLS

POINT	T1	S	T2	M2	T1	S	T2	M2
0	46037	-537	574	203	55426	1768	-81	47
1	14240	-5379	233	468	18546	-5496	-355	410
2	-4824	-6005	-555	198	-3335	-6826	-1479	127
3	-14074	-4315	-1260	-227	-14434	-4868	-2529	-331
4	-16851	-1662	-1623	-458	-18727	-1391	-3031	-517
5	-15020	1061	-1566	-350	-17667	2421	-2804	-249
6	-7899	3064	-1146	-12	-9239	5325	-1902	296
7	7270	3253	-570	235	10587	5415	-690	603
8	33851	0	-270	-0	45071	-0	-8	-0

EXTERNAL NORTH SHELL

POINT	T1	S	T2	M2	T1	S	T2	M2
0	46011	-520	579	232	55365	1741	-83	38
1	14392	-5372	240	481	18450	-5508	-361	404
2	-4543	-6036	-550	189	-3474	-6814	-1485	128
3	-13835	-4393	-1264	-260	-14564	-4826	-2529	-320
4	-16910	-1758	-1642	-513	-18741	-1333	-3021	-495
5	-15549	1014	-1598	-417	-17467	2462	-2784	-219
6	-8699	3135	-1177	-71	-8880	5305	-1879	325
7	7140	3420	-580	203	10700	5336	-678	621
8	36442	0	-259	-0	43961	-0	-8	-0

L = 72 FT. TABLE NO.N- 81 W = 34 FT.

SHELL DATA:

L = 72.00 FT.
W = 34.00 FT.
T = 2.50 IN.
G = 50.00 LB/FT*2
QA= 150.00 LB/FT
QV= 200.00 LB/FT

FOR 60DEG.GLAZING:
K = 16.75 FT.
R = 25.00 FT.
AL= 31.49 FT.
θ = 72.16 DEG.

FOR 90DEG.GLAZING:
K = 16.75 FT.
R = 40.00 FT.
AL= 39.49 FT.
θ = 56.56 DEG.

FOR 60DEG.GLAZING:

B1= 0.75 FT.
D1= 4.75 FT.
B2= 0.75 FT.
D2= 4.00 FT.

EXT.NORTH BEAM:

Z'= 2.25 FT.
Y'= 1.16 FT.
IZ= 7.93 FT*4
IY= 4.59 FT*4

FOR 90DEG.GLAZING:

ALL BEAMS:

B = 1.00 FT.
D = 4.75 FT.

BEAM RESULTS:

FOR 60DEG.GLAZING:

EXT.SOUTH BEAM
F = 2878 LB.
M1= 415478 LB.FT
M2= 13055 LB.FT
S1= -17861 LB.
S2= 487 LB.
INTERNAL BEAMS
F = 5229 LB.
P = 457 LB/FT
M1= 654549 LB.FT
M2= 1680 LB.FT
S1= -27991 LB.
S2= -32 LB.
EXT.NORTH BEAM
P = 450 LB/FT
M1= 934166 LB.FT
S2= -6563 LB.

FOR 90DEG.GLAZING:

EXT.SOUTH BEAM
F = -12718 LB.
M1= 695333 LB.FT
M2= 19453 LB.FT
S1= -31649 LB.
S2= 1114 LB.
INTERNAL BEAMS
F = -47920 LB.
P = 477 LB/FT
M1= 913018 LB.FT
M2= 26495 LB.FT
S1= -44793 LB.
S2= 2186 LB.
EXT.NORTH BEAM
P = 482 LB/FT
M1= 863214 LB.FT
S1= -37665 LB.

SHELL RESULTS:

EXTERNAL SOUTH SHELL

	SHELLS WITH 60-DEGREE GLAZING				SHELLS WITH 90-DEGREE GLAZING			
POINT	T1	S	T2	M2	T1	S	T2	M2
0	50732	-126	53	817	43084	555	-150	-60
1	14010	-5424	-295	601	13969	-5346	-535	343
2	-6436	-5873	-1117	85	-3194	-6323	-1639	89
3	-15379	-3871	-1805	-404	-12172	-4558	-2657	-343
4	-17358	-985	-2087	-586	-16436	-1411	-3161	-505
5	-14699	1831	-1902	-388	-16653	2241	-2951	-210
6	-6786	3767	-1333	26	-9704	5252	-2033	367
7	9103	3706	-622	295	9512	5536	-751	684
8	36158	0	-246	0	44707	-0	-10	-0

INTS1= -40761 LB.

POINT	T1	S	T2	M2	T1	S	T2	M2
0	49443	-228	513	24	55453	2091	-97	-34
1	16343	-5687	143	398	18151	-5514	-365	430
2	-4415	-6550	-761	190	-3254	-6872	-1564	152
3	-15308	-4736	-1586	-202	-13600	-4906	-2684	-343
4	-19043	-1699	-2008	-413	-17606	-1462	-3229	-556
5	-17015	1479	-1916	-284	-17046	2360	-3009	-276
6	-8458	3774	-1375	72	-9507	5386	-2062	314
7	9047	3872	-645	311	9873	5602	-757	656
8	38182	0	-250	0	44929	-0	-10	-0

EXTERNAL NORTH SHELL

POINT	T1	S	T2	M2	T1	S	T2	M2
0	49513	-214	515	41	55402	2060	-98	-40
1	16447	-5688	146	405	18066	-5530	-372	425
2	-4281	-6572	-760	186	-3385	-6865	-1572	152
3	-15213	-4779	-1591	-220	-13734	-4870	-2687	-335
4	-19095	-1747	-2020	-441	-17641	-1405	-3222	-536
5	-17284	1458	-1935	-318	-16877	2404	-2990	-249
6	-8840	3813	-1393	43	-9167	5372	-2039	342
7	9000	3957	-652	295	10001	5526	-745	673
8	39430	0	-246	0	43875	-0	-10	-0

DESIGN TABLES
FOR
SYMMETRICAL CYLINDRICAL SHELL ROOFS
WITH INTERNAL VALLEY BEAMS

L= 40 TABLE V- 1 W= 18

SHELL DATA:

L = 40.0 FT.
W = 18.0 FT.
R = 15.0 FT.
T = 2.5 IN.
PHI= 35.1 DEG.
G = 50.0 LB/FT2

BEAM DATA:

B = 0.75 FT.
D1 = 2.50 FT.
D2 = 2.00 FT.

EXTERNAL BEAM RESULTS:
F = 50267 LB.
M1 = 49005 LB.FT.
M2 = 1467 LB.FT.
S1 = -6801 LB.
S2 = -1370 LB.

INTERNAL BEAM RESULTS:
F = 84777 LB.
M1 = 30097 LB.FT.
S1 = -7339 LB.

EXTERNAL-HALF SHELL RESULTS:

POINT	T1	S	M2	T2
0	-950	-3948	515	-79
1	-5063	-3355	267	-674
2	-6401	-2288	-30	-1132
3	-6361	-1127	-256	-1407
4	-6185	0	-338	-1497

INTERNAL-HALF SHELL RESULTS:

POINT	T1	S	M2	T2
0	4604	-3326	-163	402
1	-1626	-3566	73	-161
2	-5837	-2863	69	-698
3	-8205	-1571	-8	-1071
4	-8957	0	-48	-1204

L= 40 TABLE V- 2 W= 20

SHELL DATA:

L = 40.0 FT.
W = 20.0 FT.
R = 15.0 FT.
T = 2.5 IN.
PHI= 39.9 DEG.
G = 50.0 LB/FT2

BEAM DATA:

B = 0.75 FT.
D1 = 2.50 FT.
D2 = 2.00 FT.

EXTERNAL BEAM RESULTS:
F = 47515 LB.
M1 = 52411 LB.FT.
M2 = 2216 LB.FT.
S1 = -6907 LB.
S2 = -1231 LB.

INTERNAL BEAM RESULTS:
F = 75569 LB.
M1 = 22512 LB.FT.
S1 = -6205 LB.

EXTERNAL-HALF SHELL RESULTS:

POINT	T1	S	M2	T2
0	-1136	-3731	543	-69
1	-5301	-2988	276	-678
2	-5586	-1824	-39	-1103
3	-4344	-799	-270	-1325
4	-3659	0	-353	-1390

INTERNAL-HALF SHELL RESULTS:

POINT	T1	S	M2	T2
0	4794	-2967	-218	446
1	-1012	-3320	88	-123
2	-4800	-2690	81	-683
3	-6787	-1474	-22	-1074
4	-7380	0	-73	-1213

L= 40 TABLE V- 3 W= 22

SHELL DATA:

L = 40.0 FT.
W = 22.0 FT.
R = 17.5 FT.
T = 2.5 IN.
PHI= 37.4 DEG.
G = 50.0 LB/FT2

BEAM DATA:

B = 0.75 FT.
D1 = 2.50 FT.
D2 = 2.00 FT.

EXTERNAL BEAM RESULTS:
F = 49478 LB.
M1 = 56657 LB.FT.
M2 = 2312 LB.FT.
S1 = -7358 LB.
S2 = -1280 LB.

INTERNAL BEAM RESULTS:
F = 77824 LB.
M1 = 22799 LB.FT.
S1 = -6362 LB.

EXTERNAL-HALF SHELL RESULTS:

POINT	T1	S	M2	T2
0	-1512	-3886	481	-97
1	-5533	-2997	259	-787
2	-5374	-1720	-36	-1243
3	-3692	-700	-253	-1463
4	-2827	0	-331	-1523

INTERNAL-HALF SHELL RESULTS:

POINT	T1	S	M2	T2
0	4999	-3056	-264	496
1	-930	-3470	100	-161
2	-4649	-2804	90	-808
3	-6472	-1527	-31	-1257
4	-6981	0	-92	-1416

```
L= 40        TABLE V-  4        W= 24
*************************************

SHELL DATA:

    L   = 40.0 FT.
    W   = 24.0 FT.
    R   = 20.0 FT.
    T   =  2.5 IN.
    PHI= 35.5 DEG.
    G   = 50.0 LB/FT2

BEAM DATA:

    B  = 0.75 FT.
    D1 = 2.50 FT.
    D2 = 2.00 FT.

EXTERNAL BEAM RESULTS:
    F  =  51514 LB.
    M1 =  60221 LB.FT.
    M2 =   2383 LB.FT.
    S1 =  -7758 LB.
    S2 =  -1333 LB.

INTERNAL BEAM RESULTS:
    F  =  79580 LB.
    M1 =  22938 LB.FT.
    S1 =  -6478 LB.

EXTERNAL-HALF SHELL RESULTS:

POINT     T1      S      M2      T2

  0     -1793   -4045    417    -120
  1     -5741   -3012    241    -897
  2     -5207   -1616    -31   -1383
  3     -3134    -596   -234   -1597
  4     -2108       0   -306   -1650

INTERNAL-HALF SHELL RESULTS:

POINT     T1      S      M2      T2

  0      5172   -3125   -312     545
  1      -888   -3596    114    -201
  2     -4518   -2890    101    -935
  3     -6133   -1560    -40   -1439
  4     -6535       0   -111   -1616
```

```
L= 40        TABLE V-  5        W= 26
*************************************

SHELL DATA:

    L   = 40.0 FT.
    W   = 26.0 FT.
    R   = 20.0 FT.
    T   =  2.5 IN.
    PHI= 39.1 DEG.
    G   = 50.0 LB/FT2

BEAM DATA:

    B  = 0.75 FT.
    D1 = 2.50 FT.
    D2 = 2.00 FT.

EXTERNAL BEAM RESULTS:
    F  =  51183 LB.
    M1 =  62343 LB.FT.
    M2 =   2898 LB.FT.
    S1 =  -7905 LB.
    S2 =  -1284 LB.

INTERNAL BEAM RESULTS:
    F  =  72921 LB.
    M1 =  18940 LB.FT.
    S1 =  -5773 LB.

EXTERNAL-HALF SHELL RESULTS:

POINT     T1      S      M2      T2

  0     -1711   -4019    398    -123
  1     -6177   -2797    219    -942
  2     -4968   -1218    -41   -1383
  3     -1998    -287   -213   -1512
  4      -587       0   -267   -1523

INTERNAL-HALF SHELL RESULTS:

POINT     T1      S      M2      T2

  0      5071   -2863   -366     551
  1      -823   -3375    132    -198
  2     -4049   -2663    114    -937
  3     -5113   -1398    -50   -1433
  4     -5252       0   -131   -1603
```

```
L= 40        TABLE V-  6        W= 28
*************************************

SHELL DATA:

    L   = 40.0 FT.
    W   = 28.0 FT.
    R   = 22.5 FT.
    T   =  2.5 IN.
    PHI= 37.3 DEG.
    G   = 50.0 LB/FT2

BEAM DATA:

    B  = 0.75 FT.
    D1 = 2.50 FT.
    D2 = 2.00 FT.

EXTERNAL BEAM RESULTS:
    F  =  53565 LB.
    M1 =  64763 LB.FT.
    M2 =   2861 LB.FT.
    S1 =  -8236 LB.
    S2 =  -1356 LB.

INTERNAL BEAM RESULTS:
    F  =  75231 LB.
    M1 =  19616 LB.FT.
    S1 =  -5963 LB.

EXTERNAL-HALF SHELL RESULTS:

POINT     T1      S      M2      T2

  0     -1866   -4207    336    -144
  1     -6303   -2850    201   -1062
  2     -4884   -1150    -38   -1535
  3     -1708    -209   -193   -1652
  4      -216       0   -240   -1653

INTERNAL-HALF SHELL RESULTS:

POINT     T1      S      M2      T2

  0      5220   -2954   -414     592
  1      -928   -3501    148    -251
  2     -4071   -2712    125   -1074
  3     -4846   -1392    -58   -1612
  4     -4829       0   -147   -1793
```

```
L= 42        TABLE V-  7          W= 18
*************************************

SHELL DATA:

     L   = 42.0 FT.
     W   = 18.0 FT.
     R   = 15.0 FT.
     T   =  2.5 IN.
     PHI= 35.1 DEG.
     G   = 50.0 LB/FT2

BEAM DATA:

     B   = 0.75 FT.
     D1  = 2.50 FT.
     D2  = 2.00 FT.

EXTERNAL BEAM RESULTS:
     F   =   55897 LB.
     M1  =   51199 LB.FT.
     M2  =    1457 LB.FT.
     S1  =   -6956 LB.
     S2  =   -1465 LB.

INTERNAL BEAM RESULTS:
     F   =   93368 LB.
     M1  =   32689 LB.FT.
     S1  =   -7661 LB.

EXTERNAL-HALF SHELL RESULTS:

POINT      T1        S       M2      T2

  0       -685    -4181     570     -73
  1      -5371    -3610     292    -679
  2      -7123    -2504     -33   -1154
  3      -7333    -1250    -278   -1444
  4      -7241        0    -368   -1540

INTERNAL-HALF SHELL RESULTS:

POINT      T1        S       M2      T2

  0       5143    -3491    -161     407
  1      -1734    -3752      76    -157
  2      -6425    -3021      72    -696
  3      -9092    -1660      -5   -1071
  4      -9947        0     -45   -1205
```

```
L= 42        TABLE V-  8          W= 20
*************************************

SHELL DATA:

     L   = 42.0 FT.
     W   = 20.0 FT.
     R   = 15.0 FT.
     T   =  2.5 IN.
     PHI= 39.9 DEG.
     G   = 50.0 LB/FT2

BEAM DATA:

     B   = 0.75 FT.
     D1  = 2.50 FT.
     D2  = 2.00 FT.

EXTERNAL BEAM RESULTS:
     F   =   52621 LB.
     M1  =   54636 LB.FT.
     M2  =    2249 LB.FT.
     S1  =   -7030 LB.
     S2  =   -1314 LB.

INTERNAL BEAM RESULTS:
     F   =   83090 LB.
     M1  =   24308 LB.FT.
     S1  =   -6463 LB.

EXTERNAL-HALF SHELL RESULTS:

POINT      T1        S       M2      T2

  0       -895    -3936     613     -60
  1      -5580    -3221     309    -679
  2      -6180    -2023     -43   -1123
  3      -5096     -914    -301   -1364
  4      -4457        0    -394   -1437

INTERNAL-HALF SHELL RESULTS:

POINT      T1        S       M2      T2

  0       5342    -3107    -218     451
  1      -1040    -3492      91    -118
  2      -5264    -2842      84    -680
  3      -7533    -1562     -20   -1075
  4      -8226        0     -72   -1215
```

```
L= 42        TABLE V-  9          W= 22
*************************************

SHELL DATA:

     L   = 42.0 FT.
     W   = 22.0 FT.
     R   = 17.5 FT.
     T   =  2.5 IN.
     PHI= 37.4 DEG.
     G   = 50.0 LB/FT2

BEAM DATA:

     B   = 0.75 FT.
     D1  = 2.50 FT.
     D2  = 2.00 FT.

EXTERNAL BEAM RESULTS:
     F   =   54677 LB.
     M1  =   59468 LB.FT.
     M2  =    2382 LB.FT.
     S1  =   -7508 LB.
     S2  =   -1360 LB.

INTERNAL BEAM RESULTS:
     F   =   85480 LB.
     M1  =   24526 LB.FT.
     S1  =   -6616 LB.

EXTERNAL-HALF SHELL RESULTS:

POINT      T1        S       M2      T2

  0      -1321    -4089     555     -88
  1      -5849    -3225     292    -788
  2      -5938    -1911     -40   -1264
  3      -4341     -808    -287   -1505
  4      -3493        0    -375   -1573

INTERNAL-HALF SHELL RESULTS:

POINT      T1        S       M2      T2

  0       5573    -3196    -265     503
  1       -930    -3650     103    -134
  2      -5086    -2967      93    -804
  3      -7197    -1624     -30   -1258
  4      -7809        0     -92   -1419
```

L= 42 TABLE V- 10 W= 24
****

SHELL DATA:

L = 42.0 FT.
W = 24.0 FT.
R = 20.0 FT.
T = 2.5 IN.
PHI= 35.5 DEG.
G = 50.0 LB/FT2

BEAM DATA:

B = 0.75 FT.
D1 = 2.50 FT.
D2 = 2.00 FT.

EXTERNAL BEAM RESULTS:
F = 56823 LB.
M1 = 63641 LB.FT.
M2 = 2488 LB.FT.
S1 = -7942 LB.
S2 = -1412 LB.

INTERNAL BEAM RESULTS:
F = 87290 LB.
M1 = 24572 LB.FT.
S1 = -6723 LB.

EXTERNAL-HALF SHELL RESULTS:

POINT	T1	S	M2	T2
0	-1657	-4250	492	-112
1	-6100	-3233	275	-897
2	-5746	-1794	-37	-1404
3	-3688	-697	-269	-1638
4	-2648	0	-351	-1699

INTERNAL-HALF SHELL RESULTS:

POINT	T1	S	M2	T2
0	5767	-3264	-315	553
1	-860	-3783	117	-191
2	-4927	-3064	104	-929
3	-6836	-1665	-40	-1441
4	-7345	0	-112	-1621

L= 42 TABLE V- 11 W= 26
****

SHELL DATA:

L = 42.0 FT.
W = 26.0 FT.
R = 20.0 FT.
T = 2.5 IN.
PHI= 39.1 DEG.
G = 50.0 LB/FT2

BEAM DATA:

B = 0.75 FT.
D1 = 2.50 FT.
D2 = 2.00 FT.

EXTERNAL BEAM RESULTS:
F = 56309 LB.
M1 = 66271 LB.FT.
M2 = 3062 LB.FT.
S1 = -8109 LB.
S2 = -1355 LB.

INTERNAL BEAM RESULTS:
F = 79696 LB.
M1 = 20107 LB.FT.
S1 = -5964 LB.

EXTERNAL-HALF SHELL RESULTS:

POINT	T1	S	M2	T2
0	-1625	-4211	472	-114
1	-6605	-2993	253	-940
2	-5472	-1362	-46	-1399
3	-2385	-361	-248	-1547
4	-903	0	-313	-1567

INTERNAL-HALF SHELL RESULTS:

POINT	T1	S	M2	T2
0	5637	-2980	-372	561
1	-751	-3547	136	-184
2	-4376	-2834	118	-929
3	-5727	-1507	-51	-1436
4	-5973	0	-134	-1611

L= 42 TABLE V- 12 W= 28
****

SHELL DATA:

L = 42.0 FT.
W = 28.0 FT.
R = 22.5 FT.
T = 2.5 IN.
PHI= 37.3 DEG.
G = 50.0 LB/FT2

BEAM DATA:

B = 0.75 FT.
D1 = 2.50 FT.
D2 = 2.00 FT.

EXTERNAL BEAM RESULTS:
F = 58908 LB.
M1 = 69223 LB.FT.
M2 = 3047 LB.FT.
S1 = -8477 LB.
S2 = -1428 LB.

INTERNAL BEAM RESULTS:
F = 82049 LB.
M1 = 20728 LB.FT.
S1 = -6144 LB.

EXTERNAL-HALF SHELL RESULTS:

POINT	T1	S	M2	T2
0	-1821	-4406	406	-136
1	-6778	-3042	234	-1061
2	-5385	-1280	-43	-1551
3	-2032	-270	-226	-1686
4	-443	0	-284	-1694

INTERNAL-HALF SHELL RESULTS:

POINT	T1	S	M2	T2
0	5800	-3068	-422	605
1	-837	-3679	152	-233
2	-4380	-2895	130	-1063
3	-5448	-1510	-59	-1616
4	-5542	0	-152	-1805

```
L= 44        TABLE V- 13          W= 20
**************************************

SHELL DATA:

     L   = 44.0 FT.
     W   = 20.0 FT.
     R   = 15.0 FT.
     T   =  2.5 IN.
     PHI= 39.9 DEG.
     G   = 50.0 LB/FT2

BEAM DATA:

     B  = 0.75 FT.
     D1 = 2.75 FT.
     D2 = 2.25 FT.

EXTERNAL BEAM RESULTS:
     F  =  56373 LB.
     M1 =  63912 LB.FT.
     M2 =   2124 LB.FT.
     S1 =  -8077 LB.
     S2 =  -1364 LB.

INTERNAL BEAM RESULTS:
     F  =  91319 LB.
     M1 =  31615 LB.FT.
     S1 =  -7945 LB.
```

EXTERNAL-HALF SHELL RESULTS:

POINT	T1	S	M2	T2
0	-1552	-4025	569	-69
1	-5770	-3273	295	-667
2	-6419	-2096	-36	-1099
3	-5557	-972	-284	-1340
4	-5028	0	-374	-1415

INTERNAL-HALF SHELL RESULTS:

POINT	T1	S	M2	T2
0	4790	-3260	-206	402
1	-1446	-3535	88	-154
2	-5620	-2848	81	-692
3	-7898	-1561	-17	-1067
4	-8602	0	-66	-1200

```
L= 44        TABLE V- 14          W= 22
**************************************

SHELL DATA:

     L   = 44.0 FT.
     W   = 22.0 FT.
     R   = 17.5 FT.
     T   =  2.5 IN.
     PHI= 37.4 DEG.
     G   = 50.0 LB/FT2

BEAM DATA:

     B  = 0.75 FT.
     D1 = 2.75 FT.
     D2 = 2.25 FT.

EXTERNAL BEAM RESULTS:
     F  =  58418 LB.
     M1 =  69029 LB.FT.
     M2 =   2243 LB.FT.
     S1 =  -8571 LB.
     S2 =  -1408 LB.

INTERNAL BEAM RESULTS:
     F  =  94080 LB.
     M1 =  31852 LB.FT.
     S1 =  -8137 LB.
```

EXTERNAL-HALF SHELL RESULTS:

POINT	T1	S	M2	T2
0	-1967	-4171	510	-96
1	-5975	-3280	279	-771
2	-6132	-1999	-33	-1237
3	-4795	-881	-271	-1482
4	-4073	0	-357	-1555

INTERNAL-HALF SHELL RESULTS:

POINT	T1	S	M2	T2
0	5035	-3358	-252	448
1	-1307	-3700	99	-194
2	-5419	-2977	90	-817
3	-7554	-1626	-27	-1250
4	-8186	0	-86	-1403

```
L= 44        TABLE V- 15          W= 24
**************************************

SHELL DATA:

     L   = 44.0 FT.
     W   = 24.0 FT.
     R   = 20.0 FT.
     T   =  2.5 IN.
     PHI= 35.5 DEG.
     G   = 50.0 LB/FT2

BEAM DATA:

     B  = 0.75 FT.
     D1 = 2.75 FT.
     D2 = 2.25 FT.

EXTERNAL BEAM RESULTS:
     F  =  60544 LB.
     M1 =  73442 LB.FT.
     M2 =   2337 LB.FT.
     S1 =  -9020 LB.
     S2 =  -1458 LB.

INTERNAL BEAM RESULTS:
     F  =  96206 LB.
     M1 =  31868 LB.FT.
     S1 =  -8273 LB.
```

EXTERNAL-HALF SHELL RESULTS:

POINT	T1	S	M2	T2
0	-2295	-4322	446	-119
1	-6169	-3289	263	-875
2	-5902	-1899	-29	-1374
3	-4135	-785	-254	-1618
4	-3237	0	-335	-1686

INTERNAL-HALF SHELL RESULTS:

POINT	T1	S	M2	T2
0	5248	-3434	-300	493
1	-1205	-3836	112	-235
2	-5235	-3080	101	-943
3	-7189	-1672	-37	-1432
4	-7729	0	-106	-1604

L= 44 TABLE V- 16 W= 26
****

SHELL DATA:

L = 44.0 FT.
W = 26.0 FT.
R = 20.0 FT.
T = 2.5 IN.
PHI= 39.1 DEG.
G = 50.0 LB/FT2

BEAM DATA:

B = 0.75 FT.
D1 = 2.75 FT.
D2 = 2.25 FT.

EXTERNAL BEAM RESULTS:
F = 59775 LB.
M1 = 76599 LB.FT.
M2 = 2907 LB.FT.
S1 = -9197 LB.
S2 = -1397 LB.

INTERNAL BEAM RESULTS:
F = 88290 LB.
M1 = 26179 LB.FT.
S1 = -7374 LB.

EXTERNAL-HALF SHELL RESULTS:

POINT	T1	S	M2	T2
0	-2355	-4267	429	-122
1	-6602	-3037	246	-912
2	-5549	-1482	-37	-1368
3	-2789	-468	-237	-1537
4	-1464	0	-305	-1568

INTERNAL-HALF SHELL RESULTS:

POINT	T1	S	M2	T2
0	5246	-3151	-359	515
1	-1001	-3617	132	-218
2	-4637	-2877	115	-938
3	-6088	-1535	-48	-1429
4	-6390	0	-129	-1599

L= 44 TABLE V- 17 W= 28
****

SHELL DATA:

L = 44.0 FT.
W = 28.0 FT.
R = 22.5 FT.
T = 2.5 IN.
PHI= 37.3 DEG.
G = 50.0 LB/FT2

BEAM DATA:

B = 0.75 FT.
D1 = 2.75 FT.
D2 = 2.25 FT.

EXTERNAL BEAM RESULTS:
F = 62386 LB.
M1 = 79716 LB.FT.
M2 = 2883 LB.FT.
S1 = -9584 LB.
S2 = -1468 LB.

INTERNAL BEAM RESULTS:
F = 90861 LB.
M1 = 26899 LB.FT.
S1 = -7567 LB.

EXTERNAL-HALF SHELL RESULTS:

POINT	T1	S	M2	T2
0	-2548	-4454	361	-143
1	-6743	-3087	228	-1026
2	-5444	-1414	-32	-1515
3	-2431	-390	-217	-1677
4	-1003	0	-278	-1701

INTERNAL-HALF SHELL RESULTS:

POINT	T1	S	M2	T2
0	5405	-3243	-409	555
1	-1060	-3749	147	-269
2	-4616	-2945	127	-1071
3	-5805	-1547	-57	-1609
4	-5967	0	-147	-1793

L= 44 TABLE V- 18 W= 30
****

SHELL DATA:

L = 44.0 FT.
W = 30.0 FT.
R = 25.0 FT.
T = 2.5 IN.
PHI= 35.8 DEG.
G = 50.0 LB/FT2

BEAM DATA:

B = 0.75 FT.
D1 = 2.75 FT.
D2 = 2.25 FT.

EXTERNAL BEAM RESULTS:
F = 64948 LB.
M1 = 82187 LB.FT.
M2 = 2852 LB.FT.
S1 = -9921 LB.
S2 = -1538 LB.

INTERNAL BEAM RESULTS:
F = 93233 LB.
M1 = 27576 LB.FT.
S1 = -7785 LB.

EXTERNAL-HALF SHELL RESULTS:

POINT	T1	S	M2	T2
0	-2660	-4637	298	-161
1	-6853	-3140	213	-1143
2	-5353	-1349	-28	-1664
3	-2122	-314	-198	-1817
4	-604	0	-252	-1832

INTERNAL-HALF SHELL RESULTS:

POINT	T1	S	M2	T2
0	5550	-3328	-460	592
1	-1159	-3868	164	-325
2	-4619	-2987	138	-1209
3	-5499	-1535	-65	-1787
4	-5495	0	-164	-1982

```
L= 46        TABLE V- 19          W= 22
*************************************

SHELL DATA:

     L   = 46.0 FT.
     W   = 22.0 FT.
     R   = 17.5 FT.
     T   =  2.5 IN.
     PHI= 37.4 DEG.
     G   = 50.0 LB/FT2

BEAM DATA:

     B  = 0.75 FT.
     D1 = 2.75 FT.
     D2 = 2.25 FT.

EXTERNAL BEAM RESULTS:
     F  =  64076 LB.
     M1 =  72093 LB.FT.
     M2 =   2291 LB.FT.
     S1 =  -8745 LB.
     S2 =  -1490 LB.

INTERNAL BEAM RESULTS:
     F  = 102580 LB.
     M1 =  34167 LB.FT.
     S1 =  -8447 LB.

EXTERNAL-HALF SHELL RESULTS:

POINT      T1        S       M2      T2

  0      -1786    -4376     578     -88
  1      -6289    -3507     310    -772
  2      -6729    -2190     -38   -1257
  3      -5523     -989    -301   -1519
  4      -4838        0    -397   -1599

INTERNAL-HALF SHELL RESULTS:

POINT      T1        S       M2      T2

  0       5569    -3502    -253     453
  1      -1348    -3875     101    -189
  2      -5894    -3131      93    -814
  3      -8311    -1715     -26   -1250
  4      -9042        0     -86   -1405
```

```
L= 46        TABLE V- 20          W= 24
*************************************

SHELL DATA:

     L   = 46.0 FT.
     W   = 24.0 FT.
     R   = 20.0 FT.
     T   =  2.5 IN.
     PHI= 35.5 DEG.
     G   = 50.0 LB/FT2

BEAM DATA:

     B  = 0.75 FT.
     D1 = 2.75 FT.
     D2 = 2.25 FT.

EXTERNAL BEAM RESULTS:
     F  =  66291 LB.
     M1 =  77131 LB.FT.
     M2 =   2417 LB.FT.
     S1 =  -9222 LB.
     S2 =  -1537 LB.

INTERNAL BEAM RESULTS:
     F  = 104799 LB.
     M1 =  34073 LB.FT.
     S1 =  -8576 LB.

EXTERNAL-HALF SHELL RESULTS:

POINT      T1        S       M2      T2

  0      -2167    -4527     516    -111
  1      -6516    -3510     294    -877
  2      -6467    -2080     -34   -1394
  3      -4767     -886    -286   -1656
  4      -3878        0    -377   -1731

INTERNAL-HALF SHELL RESULTS:

POINT      T1        S       M2      T2

  0       5807    -3578    -302     500
  1      -1217    -4018     115    -228
  2      -5682    -3244     104    -939
  3      -7923    -1769     -36   -1433
  4      -8565        0    -106   -1607
```

```
L= 46        TABLE V- 21          W= 26
*************************************

SHELL DATA:

     L   = 46.0 FT.
     W   = 26.0 FT.
     R   = 20.0 FT.
     T   =  2.5 IN.
     PHI= 39.1 DEG.
     G   = 50.0 LB/FT2

BEAM DATA:

     B  = 0.75 FT.
     D1 = 2.75 FT.
     D2 = 2.25 FT.

EXTERNAL BEAM RESULTS:
     F  =  65265 LB.
     M1 =  80837 LB.FT.
     M2 =   3044 LB.FT.
     S1 =  -9414 LB.
     S2 =  -1468 LB.

INTERNAL BEAM RESULTS:
     F  =  95926 LB.
     M1 =  27791 LB.FT.
     S1 =  -7618 LB.

EXTERNAL-HALF SHELL RESULTS:

POINT      T1        S       M2      T2

  0      -2284    -4457     501    -114
  1      -7001    -3234     279    -910
  2      -6064    -1635     -42   -1384
  3      -3251     -549    -271   -1570
  4      -1884        0    -350   -1610

INTERNAL-HALF SHELL RESULTS:

POINT      T1        S       M2      T2

  0       5791    -3275    -363     523
  1       -962    -3787     134    -208
  2      -5000    -3039     119    -933
  3      -6732    -1635     -48   -1431
  4      -7138        0    -131   -1605
```

```
L= 46        TABLE V- 22         W= 28
**********************************

SHELL DATA:

    L   = 46.0 FT.
    W   = 28.0 FT.
    R   = 22.5 FT.
    T   =  2.5 IN.
    PHI= 37.3 DEG.
    G   = 50.0 LB/FT2

BEAM DATA:

    B  = 0.75 FT.
    D1 = 2.75 FT.
    D2 = 2.25 FT.

EXTERNAL BEAM RESULTS:
    F  =  68072 LB.
    M1 =  84569 LB.FT.
    M2 =   3043 LB.FT.
    S1 =  -9838 LB.
    S2 =  -1539 LB.

INTERNAL BEAM RESULTS:
    F  =  98567 LB.
    M1 =  28443 LB.FT.
    S1 =  -7822 LB.
```

EXTERNAL-HALF SHELL RESULTS:

POINT	T1	S	M2	T2
0	-2524	-4649	430	-136
1	-7185	-3278	260	-1025
2	-5948	-1554	-37	-1530
3	-2826	-461	-249	-1709
4	-1332	0	-321	-1740

INTERNAL-HALF SHELL RESULTS:

POINT	T1	S	M2	T2
0	5967	-3365	-415	566
1	-1000	-3925	151	-255
2	-4960	-3118	131	-1063
3	-6437	-1655	-57	-1611
4	-6707	0	-150	-1801

```
L= 46        TABLE V- 23         W= 30
**********************************

SHELL DATA:

    L   = 46.0 FT.
    W   = 30.0 FT.
    R   = 25.0 FT.
    T   =  2.5 IN.
    PHI= 35.8 DEG.
    G   = 50.0 LB/FT2

BEAM DATA:

    B  = 0.75 FT.
    D1 = 2.75 FT.
    D2 = 2.25 FT.

EXTERNAL BEAM RESULTS:
    F  =  70841 LB.
    M1 =  87578 LB.FT.
    M2 =   3032 LB.FT.
    S1 = -10209 LB.
    S2 =  -1610 LB.

INTERNAL BEAM RESULTS:
    F  = 100957 LB.
    M1 =  29040 LB.FT.
    S1 =  -8007 LB.
```

EXTERNAL-HALF SHELL RESULTS:

POINT	T1	S	M2	T2
0	-2676	-4838	363	-154
1	-7334	-3327	242	-1142
2	-5849	-1477	-33	-1679
3	-2458	-374	-229	-1848
4	-853	0	-293	-1869

INTERNAL-HALF SHELL RESULTS:

POINT	T1	S	M2	T2
0	6124	-3447	-468	605
1	-1080	-4048	168	-307
2	-4943	-3171	144	-1198
3	-6118	-1654	-66	-1791
4	-6226	0	-169	-1992

```
L= 46        TABLE V- 24         W= 32
**********************************

SHELL DATA:

    L   = 46.0 FT.
    W   = 32.0 FT.
    R   = 25.0 FT.
    T   =  2.5 IN.
    PHI= 38.7 DEG.
    G   = 50.0 LB/FT2

BEAM DATA:

    B  = 0.75 FT.
    D1 = 2.75 FT.
    D2 = 2.25 FT.

EXTERNAL BEAM RESULTS:
    F  =  71159 LB.
    M1 =  88634 LB.FT.
    M2 =   3443 LB.FT.
    S1 = -10300 LB.
    S2 =  -1591 LB.

INTERNAL BEAM RESULTS:
    F  =  95105 LB.
    M1 =  25553 LB.FT.
    S1 =  -7420 LB.
```

EXTERNAL-HALF SHELL RESULTS:

POINT	T1	S	M2	T2
0	-2460	-4859	349	-158
1	-7696	-3169	214	-1202
2	-5674	-1128	-49	-1696
3	-1574	-94	-202	-1777
4	321	0	-243	-1754

INTERNAL-HALF SHELL RESULTS:

POINT	T1	S	M2	T2
0	6022	-3247	-519	598
1	-1220	-3855	185	-326
2	-4688	-2912	153	-1215
3	-5166	-1446	-75	-1779
4	-4925	0	-184	-1963

L= 48 TABLE V- 25 W= 22

SHELL DATA:

L = 48.0 FT.
W = 22.0 FT.
R = 17.5 FT.
T = 2.5 IN.
PHI= 37.4 DEG.
G = 50.0 LB/FT2

BEAM DATA:

B = 0.75 FT.
D1 = 3.00 FT.
D2 = 2.50 FT.

EXTERNAL BEAM RESULTS:
F = 68078 LB.
M1 = 82951 LB.FT.
M2 = 2160 LB.FT.
S1 = -9877 LB.
S2 = -1534 LB.

INTERNAL BEAM RESULTS:
F = 111516 LB.
M1 = 43270 LB.FT.
S1 = -10114 LB.

EXTERNAL-HALF SHELL RESULTS:

POINT	T1	S	M2	T2
0	-2336	-4455	534	-95
1	-6448	-3566	296	-758
2	-6975	-2271	-31	-1233
3	-6006	-1052	-284	-1497
4	-5435	0	-377	-1578

INTERNAL-HALF SHELL RESULTS:

POINT	T1	S	M2	T2
0	4983	-3646	-240	399
1	-1763	-3914	98	-228
2	-6242	-3132	90	-827
3	-8660	-1710	-23	-1242
4	-9401	0	-80	-1390

L= 48 TABLE V- 26 W= 24

SHELL DATA:

L = 48.0 FT.
W = 24.0 FT.
R = 20.0 FT.
T = 2.5 IN.
PHI= 35.5 DEG.
G = 50.0 LB/FT2

BEAM DATA:

B = 0.75 FT.
D1 = 3.00 FT.
D2 = 2.50 FT.

EXTERNAL BEAM RESULTS:
F = 70277 LB.
M1 = 88222 LB.FT.
M2 = 2276 LB.FT.
S1 = -10367 LB.
S2 = -1580 LB.

INTERNAL BEAM RESULTS:
F = 114105 LB.
M1 = 43118 LB.FT.
S1 = -10276 LB.

EXTERNAL-HALF SHELL RESULTS:

POINT	T1	S	M2	T2
0	-2698	-4599	471	-117
1	-6616	-3573	281	-858
2	-6673	-2177	-26	-1367
3	-5244	-963	-270	-1636
4	-4484	0	-360	-1716

INTERNAL-HALF SHELL RESULTS:

POINT	T1	S	M2	T2
0	5241	-3734	-288	442
1	-1601	-4061	111	-270
2	-6008	-3249	100	-953
3	-8269	-1768	-33	-1424
4	-8930	0	-100	-1591

L= 48 TABLE V- 27 W= 26

SHELL DATA:

L = 48.0 FT.
W = 26.0 FT.
R = 20.0 FT.
T = 2.5 IN.
PHI= 39.1 DEG.
G = 50.0 LB/FT2

BEAM DATA:

B = 0.75 FT.
D1 = 3.00 FT.
D2 = 2.50 FT.

EXTERNAL BEAM RESULTS:
F = 68964 LB.
M1 = 92495 LB.FT.
M2 = 2900 LB.FT.
S1 = -10561 LB.
S2 = -1507 LB.

INTERNAL BEAM RESULTS:
F = 105055 LB.
M1 = 35351 LB.FT.
S1 = -9177 LB.

EXTERNAL-HALF SHELL RESULTS:

POINT	T1	S	M2	T2
0	-2885	-4513	461	-121
1	-7023	-3288	271	-888
2	-6188	-1748	-34	-1357
3	-3684	-645	-260	-1559
4	-2465	0	-340	-1607

INTERNAL-HALF SHELL RESULTS:

POINT	T1	S	M2	T2
0	5362	-3437	-350	477
1	-1254	-3850	130	-242
2	-5286	-3071	116	-943
3	-7092	-1653	-45	-1423
4	-7541	0	-126	-1591

L= 48 TABLE V- 28 W= 28

****

SHELL DATA:

L = 48.0 FT.
W = 28.0 FT.
R = 22.5 FT.
T = 2.5 IN.
PHI= 37.3 DEG.
G = 50.0 LB/FT2

BEAM DATA:

B = 0.75 FT.
D1 = 3.00 FT.
D2 = 2.50 FT.

EXTERNAL BEAM RESULTS:
F = 71769 LB.
M1 = 96422 LB.FT.
M2 = 2891 LB.FT.
S1 = -11002 LB.
S2 = -1576 LB.

INTERNAL BEAM RESULTS:
F = 107949 LB.
M1 = 36087 LB.FT.
S1 = -9416 LB.

EXTERNAL-HALF SHELL RESULTS:

POINT	T1	S	M2	T2
0	-3116	-4697	389	-142
1	-7171	-3334	253	-997
2	-6051	-1680	-28	-1500
3	-3254	-568	-240	-1699
4	-1913	0	-313	-1742

INTERNAL-HALF SHELL RESULTS:

POINT	T1	S	M2	T2
0	5540	-3532	-401	516
1	-1265	-3989	146	-291
2	-5222	-3155	128	-1073
3	-6793	-1680	-55	-1604
4	-7118	0	-145	-1788

L= 48 TABLE V- 29 W= 30

****

SHELL DATA:

L = 48.0 FT.
W = 30.0 FT.
R = 25.0 FT.
T = 2.5 IN.
PHI= 35.8 DEG.
G = 50.0 LB/FT2

BEAM DATA:

B = 0.75 FT.
D1 = 3.00 FT.
D2 = 2.50 FT.

EXTERNAL BEAM RESULTS:
F = 74545 LB.
M1 = 99576 LB.FT.
M2 = 2873 LB.FT.
S1 = -11391 LB.
S2 = -1644 LB.

INTERNAL BEAM RESULTS:
F = 110548 LB.
M1 = 36748 LB.FT.
S1 = -9631 LB.

EXTERNAL-HALF SHELL RESULTS:

POINT	T1	S	M2	T2
0	-3262	-4878	320	-160
1	-7291	-3383	237	-1108
2	-5935	-1615	-22	-1645
3	-2880	-493	-220	-1840
4	-1431	0	-288	-1876

INTERNAL-HALF SHELL RESULTS:

POINT	T1	S	M2	T2
0	5699	-3617	-454	552
1	-1317	-4112	163	-344
2	-5181	-3215	140	-1207
3	-6472	-1686	-64	-1783
4	-6647	0	-164	-1980

L= 48 TABLE V- 30 W= 32

****

SHELL DATA:

L = 48.0 FT.
W = 32.0 FT.
R = 25.0 FT.
T = 2.5 IN.
PHI= 38.7 DEG.
G = 50.0 LB/FT2

BEAM DATA:

B = 0.75 FT.
D1 = 3.00 FT.
D2 = 2.50 FT.

EXTERNAL BEAM RESULTS:
F = 74717 LB.
M1 = 101087 LB.FT.
M2 = 3284 LB.FT.
S1 = -11501 LB.
S2 = -1622 LB.

INTERNAL BEAM RESULTS:
F = 104260 LB.
M1 = 32322 LB.FT.
S1 = -8930 LB.

EXTERNAL-HALF SHELL RESULTS:

POINT	T1	S	M2	T2
0	-3128	-4890	306	-165
1	-7628	-3214	213	-1163
2	-5724	-1272	-36	-1660
3	-1952	-221	-197	-1775
4	-203	0	-245	-1771

INTERNAL-HALF SHELL RESULTS:

POINT	T1	S	M2	T2
0	5663	-3411	-508	555
1	-1387	-3926	182	-353
2	-4880	-2978	152	-1218
3	-5517	-1499	-73	-1773
4	-5364	0	-181	-1957

L= 50 TABLE V- 31 W= 22

****

SHELL DATA:

L = 50.0 FT.
W = 22.0 FT.
R = 17.5 FT.
T = 2.5 IN.
PHI= 37.4 DEG.
G = 50.0 LB/FT2

BEAM DATA:

B = 0.75 FT.
D1 = 3.00 FT.
D2 = 2.50 FT.

EXTERNAL BEAM RESULTS:
F = 74193 LB.
M1 = 86337 LB.FT.
M2 = 2189 LB.FT.
S1 = -10078 LB.
S2 = -1616 LB.

INTERNAL BEAM RESULTS:
F = 120814 LB.
M1 = 46269 LB.FT.
S1 = -10480 LB.

EXTERNAL-HALF SHELL RESULTS:

POINT	T1	S	M2	T2
0	-2167	-4661	595	-88
1	-6770	-3791	324	-760
2	-7608	-2460	-35	-1251
3	-6805	-1159	-311	-1529
4	-6284	0	-413	-1617

INTERNAL-HALF SHELL RESULTS:

POINT	T1	S	M2	T2
0	5474	-3795	-240	404
1	-1844	-4084	100	-223
2	-6754	-3278	93	-825
3	-9447	-1794	-21	-1243
4	-10283	0	-78	-1391

L= 50 TABLE V- 32 W= 24

****

SHELL DATA:

L = 50.0 FT.
W = 24.0 FT.
R = 20.0 FT.
T = 2.5 IN.
PHI= 35.5 DEG.
G = 50.0 LB/FT2

BEAM DATA:

B = 0.75 FT.
D1 = 3.00 FT.
D2 = 2.50 FT.

EXTERNAL BEAM RESULTS:
F = 76468 LB.
M1 = 92219 LB.FT.
M2 = 2335 LB.FT.
S1 = -10592 LB.
S2 = -1659 LB.

INTERNAL BEAM RESULTS:
F = 123536 LB.
M1 = 45989 LB.FT.
S1 = -10636 LB.

EXTERNAL-HALF SHELL RESULTS:

POINT	T1	S	M2	T2
0	-2574	-4804	536	-110
1	-6960	-3793	311	-860
2	-7269	-2359	-31	-1386
3	-5948	-1065	-299	-1670
4	-5216	0	-398	-1756

INTERNAL-HALF SHELL RESULTS:

POINT	T1	S	M2	T2
0	5760	-3881	-289	448
1	-1652	-4238	113	-264
2	-6490	-3404	103	-950
3	-9031	-1859	-32	-1425
4	-9791	0	-100	-1593

L= 50 TABLE V- 33 W= 26

****

SHELL DATA:

L = 50.0 FT.
W = 26.0 FT.
R = 20.0 FT.
T = 2.5 IN.
PHI= 39.1 DEG.
G = 50.0 LB/FT2

BEAM DATA:

B = 0.75 FT.
D1 = 3.00 FT.
D2 = 2.50 FT.

EXTERNAL BEAM RESULTS:
F = 74836 LB.
M1 = 97039 LB.FT.
M2 = 3012 LB.FT.
S1 = -10792 LB.
S2 = -1579 LB.

INTERNAL BEAM RESULTS:
F = 113526 LB.
M1 = 37493 LB.FT.
S1 = -9475 LB.

EXTERNAL-HALF SHELL RESULTS:

POINT	T1	S	M2	T2
0	-2819	-4702	530	-113
1	-7402	-3486	302	-887
2	-6719	-1905	-38	-1372
3	-4217	-731	-292	-1591
4	-2980	0	-382	-1646

INTERNAL-HALF SHELL RESULTS:

POINT	T1	S	M2	T2
0	5882	-3566	-353	484
1	-1249	-4016	133	-234
2	-5684	-3224	119	-938
3	-7764	-1746	-45	-1425
4	-8314	0	-127	-1596

L= 50 TABLE V- 34 W= 28

SHELL DATA:

L = 50.0 FT.
W = 28.0 FT.
R = 22.5 FT.
T = 2.5 IN.
PHI= 37.3 DEG.
G = 50.0 LB/FT2

BEAM DATA:

B = 0.75 FT.
D1 = 3.00 FT.
D2 = 2.50 FT.

EXTERNAL BEAM RESULTS:
F = 77815 LB.
M1 = 101652 LB.FT.
M2 = 3028 LB.FT.
S1 = -11270 LB.
S2 = -1647 LB.

INTERNAL BEAM RESULTS:
F = 116520 LB.
M1 = 38150 LB.FT.
S1 = -9706 LB.

EXTERNAL-HALF SHELL RESULTS:

POINT	T1	S	M2	T2
0	-3100	-4886	456	-135
1	-7589	-3526	284	-996
2	-6567	-1826	-33	-1515
3	-3716	-646	-271	-1730
4	-2335	0	-355	-1779

INTERNAL-HALF SHELL RESULTS:

POINT	T1	S	M2	T2
0	6078	-3660	-406	524
1	-1238	-4161	149	-280
2	-5599	-3319	131	-1068
3	-7454	-1781	-55	-1606
4	-7883	0	-147	-1793

L= 50 TABLE V- 35 W= 30

SHELL DATA:

L = 50.0 FT.
W = 30.0 FT.
R = 25.0 FT.
T = 2.5 IN.
PHI= 35.8 DEG.
G = 50.0 LB/FT2

BEAM DATA:

B = 0.75 FT.
D1 = 3.00 FT.
D2 = 2.50 FT.

EXTERNAL BEAM RESULTS:
F = 80778 LB.
M1 = 105423 LB.FT.
M2 = 3030 LB.FT.
S1 = -11694 LB.
S2 = -1716 LB.

INTERNAL BEAM RESULTS:
F = 119162 LB.
M1 = 38715 LB.FT.
S1 = -9909 LB.

EXTERNAL-HALF SHELL RESULTS:

POINT	T1	S	M2	T2
0	-3290	-5075	384	-154
1	-7744	-3571	267	-1108
2	-6439	-1750	-27	-1659
3	-3280	-561	-250	-1869
4	-1770	0	-327	-1911

INTERNAL-HALF SHELL RESULTS:

POINT	T1	S	M2	T2
0	6253	-3743	-461	563
1	-1268	-4290	166	-330
2	-5538	-3389	145	-1199
3	-7119	-1795	-64	-1786
4	-7402	0	-167	-1988

L= 50 TABLE V- 36 W= 32

SHELL DATA:

L = 50.0 FT.
W = 32.0 FT.
R = 25.0 FT.
T = 2.5 IN.
PHI= 38.7 DEG.
G = 50.0 LB/FT2

BEAM DATA:

B = 0.75 FT.
D1 = 3.00 FT.
D2 = 2.50 FT.

EXTERNAL BEAM RESULTS:
F = 80904 LB.
M1 = 107382 LB.FT.
M2 = 3483 LB.FT.
S1 = -11825 LB.
S2 = -1691 LB.

INTERNAL BEAM RESULTS:
F = 112019 LB.
M1 = 33812 LB.FT.
S1 = -9153 LB.

EXTERNAL-HALF SHELL RESULTS:

POINT	T1	S	M2	T2
0	-3186	-5083	367	-159
1	-8140	-3385	241	-1161
2	-6213	-1381	-40	-1671
3	-2237	-270	-226	-1799
4	-381	0	-283	-1801

INTERNAL-HALF SHELL RESULTS:

POINT	T1	S	M2	T2
0	6197	-3519	-518	568
1	-1305	-4093	186	-334
2	-5178	-3152	157	-1207
3	-6100	-1613	-74	-1777
4	-6055	0	-187	-1968

L= 52 TABLE V- 37 W= 22
****

SHELL DATA:

L = 52.0 FT.
W = 22.0 FT.
R = 17.5 FT.
T = 2.5 IN.
PHI= 37.4 DEG.
G = 50.0 LB/FT2

BEAM DATA:

B = 0.75 FT.
D1 = 3.00 FT.
D2 = 2.50 FT.

EXTERNAL BEAM RESULTS:
F = 80584 LB.
M1 = 89523 LB.FT.
M2 = 2203 LB.FT.
S1 = -10268 LB.
S2 = -1699 LB.

INTERNAL BEAM RESULTS:
F = 130495 LB.
M1 = 49396 LB.FT.
S1 = -10848 LB.

EXTERNAL-HALF SHELL RESULTS:

POINT	T1	S	M2	T2
0	-1958	-4868	655	-81
1	-7082	-4020	352	-762
2	-8269	-2654	-38	-1268
3	-7663	-1270	-337	-1561
4	-7205	0	-447	-1655

INTERNAL-HALF SHELL RESULTS:

POINT	T1	S	M2	T2
0	5984	-3941	-239	408
1	-1951	-4254	103	-219
2	-7287	-3423	96	-823
3	-10264	-1878	-19	-1243
4	-11200	0	-77	-1393

L= 52 TABLE V- 38 W= 24
****

SHELL DATA:

L = 52.0 FT.
W = 24.0 FT.
R = 20.0 FT.
T = 2.5 IN.
PHI= 35.5 DEG.
G = 50.0 LB/FT2

BEAM DATA:

B = 0.75 FT.
D1 = 3.00 FT.
D2 = 2.50 FT.

EXTERNAL BEAM RESULTS:
F = 82931 LB.
M1 = 95975 LB.FT.
M2 = 2379 LB.FT.
S1 = -108[illegible]1 LB.
S2 = -1740 LB.

INTERNAL BEAM RESULTS:
F = 133360 LB.
M1 = 48987 LB.FT.
S1 = -10999 LB.

EXTERNAL-HALF SHELL RESULTS:

POINT	T1	S	M2	T2
0	-2407	-5010	600	-104
1	-7290	-4019	340	-862
2	-7890	-2548	-35	-1404
3	-6711	-1173	-328	-1704
4	-6020	0	-436	-1796

INTERNAL-HALF SHELL RESULTS:

POINT	T1	S	M2	T2
0	6300	-4028	-290	453
1	-1708	-4415	116	-259
2	-6994	-3558	106	-947
3	-9823	-1948	-30	-1426
4	-10685	0	-99	-1595

L= 52 TABLE V- 39 W= 26
****

SHELL DATA:

L = 52.0 FT.
W = 26.0 FT.
R = 20.0 FT.
T = 2.5 IN.
PHI= 39.1 DEG.
G = 50.0 LB/FT2

BEAM DATA:

B = 0.75 FT.
D1 = 3.00 FT.
D2 = 2.50 FT.

EXTERNAL BEAM RESULTS:
F = 80940 LB.
M1 = 101301 LB.FT.
M2 = 3108 LB.FT.
S1 = -11002 LB.
S2 = -1651 LB.

INTERNAL BEAM RESULTS:
F = 122354 LB.
M1 = 39735 LB.FT.
S1 = -9778 LB.

EXTERNAL-HALF SHELL RESULTS:

POINT	T1	S	M2	T2
0	-2709	-4890	601	-105
1	-7761	-3689	335	-886
2	-7270	-2072	-43	-1388
3	-4804	-823	-325	-1623
4	-3564	0	-426	-1685

INTERNAL-HALF SHELL RESULTS:

POINT	T1	S	M2	T2
0	6423	-3696	-355	489
1	-1247	-4182	136	-227
2	-6099	-3376	122	-935
3	-8463	-1837	-44	-1426
4	-9115	0	-127	-1599

L= 52 TABLE V- 40 W= 28

SHELL DATA:

L = 52.0 FT.
W = 28.0 FT.
R = 22.5 FT.
T = 2.5 IN.
PHI= 37.3 DEG.
G = 50.0 LB/FT2

BEAM DATA:

B = 0.75 FT.
D1 = 3.00 FT.
D2 = 2.50 FT.

EXTERNAL BEAM RESULTS:
F = 84090 LB.
M1 = 106630 LB.FT.
M2 = 3151 LB.FT.
S1 = -11516 LB.
S2 = -1719 LB.

INTERNAL BEAM RESULTS:
F = 125454 LB.
M1 = 40313 LB.FT.
S1 = -10002 LB.

EXTERNAL-HALF SHELL RESULTS:

POINT	T1	S	M2	T2
0	-3043	-5080	526	-128
1	-7989	-3725	316	-995
2	-7101	-1982	-37	-1530
3	-4228	-730	-304	-1762
4	-2819	0	-398	-1818

INTERNAL-HALF SHELL RESULTS:

POINT	T1	S	M2	T2
0	6639	-3789	-409	532
1	-1213	-4333	152	-271
2	-5994	-3481	135	-1062
3	-8139	-1879	-54	-1607
4	-8674	0	-149	-1798

L= 52 TABLE V- 41 W= 30

SHELL DATA:

L = 52.0 FT.
W = 30.0 FT.
R = 25.0 FT.
T = 2.5 IN.
PHI= 35.8 DEG.
G = 50.0 LB/FT2

BEAM DATA:

B = 0.75 FT.
D1 = 3.00 FT.
D2 = 2.50 FT.

EXTERNAL BEAM RESULTS:
F = 87237 LB.
M1 = 111063 LB.FT.
M2 = 3176 LB.FT.
S1 = -11974 LB.
S2 = -1788 LB.

INTERNAL BEAM RESULTS:
F = 128143 LB.
M1 = 40779 LB.FT.
S1 = -10194 LB.

EXTERNAL-HALF SHELL RESULTS:

POINT	T1	S	M2	T2
0	-3280	-5270	452	-147
1	-8183	-3765	298	-1107
2	-6960	-1895	-32	-1674
3	-3724	-637	-282	-1900
4	-2164	0	-370	-1948

INTERNAL-HALF SHELL RESULTS:

POINT	T1	S	M2	T2
0	6830	-3870	-466	573
1	-1220	-4468	169	-318
2	-5911	-3561	148	-1192
3	-7790	-1902	-65	-1788
4	-8184	0	-170	-1995

L= 52 TABLE V- 42 W= 32

SHELL DATA:

L = 52.0 FT.
W = 32.0 FT.
R = 25.0 FT.
T = 2.5 IN.
PHI= 38.7 DEG.
G = 50.0 LB/FT2

BEAM DATA:

B = 0.75 FT.
D1 = 3.00 FT.
D2 = 2.50 FT.

EXTERNAL BEAM RESULTS:
F = 87293 LB.
M1 = 113531 LB.FT.
M2 = 3672 LB.FT.
S1 = -12127 LB.
S2 = -1760 LB.

INTERNAL BEAM RESULTS:
F = 120107 LB.
M1 = 35377 LB.FT.
S1 = -9383 LB.

EXTERNAL-HALF SHELL RESULTS:

POINT	T1	S	M2	T2
0	-3212	-5273	432	-152
1	-8640	-3562	271	-1158
2	-6716	-1500	-45	-1683
3	-2558	-325	-257	-1825
4	-605	0	-324	-1834

INTERNAL-HALF SHELL RESULTS:

POINT	T1	S	M2	T2
0	6753	-3628	-527	580
1	-1223	-4259	190	-318
2	-5489	-3324	163	-1198
3	-6705	-1724	-75	-1780
4	-6772	0	-192	-1978

L= 54 TABLE V- 43 W= 24

SHELL DATA:

L = 54.0 FT.
W = 24.0 FT.
R = 20.0 FT.
T = 2.5 IN.
PHI= 35.5 DEG.
G = 50.0 LB/FT2

BEAM DATA:

B = 0.75 FT.
D1 = 3.25 FT.
D2 = 2.75 FT.

EXTERNAL BEAM RESULTS:
F = 87372 LB.
M1 = 109093 LB.FT.
M2 = 2241 LB.FT.
S1 = -12058 LB.
S2 = -1780 LB.

INTERNAL BEAM RESULTS:
F = 143375 LB.
M1 = 60633 LB.FT.
S1 = -12894 LB.

EXTERNAL-HALF SHELL RESULTS:

POINT	T1	S	M2	T2
0	-2916	-5083	551	-110
1	-7441	-4078	324	-847
2	-8149	-2629	-28	-1379
3	-7221	-1235	-309	-1680
4	-6647	0	-414	-1774

INTERNAL-HALF SHELL RESULTS:

POINT	T1	S	M2	T2
0	5630	-4170	-276	396
1	-2159	-4444	111	-300
2	-7345	-3548	102	-960
3	-10156	-1937	-27	-1417
4	-11021	0	-93	-1579

L= 54 TABLE V- 44 W= 26

SHELL DATA:

L = 54.0 FT.
W = 26.0 FT.
R = 20.0 FT.
T = 2.5 IN.
PHI= 39.1 DEG.
G = 50.0 LB/FT2

BEAM DATA:

B = 0.75 FT.
D1 = 3.25 FT.
D2 = 2.75 FT.

EXTERNAL BEAM RESULTS:
F = 85073 LB.
M1 = 114973 LB.FT.
M2 = 2967 LB.FT.
S1 = -12250 LB.
S2 = -1688 LB.

INTERNAL BEAM RESULTS:
F = 132392 LB.
M1 = 49507 LB.FT.
S1 = -11530 LB.

EXTERNAL-HALF SHELL RESULTS:

POINT	T1	S	M2	T2
0	-3259	-4949	557	-111
1	-7818	-3746	324	-868
2	-7442	-2172	-35	-1365
3	-5277	-906	-311	-1608
4	-4180	0	-410	-1676

INTERNAL-HALF SHELL RESULTS:

POINT	T1	S	M2	T2
0	5907	-3851	-341	443
1	-1606	-4234	131	-262
2	-6420	-3392	118	-946
3	-8820	-1843	-41	-1419
4	-9499	0	-121	-1586

L= 54 TABLE V- 45 W= 28

SHELL DATA:

L = 54.0 FT.
W = 28.0 FT.
R = 22.5 FT.
T = 2.5 IN.
PHI= 37.3 DEG.
G = 50.0 LB/FT2

BEAM DATA:

B = 0.75 FT.
D1 = 3.25 FT.
D2 = 2.75 FT.

EXTERNAL BEAM RESULTS:
F = 88197 LB.
M1 = 120548 LB.FT.
M2 = 3001 LB.FT.
S1 = -12779 LB.
S2 = -1754 LB.

INTERNAL BEAM RESULTS:
F = 135811 LB.
M1 = 50143 LB.FT.
S1 = -11793 LB.

EXTERNAL-HALF SHELL RESULTS:

POINT	T1	S	M2	T2
0	-3578	-5131	481	-133
1	-8001	-3783	306	-973
2	-7245	-2096	-29	-1504
3	-4697	-824	-291	-1749
4	-3442	0	-385	-1813

INTERNAL-HALF SHELL RESULTS:

POINT	T1	S	M2	T2
0	6130	-3950	-395	482
1	-1547	-4387	147	-308
2	-6293	-3501	131	-1074
3	-8493	-1890	-52	-1600
4	-9065	0	-142	-1784

L= 54 TABLE V- 46 W= 30
****

SHELL DATA:

```
L   = 54.0 FT.
W   = 30.0 FT.
R   = 25.0 FT.
T   =  2.5 IN.
PHI= 35.8 DEG.
G   = 50.0 LB/FT2
```

BEAM DATA:

```
B  = 0.75 FT.
D1 = 3.25 FT.
D2 = 2.75 FT.
```

EXTERNAL BEAM RESULTS:

```
F  =  91322 LB.
M1 = 125174 LB.FT.
M2 =   3018 LB.FT.
S1 = -13254 LB.
S2 =  -1820 LB.
```

INTERNAL BEAM RESULTS:

```
F  = 138771 LB.
M1 =  50635 LB.FT.
S1 = -12016 LB.
```

EXTERNAL-HALF SHELL RESULTS:

POINT	T1	S	M2	T2
0	-3804	-5312	407	-152
1	-8156	-3824	289	-1080
2	-7080	-2022	-22	-1645
3	-4189	-743	-270	-1889
4	-2791	0	-359	-1948

INTERNAL-HALF SHELL RESULTS:

POINT	T1	S	M2	T2
0	6329	-4036	-450	519
1	-1526	-4523	164	-357
2	-6187	-3587	144	-1204
3	-8142	-1918	-62	-1780
4	-8584	0	-164	-1980

L= 54 TABLE V- 47 W= 32
****

SHELL DATA:

```
L   = 54.0 FT.
W   = 32.0 FT.
R   = 25.0 FT.
T   =  2.5 IN.
PHI= 38.7 DEG.
G   = 50.0 LB/FT2
```

BEAM DATA:

```
B  = 0.75 FT.
D1 = 3.25 FT.
D2 = 2.75 FT.
```

EXTERNAL BEAM RESULTS:

```
F  =  91147 LB.
M1 = 128235 LB.FT.
M2 =   3515 LB.FT.
S1 = -13420 LB.
S2 =  -1788 LB.
```

INTERNAL BEAM RESULTS:

```
F  = 130379 LB.
M1 =  43975 LB.FT.
S1 = -11081 LB.
```

EXTERNAL-HALF SHELL RESULTS:

POINT	T1	S	M2	T2
0	-3814	-5302	388	-158
1	-8571	-3611	267	-1127
2	-6784	-1636	-33	-1652
3	-2982	-443	-249	-1820
4	-1193	0	-320	-1843

INTERNAL-HALF SHELL RESULTS:

POINT	T1	S	M2	T2
0	6339	-3792	-513	536
1	-1452	-4325	186	-349
2	-5722	-3371	160	-1205
3	-7062	-1759	-73	-1774
4	-7200	0	-187	-1968

L= 54 TABLE V- 48 W= 34
****

SHELL DATA:

```
L   = 54.0 FT.
W   = 34.0 FT.
R   = 27.5 FT.
T   =  2.5 IN.
PHI= 37.2 DEG.
G   = 50.0 LB/FT2
```

BEAM DATA:

```
B  = 0.75 FT.
D1 = 3.25 FT.
D2 = 2.75 FT.
```

EXTERNAL BEAM RESULTS:

```
F  =  94503 LB.
M1 = 131357 LB.FT.
M2 =   3446 LB.FT.
S1 = -13822 LB.
S2 =  -1864 LB.
```

INTERNAL BEAM RESULTS:

```
F  = 133960 LB.
M1 =  45254 LB.FT.
S1 = -11390 LB.
```

EXTERNAL-HALF SHELL RESULTS:

POINT	T1	S	M2	T2
0	-3915	-5498	317	-175
1	-8656	-3680	252	-1240
2	-6691	-1589	-28	-1801
3	-2691	-379	-230	-1966
4	-820	0	-294	-1981

INTERNAL-HALF SHELL RESULTS:

POINT	T1	S	M2	T2
0	6505	-3896	-569	568
1	-1572	-4458	204	-409
2	-5753	-3422	172	-1343
3	-6758	-1751	-82	-1952
4	-6715	0	-206	-2156

L= 56 TABLE V- 49 W= 24

SHELL DATA:

L = 56.0 FT.
W = 24.0 FT.
R = 20.0 FT.
T = 2.5 IN.
PHI= 35.5 DEG.
G = 50.0 LB/FT2

BEAM DATA:

B = 0.75 FT.
D1 = 3.25 FT.
D2 = 2.75 FT.

EXTERNAL BEAM RESULTS:
F = 94284 LB.
M1 = 113256 LB.FT.
M2 = 2270 LB.FT.
S1 = -12296 LB.
S2 = -1861 LB.

INTERNAL BEAM RESULTS:
F = 153994 LB.
M1 = 64414 LB.FT.
S1 = -13313 LB.

EXTERNAL-HALF SHELL RESULTS:

POINT	T1	S	M2	T2
0	-2761	-5289	610	-104
1	-7780	-4301	351	-849
2	-8804	-2815	-31	-1396
3	-8045	-1341	-335	-1710
4	-7525	0	-448	-1810

INTERNAL-HALF SHELL RESULTS:

POINT	T1	S	M2	T2
0	6125	-4319	-276	400
1	-2254	-4616	114	-296
2	-7880	-3695	105	-958
3	-10973	-2021	-26	-1418
4	-11936	0	-92	-1581

L= 56 TABLE V- 50 W= 26

SHELL DATA:

L = 56.0 FT.
W = 26.0 FT.
R = 20.0 FT.
T = 2.5 IN.
PHI= 39.1 DEG.
G = 50.0 LB/FT2

BEAM DATA:

B = 0.75 FT.
D1 = 3.25 FT.
D2 = 2.75 FT.

EXTERNAL BEAM RESULTS:
F = 91584 LB.
M1 = 119559 LB.FT.
M2 = 3042 LB.FT.
S1 = -12479 LB.
S2 = -1762 LB.

INTERNAL BEAM RESULTS:
F = 142026 LB.
M1 = 52383 LB.FT.
S1 = -11886 LB.

EXTERNAL-HALF SHELL RESULTS:

POINT	T1	S	M2	T2
0	-3152	-5137	624	-104
1	-8169	-3950	355	-867
2	-8016	-2340	-39	-1380
3	-5929	-1000	-342	-1638
4	-4847	0	-451	-1712

INTERNAL-HALF SHELL RESULTS:

POINT	T1	S	M2	T2
0	6416	-3983	-342	448
1	-1637	-4397	134	-256
2	-6867	-3537	121	-943
3	-9544	-1929	-40	-1419
4	-10322	0	-121	-1588

L= 56 TABLE V- 51 W= 28

SHELL DATA:

L = 56.0 FT.
W = 28.0 FT.
R = 22.5 FT.
T = 2.5 IN.
PHI= 37.3 DEG.
G = 50.0 LB/FT2

BEAM DATA:

B = 0.75 FT.
D1 = 3.25 FT.
D2 = 2.75 FT.

EXTERNAL BEAM RESULTS:
F = 94857 LB.
M1 = 125892 LB.FT.
M2 = 3103 LB.FT.
S1 = -13042 LB.
S2 = -1826 LB.

INTERNAL BEAM RESULTS:
F = 145586 LB.
M1 = 52929 LB.FT.
S1 = -12143 LB.

EXTERNAL-HALF SHELL RESULTS:

POINT	T1	S	M2	T2
0	-3522	-5321	548	-127
1	-8386	-3982	336	-973
2	-7798	-2255	-33	-1519
3	-5273	-912	-322	-1779
4	-4010	0	-426	-1849

INTERNAL-HALF SHELL RESULTS:

POINT	T1	S	M2	T2
0	6661	-4083	-397	488
1	-1555	-4556	150	-301
2	-6720	-3655	134	-1070
3	-9202	-1982	-51	-1601
4	-9877	0	-143	-1788

```
L= 56        TABLE V- 52          W= 30
**********************************

SHELL DATA:

     L   = 56.0 FT.
     W   = 30.0 FT.
     R   = 25.0 FT.
     T   =  2.5 IN.
     PHI= 35.8 DEG.
     G   = 50.0 LB/FT2

BEAM DATA:

     B  = 0.75 FT.
     D1 = 3.25 FT.
     D2 = 2.75 FT.

EXTERNAL BEAM RESULTS:
     F  =   98145 LB.
     M1 =  131241 LB.FT.
     M2 =    3144 LB.FT.
     S1 =  -13551 LB.
     S2 =   -1892 LB.

INTERNAL BEAM RESULTS:
     F  =  148623 LB.
     M1 =   53307 LB.FT.
     S1 =  -12357 LB.

EXTERNAL-HALF SHELL RESULTS:

POINT     T1       S      M2      T2

  0     -3798   -5505    472    -146
  1     -8575   -4018    319   -1079
  2     -7615   -2172    -27   -1659
  3     -4695    -823   -301   -1918
  4     -3268       0   -399   -1983

INTERNAL-HALF SHELL RESULTS:

POINT     T1       S      M2      T2

  0      6879   -4168   -454     527
  1     -1511   -4697    167    -347
  2     -6592   -3750    148   -1198
  3     -8837   -2018    -61   -1782
  4     -9386       0   -166   -1985
```

```
L= 56        TABLE V- 53          W= 32
**********************************

SHELL DATA:

     L   = 56.0 FT.
     W   = 32.0 FT.
     R   = 25.0 FT.
     T   =  2.5 IN.
     PHI= 38.7 DEG.
     G   = 50.0 LB/FT2

BEAM DATA:

     B  = 0.75 FT.
     D1 = 3.25 FT.
     D2 = 2.75 FT.

EXTERNAL BEAM RESULTS:
     F  =   97845 LB.
     M1 =  134900 LB.FT.
     M2 =    3685 LB.FT.
     S1 =  -13737 LB.
     S2 =   -1856 LB.

INTERNAL BEAM RESULTS:
     F  =  139315 LB.
     M1 =   46032 LB.FT.
     S1 =  -11363 LB.

EXTERNAL-HALF SHELL RESULTS:

POINT     T1       S      M2      T2

  0     -3853   -5489    452    -152
  1     -9043   -3788    297   -1125
  2     -7291   -1761    -38   -1663
  3     -3360    -506   -280   -1846
  4     -1499       0   -360   -1874

INTERNAL-HALF SHELL RESULTS:

POINT     T1       S      M2      T2

  0      6878   -3907   -520     546
  1     -1396   -4489    189    -336
  2     -6062   -3534    164   -1197
  3     -7691   -1862    -73   -1776
  4     -7939       0   -190   -1975
```

```
L= 56        TABLE V- 54          W= 34
**********************************

SHELL DATA:

     L   = 56.0 FT.
     W   = 34.0 FT.
     R   = 27.5 FT.
     T   =  2.5 IN.
     PHI= 37.2 DEG.
     G   = 50.0 LB/FT2

BEAM DATA:

     B  = 0.75 FT.
     D1 = 3.25 FT.
     D2 = 2.75 FT.

EXTERNAL BEAM RESULTS:
     F  =  101433 LB.
     M1 =  138613 LB.FT.
     M2 =    3630 LB.FT.
     S1 =  -14172 LB.
     S2 =   -1934 LB.

INTERNAL BEAM RESULTS:
     F  =  142956 LB.
     M1 =   47234 LB.FT.
     S1 =  -11661 LB.

EXTERNAL-HALF SHELL RESULTS:

POINT     T1       S      M2      T2

  0     -3989   -5690    378    -170
  1     -9159   -3854    280   -1239
  2     -7191   -1706    -32   -1812
  3     -3023    -434   -259   -1989
  4     -1063       0   -333   -2010

INTERNAL-HALF SHELL RESULTS:

POINT     T1       S      M2      T2

  0      7058   -4009   -578     580
  1     -1499   -4628    208    -392
  2     -6078   -3595    178   -1334
  3     -7379   -1862    -83   -1955
  4     -7449       0   -211   -2166
```

```
L= 58        TABLE V- 55          W= 26
***************************************

SHELL DATA:

     L   = 58.0 FT.
     W   = 26.0 FT.
     R   = 20.0 FT.
     T   =  2.5 IN.
     PHI= 39.1 DEG.
     G   = 50.0 LB/FT2

BEAM DATA:

     B  = 0.75 FT.
     D1 = 3.50 FT.
     D2 = 3.00 FT.

EXTERNAL BEAM RESULTS:
     F  =  96007 LB.
     M1 = 134773 LB.FT.
     M2 =   2908 LB.FT.
     S1 = -13793 LB.
     S2 =  -1798 LB.

INTERNAL BEAM RESULTS:
     F  = 152423 LB.
     M1 =  64131 LB.FT.
     S1 = -13777 LB.

EXTERNAL-HALF SHELL RESULTS:

POINT      T1        S      M2      T2

  0     -3631    -5200     582    -110
  1     -8259    -4011     342    -853
  2     -8233    -2434     -32   -1359
  3     -6426    -1074    -327   -1623
  4     -5475        0    -434   -1701

INTERNAL-HALF SHELL RESULTS:

POINT      T1        S      M2      T2

  0      5864    -4128    -329     402
  1     -2027    -4441     130    -290
  2     -7203    -3546     118    -953
  3     -9897    -1930     -37   -1413
  4    -10693        0    -115   -1575
```

```
L= 58        TABLE V- 56          W= 28
***************************************

SHELL DATA:

     L   = 58.0 FT.
     W   = 28.0 FT.
     R   = 22.5 FT.
     T   =  2.5 IN.
     PHI= 37.3 DEG.
     G   = 50.0 LB/FT2

BEAM DATA:

     B  = 0.75 FT.
     D1 = 3.50 FT.
     D2 = 3.00 FT.

EXTERNAL BEAM RESULTS:
     F  =  99256 LB.
     M1 = 141357 LB.FT.
     M2 =   2961 LB.FT.
     S1 = -14370 LB.
     S2 =  -1860 LB.

INTERNAL BEAM RESULTS:
     F  = 156338 LB.
     M1 =  64718 LB.FT.
     S1 = -14076 LB.

EXTERNAL-HALF SHELL RESULTS:

POINT      T1        S      M2      T2

  0     -3979    -5376     506    -132
  1     -8428    -4045     325    -955
  2     -7985    -2362     -26   -1496
  3     -5766     -997    -308   -1765
  4     -4646        0    -411   -1841

INTERNAL-HALF SHELL RESULTS:

POINT      T1        S      M2      T2

  0      6118    -4234    -382     439
  1     -1921    -4602     145    -337
  2     -7035    -3667     130   -1081
  3     -9551    -1986     -48   -1593
  4    -10253        0    -137   -1774
```

```
L= 58        TABLE V- 57          W= 30
***************************************

SHELL DATA:

     L   = 58.0 FT.
     W   = 30.0 FT.
     R   = 25.0 FT.
     T   =  2.5 IN.
     PHI= 35.8 DEG.
     G   = 50.0 LB/FT2

BEAM DATA:

     B  = 0.75 FT.
     D1 = 3.50 FT.
     D2 = 3.00 FT.

EXTERNAL BEAM RESULTS:
     F  = 102521 LB.
     M1 = 146910 LB.FT.
     M2 =   2994 LB.FT.
     S1 = -14893 LB.
     S2 =  -1924 LB.

INTERNAL BEAM RESULTS:
     F  = 159688 LB.
     M1 =  65098 LB.FT.
     S1 = -14325 LB.

EXTERNAL-HALF SHELL RESULTS:

POINT      T1        S      M2      T2

  0     -4238    -5553     429    -151
  1     -8577    -4084     309   -1057
  2     -7776    -2290     -18   -1634
  3     -5184     -919    -288   -1906
  4     -3908        0    -387   -1979

INTERNAL-HALF SHELL RESULTS:

POINT      T1        S      M2      T2

  0      6347    -4324    -439     474
  1     -1850    -4745     162    -386
  2     -6887    -3766     144   -1210
  3     -9184    -2026     -59   -1774
  4     -9770        0    -159   -1971
```

```
L= 58        TABLE V- 58         W= 32
**************************************

SHELL DATA:

    L   = 58.0 FT.
    W   = 32.0 FT.
    R   = 25.0 FT.
    T   =  2.5 IN.
    PHI= 38.7 DEG.
    G   = 50.0 LB/FT2

BEAM DATA:

    B  = 0.75 FT.
    D1 = 3.50 FT.
    D2 = 3.00 FT.

EXTERNAL BEAM RESULTS:
    F  = 101949 LB.
    M1 = 151244 LB.FT.
    M2 =   3537 LB.FT.
    S1 = -15089 LB.
    S2 =  -1883 LB.

INTERNAL BEAM RESULTS:
    F  = 150127 LB.
    M1 =  56319 LB.FT.
    S1 = -13203 LB.
```

EXTERNAL-HALF SHELL RESULTS:

POINT	T1	S	M2	T2
0	-4360	-5522	411	-157
1	-8997	-3845	291	-1098
2	-7395	-1890	-28	-1636
3	-3807	-614	-271	-1840
4	-2104	0	-355	-1880

INTERNAL-HALF SHELL RESULTS:

POINT	T1	S	M2	T2
0	6441	-4065	-506	502
1	-1659	-4550	185	-367
2	-6318	-3572	161	-1206
3	-8047	-1888	-71	-1770
4	-8355	0	-185	-1964

```
L= 58        TABLE V- 59         W= 34
**************************************

SHELL DATA:

    L   = 58.0 FT.
    W   = 34.0 FT.
    R   = 27.5 FT.
    T   =  2.5 IN.
    PHI= 37.2 DEG.
    G   = 50.0 LB/FT2

BEAM DATA:

    B  = 0.75 FT.
    D1 = 3.50 FT.
    D2 = 3.00 FT.

EXTERNAL BEAM RESULTS:
    F  = 105549 LB.
    M1 = 155080 LB.FT.
    M2 =   3476 LB.FT.
    S1 = -15541 LB.
    S2 =  -1959 LB.

INTERNAL BEAM RESULTS:
    F  = 153995 LB.
    M1 =  57603 LB.FT.
    S1 = -13539 LB.
```

EXTERNAL-HALF SHELL RESULTS:

POINT	T1	S	M2	T2
0	-4486	-5717	335	-174
1	-9090	-3912	276	-1208
2	-7282	-1844	-21	-1783
3	-3465	-552	-251	-1985
4	-1664	0	-329	-2020

INTERNAL-HALF SHELL RESULTS:

POINT	T1	S	M2	T2
0	6617	-4170	-564	534
1	-1739	-4688	203	-424
2	-6311	-3638	174	-1341
3	-7733	-1895	-81	-1948
4	-7874	0	-206	-2155

```
L= 58        TABLE V- 60         W= 36
**************************************

SHELL DATA:

    L   = 58.0 FT.
    W   = 36.0 FT.
    R   = 30.0 FT.
    T   =  2.5 IN.
    PHI= 36.0 DEG.
    G   = 50.0 LB/FT2

BEAM DATA:

    B  = 0.75 FT.
    D1 = 3.50 FT.
    D2 = 3.00 FT.

EXTERNAL BEAM RESULTS:
    F  = 109065 LB.
    M1 = 158134 LB.FT.
    M2 =   3414 LB.FT.
    S1 = -15945 LB.
    S2 =  -2033 LB.

INTERNAL BEAM RESULTS:
    F  = 157673 LB.
    M1 =  58977 LB.FT.
    S1 = -13860 LB.
```

EXTERNAL-HALF SHELL RESULTS:

POINT	T1	S	M2	T2
0	-4543	-5907	263	-189
1	-9162	-3981	263	-1319
2	-7185	-1800	-15	-1931
3	-3158	-488	-232	-2131
4	-1266	0	-305	-2159

INTERNAL-HALF SHELL RESULTS:

POINT	T1	S	M2	T2
0	6782	-4270	-621	562
1	-1858	-4816	222	-487
2	-6329	-3681	188	-1480
3	-7395	-1881	-90	-2125
4	-7344	0	-225	-2342

L= 60 TABLE V- 61 W= 26

SHELL DATA:

L = 60.0 FT.
W = 26.0 FT.
R = 20.0 FT.
T = 2.5 IN.
PHI= 39.1 DEG.
G = 50.0 LB/FT2

BEAM DATA:

B = 0.75 FT.
D1 = 3.50 FT.
D2 = 3.00 FT.

EXTERNAL BEAM RESULTS:
F = 102932 LB.
M1 = 139724 LB.FT.
M2 = 2965 LB.FT.
S1 = -14044 LB.
S2 = -1872 LB.

INTERNAL BEAM RESULTS:
F = 162830 LB.
M1 = 67733 LB.FT.
S1 = -14185 LB.

EXTERNAL-HALF SHELL RESULTS:

POINT	T1	S	M2	T2
0	-3526	-5386	645	-103
1	-8608	-4214	371	-853
2	-8833	-2603	-36	-1373
3	-7138	-1166	-355	-1651
4	-6217	0	-472	-1734

INTERNAL-HALF SHELL RESULTS:

POINT	T1	S	M2	T2
0	6340	-4262	-329	407
1	-2092	-4600	132	-285
2	-7680	-3684	121	-951
3	-10645	-2010	-35	-1413
4	-11536	0	-114	-1577

L= 60 TABLE V- 62 W= 28

SHELL DATA:

L = 60.0 FT.
W = 28.0 FT.
R = 22.5 FT.
T = 2.5 IN.
PHI= 37.3 DEG.
G = 50.0 LB/FT2

BEAM DATA:

B = 0.75 FT.
D1 = 3.50 FT.
D2 = 3.00 FT.

EXTERNAL BEAM RESULTS:
F = 106314 LB.
M1 = 147091 LB.FT.
M2 = 3044 LB.FT.
S1 = -14653 LB.
S2 = -1933 LB.

INTERNAL BEAM RESULTS:
F = 166920 LB.
M1 = 68219 LB.FT.
S1 = -14481 LB.

EXTERNAL-HALF SHELL RESULTS:

POINT	T1	S	M2	T2
0	-3923	-5566	569	-125
1	-8805	-4245	354	-955
2	-8559	-2523	-30	-1510
3	-6402	-1087	-337	-1793
4	-5291	0	-450	-1875

INTERNAL-HALF SHELL RESULTS:

POINT	T1	S	M2	T2
0	6618	-4369	-384	444
1	-1961	-4767	148	-331
2	-7492	-3814	133	-1078
3	-10284	-2073	-47	-1594
4	-11084	0	-137	-1776

L= 60 TABLE V- 63 W= 30

SHELL DATA:

L = 60.0 FT.
W = 30.0 FT.
R = 25.0 FT.
T = 2.5 IN.
PHI= 35.8 DEG.
G = 50.0 LB/FT2

BEAM DATA:

B = 0.75 FT.
D1 = 3.50 FT.
D2 = 3.00 FT.

EXTERNAL BEAM RESULTS:
F = 109722 LB.
M1 = 153413 LB.FT.
M2 = 3102 LB.FT.
S1 = -15207 LB.
S2 = -1996 LB.

INTERNAL BEAM RESULTS:
F = 170379 LB.
M1 = 68469 LB.FT.
S1 = -14722 LB.

EXTERNAL-HALF SHELL RESULTS:

POINT	T1	S	M2	T2
0	-4231	-5745	492	-145
1	-8982	-4278	338	-1057
2	-8329	-2443	-23	-1648
3	-5748	-1003	-318	-1933
4	-4461	0	-426	-2012

INTERNAL-HALF SHELL RESULTS:

POINT	T1	S	M2	T2
0	6868	-4460	-442	481
1	-1867	-4916	165	-378
2	-7321	-3921	147	-1206
3	-9902	-2120	-58	-1775
4	-10590	0	-160	-1975

L= 60 TABLE V- 64 W= 32

SHELL DATA:

```
L   = 60.0 FT.
W   = 32.0 FT.
R   = 25.0 FT.
T   =  2.5 IN.
PHI= 38.7 DEG.
G   = 50.0 LB/FT2
```

BEAM DATA:

```
B  = 0.75 FT.
D1 = 3.50 FT.
D2 = 3.00 FT.
```

EXTERNAL BEAM RESULTS:

```
F  = 108972 LB.
M1 = 158412 LB.FT.
M2 =   3689 LB.FT.
S1 = -15420 LB.
S2 =  -1951 LB.
```

INTERNAL BEAM RESULTS:

```
F  = 159892 LB.
M1 =  58952 LB.FT.
S1 = -13539 LB.
```

EXTERNAL-HALF SHELL RESULTS:

POINT	T1	S	M2	T2
0	-4403	-5705	474	-151
1	-9448	-4022	321	-1097
2	-7910	-2021	-32	-1648
3	-4240	-682	-301	-1864
4	-2487	0	-394	-1910

INTERNAL-HALF SHELL RESULTS:

POINT	T1	S	M2	T2
0	6960	-4185	-511	511
1	-1632	-4712	188	-357
2	-6686	-3728	165	-1200
3	-8699	-1985	-71	-1772
4	-9113	0	-187	-1970

L= 60 TABLE V- 65 W= 34

SHELL DATA:

```
L   = 60.0 FT.
W   = 34.0 FT.
R   = 27.5 FT.
T   =  2.5 IN.
PHI= 37.2 DEG.
G   = 50.0 LB/FT2
```

BEAM DATA:

```
B  = 0.75 FT.
D1 = 3.50 FT.
D2 = 3.00 FT.
```

EXTERNAL BEAM RESULTS:

```
F  = 112788 LB.
M1 = 162918 LB.FT.
M2 =   3644 LB.FT.
S1 = -15906 LB.
S2 =  -2027 LB.
```

INTERNAL BEAM RESULTS:

```
F  = 163843 LB.
M1 =  60206 LB.FT.
S1 = -13864 LB.
```

EXTERNAL-HALF SHELL RESULTS:

POINT	T1	S	M2	T2
0	-4567	-5905	395	-169
1	-9569	-4086	304	-1206
2	-7788	-1967	-25	-1793
3	-3849	-613	-279	-2008
4	-1982	0	-367	-2048

INTERNAL-HALF SHELL RESULTS:

POINT	T1	S	M2	T2
0	7152	-4289	-571	544
1	-1692	-4856	207	-411
2	-6663	-3803	179	-1334
3	-8376	-1999	-81	-1951
4	-8627	0	-209	-2163

L= 60 TABLE V- 66 W= 36

SHELL DATA:

```
L   = 60.0 FT.
W   = 36.0 FT.
R   = 30.0 FT.
T   =  2.5 IN.
PHI= 36.0 DEG.
G   = 50.0 LB/FT2
```

BEAM DATA:

```
B  = 0.75 FT.
D1 = 3.50 FT.
D2 = 3.00 FT.
```

EXTERNAL BEAM RESULTS:

```
F  = 116524 LB.
M1 = 166559 LB.FT.
M2 =   3594 LB.FT.
S1 = -16342 LB.
S2 =  -2103 LB.
```

INTERNAL BEAM RESULTS:

```
F  = 167563 LB.
M1 =  61416 LB.FT.
S1 = -14172 LB.
```

EXTERNAL-HALF SHELL RESULTS:

POINT	T1	S	M2	T2
0	-4656	-6101	320	-185
1	-9667	-4153	289	-1318
2	-7682	-1915	-19	-1941
3	-3499	-543	-260	-2153
4	-1526	0	-341	-2186

INTERNAL-HALF SHELL RESULTS:

POINT	T1	S	M2	T2
0	7330	-4386	-631	575
1	-1794	-4987	226	-470
2	-6665	-3856	193	-1471
3	-8029	-1993	-91	-2129
4	-8091	0	-230	-2351

```
L= 62       TABLE V- 67         W= 26
*************************************

SHELL DATA:

    L   = 62.0 FT.
    W   = 26.0 FT.
    R   = 20.0 FT.
    T   =  2.5 IN.
    PHI= 39.1 DEG.
    G   = 50.0 LB/FT2

BEAM DATA:

    B   = 0.75 FT.
    D1  = 3.75 FT.
    D2  = 3.00 FT.

EXTERNAL BEAM RESULTS:
    F   = 111858 LB.
    M1  = 189628 LB.FT.
    M2  =   3454 LB.FT.
    S1  = -15975 LB.
    S2  =  -1957 LB.

INTERNAL BEAM RESULTS:
    F   = 185991 LB.
    M1  =  93245 LB.FT.
    S1  = -15306 LB.
```

EXTERNAL-HALF SHELL RESULTS:

POINT	T1	S	M2	T2
0	-3151	-5667	769	-88
1	-9121	-4509	425	-861
2	-9682	-2825	-46	-1407
3	-8033	-1284	-409	-1704
4	-7098	0	-542	-1794

INTERNAL-HALF SHELL RESULTS:

POINT	T1	S	M2	T2
0	7146	-4712	-357	537
1	-2378	-5079	146	-211
2	-8739	-4072	134	-929
3	-12158	-2225	-37	-1427
4	-13199	0	-123	-1604

```
L= 62       TABLE V- 68         W= 28
*************************************

SHELL DATA:

    L   = 62.0 FT.
    W   = 28.0 FT.
    R   = 22.5 FT.
    T   =  2.5 IN.
    PHI= 37.3 DEG.
    G   = 50.0 LB/FT2

BEAM DATA:

    B   = 0.75 FT.
    D1  = 3.75 FT.
    D2  = 3.00 FT.

EXTERNAL BEAM RESULTS:
    F   = 114962 LB.
    M1  = 199594 LB.FT.
    M2  =   3557 LB.FT.
    S1  = -16659 LB.
    S2  =  -2010 LB.

INTERNAL BEAM RESULTS:
    F   = 190180 LB.
    M1  =  93492 LB.FT.
    S1  = -15560 LB.
```

EXTERNAL-HALF SHELL RESULTS:

POINT	T1	S	M2	T2
0	-3569	-5825	688	-112
1	-9314	-4521	405	-962
2	-9344	-2728	-40	-1542
3	-7177	-1191	-389	-1843
4	-6029	0	-517	-1931

INTERNAL-HALF SHELL RESULTS:

POINT	T1	S	M2	T2
0	7439	-4818	-416	583
1	-2212	-5251	162	-252
2	-8494	-4207	146	-1054
3	-11725	-2291	-50	-1610
4	-12669	0	-148	-1806

```
L= 62       TABLE V- 69         W= 30
*************************************

SHELL DATA:

    L   = 62.0 FT.
    W   = 30.0 FT.
    R   = 25.0 FT.
    T   =  2.5 IN.
    PHI= 35.8 DEG.
    G   = 50.0 LB/FT2

BEAM DATA:

    B   = 0.75 FT.
    D1  = 3.75 FT.
    D2  = 3.00 FT.

EXTERNAL BEAM RESULTS:
    F   = 118178 LB.
    M1  = 208241 LB.FT.
    M2  =   3635 LB.FT.
    S1  = -17281 LB.
    S2  =  -2066 LB.

INTERNAL BEAM RESULTS:
    F   = 193634 LB.
    M1  =  93412 LB.FT.
    S1  = -15755 LB.
```

EXTERNAL-HALF SHELL RESULTS:

POINT	T1	S	M2	T2
0	-3899	-5988	604	-133
1	-9496	-4538	386	-1064
2	-9063	-2631	-32	-1677
3	-6420	-1096	-367	-1980
4	-5075	0	-490	-2064

INTERNAL-HALF SHELL RESULTS:

POINT	T1	S	M2	T2
0	7700	-4905	-477	629
1	-2085	-5402	180	-293
2	-8270	-4319	160	-1180
3	-11273	-2341	-62	-1792
4	-12098	0	-173	-2006

```
L= 62        TABLE V- 70         W= 32
*************************************

SHELL DATA:

     L   = 62.0 FT.
     W   = 32.0 FT.
     R   = 25.0 FT.
     T   =  2.5 IN.
     PHI= 38.7 DEG.
     G   = 50.0 LB/FT2

BEAM DATA:

     B  = 0.75 FT.
     D1 = 3.75 FT.
     D2 = 3.00 FT.

EXTERNAL BEAM RESULTS:
     F  = 116831 LB.
     M1 = 214746 LB.FT.
     M2 =   4324 LB.FT.
     S1 = -17534 LB.
     S2 =  -2006 LB.

INTERNAL BEAM RESULTS:
     F  = 181257 LB.
     M1 =  79806 LB.FT.
     S1 = -14362 LB.

EXTERNAL-HALF SHELL RESULTS:

POINT     T1      S      M2      T2

  0     -4053   -5910   580    -139
  1     -9987   -4253   367   -1100
  2     -8589   -2177   -41   -1672
  3     -4746    -754  -348   -1903
  4     -2886       0  -456   -1954

INTERNAL-HALF SHELL RESULTS:

POINT     T1      S      M2      T2

  0      7751   -4592  -553     657
  1     -1797   -5163   205    -272
  2     -7508   -4096   180   -1174
  3     -9892   -2192   -76   -1789
  4    -10423       0  -203   -2004
```

```
L= 62        TABLE V- 71         W= 34
*************************************

SHELL DATA:

     L   = 62.0 FT.
     W   = 34.0 FT.
     R   = 27.5 FT.
     T   =  2.5 IN.
     PHI= 37.2 DEG.
     G   = 50.0 LB/FT2

BEAM DATA:

     B  = 0.75 FT.
     D1 = 3.75 FT.
     D2 = 3.00 FT.

EXTERNAL BEAM RESULTS:
     F  = 120680 LB.
     M1 = 221122 LB.FT.
     M2 =   4285 LB.FT.
     S1 = -18077 LB.
     S2 =  -2080 LB.

INTERNAL BEAM RESULTS:
     F  = 185239 LB.
     M1 =  81156 LB.FT.
     S1 = -14658 LB.

EXTERNAL-HALF SHELL RESULTS:

POINT     T1      S      M2      T2

  0     -4245   -6114   494    -159
  1    -10128   -4309   347   -1209
  2     -8443   -2110   -34   -1816
  3     -4289    -674  -324   -2044
  4     -2298       0  -424   -2089

INTERNAL-HALF SHELL RESULTS:

POINT     T1      S      M2      T2

  0      7950   -4693  -616     698
  1     -1839   -5309   225    -321
  2     -7451   -4179   195   -1306
  3     -9519   -2210   -88   -1969
  4     -9881       0  -227   -2197
```

```
L= 62        TABLE V- 72         W= 36
*************************************

SHELL DATA:

     L   = 62.0 FT.
     W   = 36.0 FT.
     R   = 30.0 FT.
     T   =  2.5 IN.
     PHI= 36.0 DEG.
     G   = 50.0 LB/FT2

BEAM DATA:

     B  = 0.75 FT.
     D1 = 3.75 FT.
     D2 = 3.00 FT.

EXTERNAL BEAM RESULTS:
     F  = 124508 LB.
     M1 = 226368 LB.FT.
     M2 =   4240 LB.FT.
     S1 = -18562 LB.
     S2 =  -2155 LB.

INTERNAL BEAM RESULTS:
     F  = 188933 LB.
     M1 =  82439 LB.FT.
     S1 = -14935 LB.

EXTERNAL-HALF SHELL RESULTS:

POINT     T1      S      M2      T2

  0     -4361   -6308   412    -175
  1    -10248   -4368   329   -1321
  2     -8320   -2046   -27   -1962
  3     -3882    -595  -301   -2186
  4     -1770       0  -395   -2223

INTERNAL-HALF SHELL RESULTS:

POINT     T1      S      M2      T2

  0      8132   -4786  -680     736
  1     -1925   -5442   245    -375
  2     -7421   -4235   210   -1441
  3     -9123   -2207   -98   -2148
  4     -9289       0  -250   -2388
```

L= 64 TABLE V- 73 W= 28

SHELL DATA:

L = 64.0 FT.
W = 28.0 FT.
R = 22.5 FT.
T = 2.5 IN.
PHI= 37.3 DEG.
G = 50.0 LB/FT2

BEAM DATA:

B = 0.75 FT.
D1 = 4.00 FT.
D2 = 3.25 FT.

EXTERNAL BEAM RESULTS:
F = 120055 LB.
M1 = 219918 LB.FT.
M2 = 3388 LB.FT.
S1 = -18154 LB.
S2 = -2049 LB.

INTERNAL BEAM RESULTS:
F = 201956 LB.
M1 = 111021 LB.FT.
S1 = -17823 LB.

EXTERNAL-HALF SHELL RESULTS:

POINT	T1	S	M2	T2
0	-4097	-5893	641	-118
1	-9425	-4583	391	-948
2	-9578	-2820	-32	-1521
3	-7697	-1263	-372	-1826
4	-6685	0	-498	-1918

INTERNAL-HALF SHELL RESULTS:

POINT	T1	S	M2	T2
0	6796	-4956	-400	528
1	-2643	-5281	157	-290
2	-8841	-4202	142	-1067
3	-12076	-2284	-47	-1602
4	-13034	0	-141	-1790

L= 64 TABLE V- 74 W= 30

SHELL DATA:

L = 64.0 FT.
W = 30.0 FT.
R = 25.0 FT.
T = 2.5 IN.
PHI= 35.8 DEG.
G = 50.0 LB/FT2

BEAM DATA:

B = 0.75 FT.
D1 = 4.00 FT.
D2 = 3.25 FT.

EXTERNAL BEAM RESULTS:
F = 123274 LB.
M1 = 228769 LB.FT.
M2 = 3460 LB.FT.
S1 = -18786 LB.
S2 = -2104 LB.

INTERNAL BEAM RESULTS:
F = 205795 LB.
M1 = 110853 LB.FT.
S1 = -18052 LB.

EXTERNAL-HALF SHELL RESULTS:

POINT	T1	S	M2	T2
0	-4410	-6051	557	-139
1	-9564	-4603	373	-1046
2	-9268	-2736	-24	-1654
3	-6936	-1179	-351	-1965
4	-5737	0	-473	-2055

INTERNAL-HALF SHELL RESULTS:

POINT	T1	S	M2	T2
0	7074	-5050	-460	570
1	-2490	-5436	174	-334
2	-8597	-4318	155	-1193
3	-11622	-2337	-59	-1783
4	-12472	0	-166	-1990

L= 64 TABLE V- 75 W= 32

SHELL DATA:

L = 64.0 FT.
W = 32.0 FT.
R = 25.0 FT.
T = 2.5 IN.
PHI= 38.7 DEG.
G = 50.0 LB/FT2

BEAM DATA:

B = 0.75 FT.
D1 = 4.00 FT.
D2 = 3.25 FT.

EXTERNAL BEAM RESULTS:
F = 121607 LB.
M1 = 236069 LB.FT.
M2 = 4147 LB.FT.
S1 = -19043 LB.
S2 = -2040 LB.

INTERNAL BEAM RESULTS:
F = 193332 LB.
M1 = 94997 LB.FT.
S1 = -16511 LB.

EXTERNAL-HALF SHELL RESULTS:

POINT	T1	S	M2	T2
0	-4619	-5969	537	-145
1	-9991	-4312	358	-1079
2	-8726	-2294	-32	-1649
3	-5222	-850	-337	-1894
4	-3522	0	-445	-1954

INTERNAL-HALF SHELL RESULTS:

POINT	T1	S	M2	T2
0	7241	-4745	-536	608
1	-2121	-5214	199	-307
2	-7803	-4122	175	-1184
3	-10261	-2207	-73	-1781
4	-10840	0	-197	-1989

L= 64 TABLE V- 76 W= 34

SHELL DATA:

L = 64.0 FT.
W = 34.0 FT.
R = 27.5 FT.
T = 2.5 IN.
PHI= 37.2 DEG.
G = 50.0 LB/FT2

BEAM DATA:

B = 0.75 FT.
D1 = 4.00 FT.
D2 = 3.25 FT.

EXTERNAL BEAM RESULTS:
F = 125472 LB.
M1 = 242603 LB.FT.
M2 = 4102 LB.FT.
S1 = -19601 LB.
S2 = -2112 LB.

INTERNAL BEAM RESULTS:
F = 197587 LB.
M1 = 96446 LB.FT.
S1 = -16845 LB.

EXTERNAL-HALF SHELL RESULTS:

POINT	T1	S	M2	T2
0	-4801	-6159	450	-163
1	-10105	-4368	339	-1185
2	-8563	-2237	-24	-1790
3	-4761	-780	-313	-2037
4	-2934	0	-416	-2092

INTERNAL-HALF SHELL RESULTS:

POINT	T1	S	M2	T2
0	7442	-4849	-599	646
1	-2139	-5361	219	-358
2	-7724	-4207	191	-1316
3	-9886	-2230	-84	-1961
4	-10307	0	-220	-2184

L= 64 TABLE V- 77 W= 36

SHELL DATA:

L = 64.0 FT.
W = 36.0 FT.
R = 30.0 FT.
T = 2.5 IN.
PHI= 36.0 DEG.
G = 50.0 LB/FT2

BEAM DATA:

B = 0.75 FT.
D1 = 4.00 FT.
D2 = 3.25 FT.

EXTERNAL BEAM RESULTS:
F = 129321 LB.
M1 = 247960 LB.FT.
M2 = 4052 LB.FT.
S1 = -20101 LB.
S2 = -2185 LB.

INTERNAL BEAM RESULTS:
F = 201512 LB.
M1 = 97806 LB.FT.
S1 = -17153 LB.

EXTERNAL-HALF SHELL RESULTS:

POINT	T1	S	M2	T2
0	-4911	-6348	368	-180
1	-10201	-4429	323	-1293
2	-8426	-2182	-16	-1934
3	-4348	-710	-292	-2180
4	-2403	0	-388	-2230

INTERNAL-HALF SHELL RESULTS:

POINT	T1	S	M2	T2
0	7626	-4945	-663	681
1	-2198	-5494	239	-412
2	-7671	-4268	206	-1450
3	-9489	-2233	-95	-2140
4	-9726	0	-243	-2376

L= 64 TABLE V- 78 W= 38

SHELL DATA:

L = 64.0 FT.
W = 38.0 FT.
R = 32.5 FT.
T = 2.5 IN.
PHI= 35.0 DEG.
G = 50.0 LB/FT2

BEAM DATA:

B = 0.75 FT.
D1 = 4.00 FT.
D2 = 3.25 FT.

EXTERNAL BEAM RESULTS:
F = 133142 LB.
M1 = 252323 LB.FT.
M2 = 3998 LB.FT.
S1 = -20550 LB.
S2 = -2258 LB.

INTERNAL BEAM RESULTS:
F = 205277 LB.
M1 = 99211 LB.FT.
S1 = -17454 LB.

EXTERNAL-HALF SHELL RESULTS:

POINT	T1	S	M2	T2
0	-4960	-6535	290	-194
1	-10283	-4494	309	-1402
2	-8310	-2129	-9	-2081
3	-3979	-639	-272	-2324
4	-1921	0	-363	-2368

INTERNAL-HALF SHELL RESULTS:

POINT	T1	S	M2	T2
0	7796	-5038	-727	714
1	-2303	-5617	260	-471
2	-7649	-4308	221	-1588
3	-9076	-2217	-106	-2318
4	-9103	0	-265	-2563

L= 66 TABLE V- 79 W= 28

SHELL DATA:

L = 66.0 FT.
W = 28.0 FT.
R = 22.5 FT.
T = 2.5 IN.
PHI= 37.3 DEG.
G = 50.0 LB/FT2

BEAM DATA:

B = 0.75 FT.
D1 = 4.00 FT.
D2 = 3.25 FT.

EXTERNAL BEAM RESULTS:
F = 128140 LB.
M1 = 228038 LB.FT.
M2 = 3464 LB.FT.
S1 = -18470 LB.
S2 = -2128 LB.

INTERNAL BEAM RESULTS:
F = 214510 LB.
M1 = 116721 LB.FT.
S1 = -18299 LB.

EXTERNAL-HALF SHELL RESULTS:

POINT	T1	S	M2	T2
0	-4022	-6099	709	-111
1	-9824	-4797	422	-949
2	-10228	-2994	-37	-1537
3	-8454	-1360	-403	-1855
4	-7469	0	-538	-1952

INTERNAL-HALF SHELL RESULTS:

POINT	T1	S	M2	T2
0	7297	-5105	-400	533
1	-2730	-5454	160	-286
2	-9374	-4351	145	-1564
3	-12902	-2370	-45	-1602
4	-13963	0	-140	-1793

L= 66 TABLE V- 80 W= 30

SHELL DATA:

L = 66.0 FT.
W = 30.0 FT.
R = 25.0 FT.
T = 2.5 IN.
PHI= 35.8 DEG.
G = 50.0 LB/FT2

BEAM DATA:

B = 0.75 FT.
D1 = 4.00 FT.
D2 = 3.25 FT.

EXTERNAL BEAM RESULTS:
F = 131440 LB.
M1 = 237889 LB.FT.
M2 = 3562 LB.FT.
S1 = -19136 LB.
S2 = -2181 LB.

INTERNAL BEAM RESULTS:
F = 218491 LB.
M1 = 116347 LB.FT.
S1 = -18520 LB.

EXTERNAL-HALF SHELL RESULTS:

POINT	T1	S	M2	T2
0	-4380	-6256	626	-132
1	-9983	-4812	404	-1048
2	-9887	-2901	-29	-1670
3	-7612	-1270	-383	-1994
4	-6421	0	-514	-2089

INTERNAL-HALF SHELL RESULTS:

POINT	T1	S	M2	T2
0	7598	-5200	-461	576
1	-2551	-5614	177	-328
2	-9106	-4475	158	-1190
3	-12429	-2430	-58	-1784
4	-13385	0	-166	-1993

L= 66 TABLE V- 81 W= 32

SHELL DATA:

L = 66.0 FT.
W = 32.0 FT.
R = 25.0 FT.
T = 2.5 IN.
PHI= 38.7 DEG.
G = 50.0 LB/FT2

BEAM DATA:

B = 0.75 FT.
D1 = 4.00 FT.
D2 = 3.25 FT.

EXTERNAL BEAM RESULTS:
F = 129462 LB.
M1 = 246045 LB.FT.
M2 = 4298 LB.FT.
S1 = -19407 LB.
S2 = -2112 LB.

INTERNAL BEAM RESULTS:
F = 205003 LB.
M1 = 99330 LB.FT.
S1 = -16910 LB.

EXTERNAL-HALF SHELL RESULTS:

POINT	T1	S	M2	T2
0	-4639	-6162	607	-138
1	-10443	-4501	390	-1079
2	-9291	-2438	-37	-1662
3	-5754	-927	-369	-1921
4	-4022	0	-488	-1985

INTERNAL-HALF SHELL RESULTS:

POINT	T1	S	M2	T2
0	7769	-4879	-540	615
1	-2131	-5384	202	-299
2	-8238	-4279	179	-1180
3	-10993	-2301	-73	-1783
4	-11681	0	-198	-1994

L= 66 TABLE V- 82 W= 34

****

SHELL DATA:

L = 66.0 FT.
W = 34.0 FT.
R = 27.5 FT.
T = 2.5 IN.
PHI= 37.2 DEG.
G = 50.0 LB/FT2

BEAM DATA:

B = 0.75 FT.
D1 = 4.00 FT.
D2 = 3.25 FT.

EXTERNAL BEAM RESULTS:
F = 133494 LB.
M1 = 253533 LB.FT.
M2 = 4272 LB.FT.
S1 = -20004 LB.
S2 = -2184 LB.

INTERNAL BEAM RESULTS:
F = 209364 LB.
M1 = 100634 LB.FT.
S1 = -17233 LB.

EXTERNAL-HALF SHELL RESULTS:

POINT	T1	S	M2	T2
0	-4861	-6354	518	-158
1	-10583	-4553	370	-1185
2	-9113	-2372	-29	-1803
3	-5234	-850	-345	-2062
4	-3359	0	-457	-2122

INTERNAL-HALF SHELL RESULTS:

POINT	T1	S	M2	T2
0	7987	-4982	-604	654
1	-2127	-5536	222	-347
2	-8141	-4372	194	-1310
3	-10607	-2331	-84	-1963
4	-11141	0	-223	-2190

L= 66 TABLE V- 83 W= 36

****

SHELL DATA:

L = 66.0 FT.
W = 36.0 FT.
R = 30.0 FT.
T = 2.5 IN.
PHI= 36.0 DEG.
G = 50.0 LB/FT2

BEAM DATA:

B = 0.75 FT.
D1 = 4.00 FT.
D2 = 3.25 FT.

EXTERNAL BEAM RESULTS:
F = 137522 LB.
M1 = 259741 LB.FT.
M2 = 4238 LB.FT.
S1 = -20540 LB.
S2 = -2257 LB.

INTERNAL BEAM RESULTS:
F = 213346 LB.
M1 = 101829 LB.FT.
S1 = -17528 LB.

EXTERNAL-HALF SHELL RESULTS:

POINT	T1	S	M2	T2
0	-5005	-6546	432	-175
1	-10703	-4610	352	-1292
2	-8962	-2309	-21	-1946
3	-4771	-774	-322	-2204
4	-2761	0	-428	-2258

INTERNAL-HALF SHELL RESULTS:

POINT	T1	S	M2	T2
0	8186	-5077	-670	692
1	-2166	-5674	243	-399
2	-8068	-4442	210	-1443
3	-10198	-2342	-96	-2143
4	-10553	0	-247	-2383

L= 66 TABLE V- 84 W= 38

****

SHELL DATA:

L = 66.0 FT.
W = 38.0 FT.
R = 32.5 FT.
T = 2.5 IN.
PHI= 35.0 DEG.
G = 50.0 LB/FT2

BEAM DATA:

B = 0.75 FT.
D1 = 4.00 FT.
D2 = 3.25 FT.

EXTERNAL BEAM RESULTS:
F = 141533 LB.
M1 = 264854 LB.FT.
M2 = 4196 LB.FT.
S1 = -21022 LB.
S2 = -2330 LB.

INTERNAL BEAM RESULTS:
F = 217128 LB.
M1 = 103055 LB.FT.
S1 = -17812 LB.

EXTERNAL-HALF SHELL RESULTS:

POINT	T1	S	M2	T2
0	-5083	-6736	350	-190
1	-10806	-4671	337	-1402
2	-8834	-2248	-14	-2091
3	-4356	-697	-300	-2346
4	-2221	0	-401	-2395

INTERNAL-HALF SHELL RESULTS:

POINT	T1	S	M2	T2
0	8369	-5167	-736	727
1	-2251	-5800	265	-455
2	-8027	-4491	226	-1579
3	-9773	-2333	-107	-2321
4	-9921	0	-270	-2573

L= 68 TABLE V- 85 W= 30
****

SHELL DATA:

L = 68.0 FT.
W = 30.0 FT.
R = 25.0 FT.
T = 2.5 IN.
PHI= 35.8 DEG.
G = 50.0 LB/FT2

BEAM DATA:

B = 0.75 FT.
D1 = 4.25 FT.
D2 = 3.50 FT.

EXTERNAL BEAM RESULTS:
F = 136806 LB.
M1 = 260364 LB.FT.
M2 = 3393 LB.FT.
S1 = -20712 LB.
S2 = -2218 LB.

INTERNAL BEAM RESULTS:
F = 230914 LB.
M1 = 136506 LB.FT.
S1 = -20957 LB.

EXTERNAL-HALF SHELL RESULTS:

POINT	T1	S	M2	T2
0	-4829	-6320	580	-137
1	-10075	-4879	390	-1033
2	-10129	-3000	-22	-1649
3	-8150	-1346	-366	-1978
4	-7095	0	-495	-2077

INTERNAL-HALF SHELL RESULTS:

POINT	T1	S	M2	T2
0	6942	-5334	-444	519
1	-2980	-5640	171	-368
2	-9444	-4467	154	-1203
3	-12770	-2421	-54	-1776
4	-13741	0	-159	-1977

L= 68 TABLE V- 86 W= 32
****

SHELL DATA:

L = 68.0 FT.
W = 32.0 FT.
R = 25.0 FT.
T = 2.5 IN.
PHI= 38.7 DEG.
G = 50.0 LB/FT2

BEAM DATA:

B = 0.75 FT.
D1 = 4.25 FT.
D2 = 3.50 FT.

EXTERNAL BEAM RESULTS:
F = 134499 LB.
M1 = 269342 LB.FT.
M2 = 4129 LB.FT.
S1 = -20980 LB.
S2 = -2144 LB.

INTERNAL BEAM RESULTS:
F = 217477 LB.
M1 = 116949 LB.FT.
S1 = -19202 LB.

EXTERNAL-HALF SHELL RESULTS:

POINT	T1	S	M2	T2
0	-5131	-6213	566	-143
1	-10469	-4564	380	-1061
2	-9464	-2549	-29	-1641
3	-6252	-1017	-357	-1911
4	-4672	0	-476	-1982

INTERNAL-HALF SHELL RESULTS:

POINT	T1	S	M2	T2
0	7231	-5023	-523	566
1	-2484	-5429	197	-333
2	-8549	-4294	174	-1190
3	-11357	-2310	-70	-1776
4	-12083	0	-191	-1980

L= 68 TABLE V- 87 W= 34
****

SHELL DATA:

L = 68.0 FT.
W = 34.0 FT.
R = 27.5 FT.
T = 2.5 IN.
PHI= 37.2 DEG.
G = 50.0 LB/FT2

BEAM DATA:

B = 0.75 FT.
D1 = 4.25 FT.
D2 = 3.50 FT.

EXTERNAL BEAM RESULTS:
F = 138542 LB.
M1 = 277001 LB.FT.
M2 = 4098 LB.FT.
S1 = -21592 LB.
S2 = -2215 LB.

INTERNAL BEAM RESULTS:
F = 222143 LB.
M1 = 118327 LB.FT.
S1 = -19564 LB.

EXTERNAL-HALF SHELL RESULTS:

POINT	T1	S	M2	T2
0	-5340	-6400	476	-162
1	-10579	-4618	362	-1164
2	-9267	-2493	-20	-1780
3	-5728	-948	-334	-2054
4	-4010	0	-447	-2122

INTERNAL-HALF SHELL RESULTS:

POINT	T1	S	M2	T2
0	7454	-5131	-587	603
1	-2457	-5582	216	-383
2	-8431	-4391	190	-1321
3	-10970	-2344	-81	-1956
4	-11552	0	-216	-2176

```
L= 68        TABLE V- 88          W= 36
*************************************

SHELL DATA:

     L   = 68.0 FT.
     W   = 36.0 FT.
     R   = 30.0 FT.
     T   =  2.5 IN.
     PHI= 36.0 DEG.
     G   = 50.0 LB/FT2

BEAM DATA:

     B  = 0.75 FT.
     D1 = 4.25 FT.
     D2 = 3.50 FT.

EXTERNAL BEAM RESULTS:
     F  = 142584 LB.
     M1 = 283336 LB.FT.
     M2 =   4059 LB.FT.
     S1 = -22142 LB.
     S2 =  -2286 LB.

INTERNAL BEAM RESULTS:
     F  = 226389 LB.
     M1 = 119568 LB.FT.
     S1 = -19892 LB.

EXTERNAL-HALF SHELL RESULTS:

POINT      T1       S      M2      T2

  0     -5474   -6587     390    -179
  1    -10674   -4676     345   -1268
  2     -9099   -2439     -11   -1921
  3     -5258    -880    -312   -2197
  4     -3410       0    -420   -2262

INTERNAL-HALF SHELL RESULTS:

POINT      T1       S      M2      T2

  0      7657   -5229    -651     638
  1     -2470   -5721     237    -436
  2     -8337   -4466     206   -1453
  3    -10560   -2360     -93   -2135
  4    -10974       0    -240   -2370
```

```
L= 68        TABLE V- 89          W= 38
*************************************

SHELL DATA:

     L   = 68.0 FT.
     W   = 38.0 FT.
     R   = 32.5 FT.
     T   =  2.5 IN.
     PHI= 35.0 DEG.
     G   = 50.0 LB/FT2

BEAM DATA:

     B  = 0.75 FT.
     D1 = 4.25 FT.
     D2 = 3.50 FT.

EXTERNAL BEAM RESULTS:
     F  = 146614 LB.
     M1 = 288537 LB.FT.
     M2 =   4014 LB.FT.
     S1 = -22638 LB.
     S2 =  -2358 LB.

INTERNAL BEAM RESULTS:
     F  = 230397 LB.
     M1 = 120835 LB.FT.
     S1 = -20206 LB.

EXTERNAL-HALF SHELL RESULTS:

POINT      T1       S      M2      T2

  0     -5545   -6773     307    -193
  1    -10755   -4739     332   -1374
  2     -8958   -2387      -3   -2065
  3     -4837    -811    -292   -2341
  4     -2866       0    -395   -2402

INTERNAL-HALF SHELL RESULTS:

POINT      T1       S      M2      T2

  0      7844   -5322    -717     670
  1     -2529   -5848     258    -493
  2     -8274   -4521     222   -1589
  3    -10135   -2358    -104   -2313
  4    -10355       0    -263   -2560
```

```
L= 68        TABLE V- 90          W= 40
*************************************

SHELL DATA:

     L   = 68.0 FT.
     W   = 40.0 FT.
     R   = 32.5 FT.
     T   =  2.5 IN.
     PHI= 37.1 DEG.
     G   = 50.0 LB/FT2

BEAM DATA:

     B  = 0.75 FT.
     D1 = 4.25 FT.
     D2 = 3.50 FT.

EXTERNAL BEAM RESULTS:
     F  = 146636 LB.
     M1 = 291911 LB.FT.
     M2 =   4465 LB.FT.
     S1 = -22796 LB.
     S2 =  -2338 LB.

INTERNAL BEAM RESULTS:
     F  = 220907 LB.
     M1 = 109719 LB.FT.
     S1 = -19091 LB.

EXTERNAL-HALF SHELL RESULTS:

POINT      T1       S      M2      T2

  0     -5509   -6774     289    -200
  1    -11110   -4546     310   -1425
  2     -8667   -2027     -16   -2081
  3     -3709    -530    -270   -2286
  4     -1380       0    -354   -2312

INTERNAL-HALF SHELL RESULTS:

POINT      T1       S      M2      T2

  0      7887   -5102    -788     685
  1     -2577   -5655     283    -495
  2     -7904   -4272     238   -1597
  3     -8977   -2160    -116   -2304
  4     -8796       0    -287   -2539
```

L= 70 TABLE V- 91 W= 30

SHELL DATA:

L = 70.0 FT.
W = 30.0 FT.
R = 25.0 FT.
T = 2.5 IN.
PHI= 35.8 DEG.
G = 50.0 LB/FT2

BEAM DATA:

B = 0.75 FT.
D1 = 4.25 FT.
D2 = 3.50 FT.

EXTERNAL BEAM RESULTS:
F = 145401 LB.
M1 = 270030 LB.FT.
M2 = 3478 LB.FT.
S1 = -21083 LB.
S2 = -2296 LB.

INTERNAL BEAM RESULTS:
F = 244403 LB.
M1 = 143077 LB.FT.
S1 = -21484 LB.

EXTERNAL-HALF SHELL RESULTS:

POINT	T1	S	M2	T2
0	-4800	-6525	645	-132
1	-10493	-5088	419	-1035
2	-10772	-3167	-27	-1664
3	-8880	-1438	-396	-2005
4	-7846	0	-534	-2109

INTERNAL-HALF SHELL RESULTS:

POINT	T1	S	M2	T2
0	7431	-5484	-446	524
1	-3073	-5814	174	-363
2	-9979	-4618	157	-1200
3	-13595	-2509	-53	-1777
4	-14669	0	-158	-1980

L= 70 TABLE V- 92 W= 32

SHELL DATA:

L = 70.0 FT.
W = 32.0 FT.
R = 25.0 FT.
T = 2.5 IN.
PHI= 38.7 DEG.
G = 50.0 LB/FT2

BEAM DATA:

B = 0.75 FT.
D1 = 4.25 FT.
D2 = 3.50 FT.

EXTERNAL BEAM RESULTS:
F = 142740 LB.
M1 = 279860 LB.FT.
M2 = 4261 LB.FT.
S1 = -21361 LB.
S2 = -2217 LB.

INTERNAL BEAM RESULTS:
F = 229956 LB.
M1 = 122193 LB.FT.
S1 = -19657 LB.

EXTERNAL-HALF SHELL RESULTS:

POINT	T1	S	M2	T2
0	-5151	-6406	634	-137
1	-10910	-4754	411	-1061
2	-10047	-2696	-34	-1654
3	-6837	-1096	-389	-1936
4	-5241	0	-518	-2013

INTERNAL-HALF SHELL RESULTS:

POINT	T1	S	M2	T2
0	7734	-5160	-526	572
1	-2522	-5596	199	-327
2	-9011	-4444	178	-1187
3	-12109	-2399	-69	-1777
4	-12940	0	-192	-1984

L= 70 TABLE V- 93 W= 34

SHELL DATA:

L = 70.0 FT.
W = 34.0 FT.
R = 27.5 FT.
T = 2.5 IN.
PHI= 37.2 DEG.
G = 50.0 LB/FT2

BEAM DATA:

B = 0.75 FT.
D1 = 4.25 FT.
D2 = 3.50 FT.

EXTERNAL BEAM RESULTS:
F = 146936 LB.
M1 = 288544 LB.FT.
M2 = 4249 LB.FT.
S1 = -22010 LB.
S2 = -2287 LB.

INTERNAL BEAM RESULTS:
F = 234758 LB.
M1 = 123413 LB.FT.
S1 = -20010 LB.

EXTERNAL-HALF SHELL RESULTS:

POINT	T1	S	M2	T2
0	-5401	-6594	542	-156
1	-11045	-4804	392	-1164
2	-9831	-2632	-25	-1793
3	-6252	-1022	-364	-2078
4	-4502	0	-488	-2152

INTERNAL-HALF SHELL RESULTS:

POINT	T1	S	M2	T2
0	7975	-5267	-591	610
1	-2474	-5754	219	-375
2	-8874	-4549	193	-1316
3	-11709	-2439	-81	-1957
4	-12401	0	-217	-2181

L= 70 TABLE V- 94 W= 36

SHELL DATA:

L = 70.0 FT.
W = 36.0 FT.
R = 30.0 FT.
T = 2.5 IN.
PHI= 36.0 DEG.
G = 50.0 LB/FT2

BEAM DATA:

B = 0.75 FT.
D1 = 4.25 FT.
D2 = 3.50 FT.

EXTERNAL BEAM RESULTS:
F = 151142 LB.
M1 = 295808 LB.FT.
M2 = 4227 LB.FT.
S1 = -22596 LB.
S2 = -2358 LB.

INTERNAL BEAM RESULTS:
F = 239086 LB.
M1 = 124467 LB.FT.
S1 = -20326 LB.

EXTERNAL-HALF SHELL RESULTS:

POINT	T1	S	M2	T2
0	-5572	-6783	453	-174
1	-11162	-4858	374	-1268
2	-9646	-2571	-16	-1933
3	-5729	-948	-341	-2220
4	-3834	0	-459	-2290

INTERNAL-HALF SHELL RESULTS:

POINT	T1	S	M2	T2
0	8195	-5365	-657	647
1	-2466	-5897	240	-426
2	-8760	-4633	210	-1447
3	-11288	-2462	-93	-2137
4	-11816	0	-242	-2376

L= 70 TABLE V- 95 W= 38

SHELL DATA:

L = 70.0 FT.
W = 38.0 FT.
R = 32.5 FT.
T = 2.5 IN.
PHI= 35.0 DEG.
G = 50.0 LB/FT2

BEAM DATA:

B = 0.75 FT.
D1 = 4.25 FT.
D2 = 3.50 FT.

EXTERNAL BEAM RESULTS:
F = 155346 LB.
M1 = 301839 LB.FT.
M2 = 4195 LB.FT.
S1 = -23127 LB.
S2 = -2430 LB.

INTERNAL BEAM RESULTS:
F = 243133 LB.
M1 = 125531 LB.FT.
S1 = -20624 LB.

EXTERNAL-HALF SHELL RESULTS:

POINT	T1	S	M2	T2
0	-5674	-6971	367	-189
1	-11263	-4916	359	-1374
2	-9491	-2512	-7	-2076
3	-5261	-874	-320	-2363
4	-3229	0	-433	-2428

INTERNAL-HALF SHELL RESULTS:

POINT	T1	S	M2	T2
0	8396	-5455	-725	682
1	-2504	-6029	262	-480
2	-8678	-4696	226	-1581
3	-10851	-2468	-105	-2316
4	-11189	0	-267	-2567

L= 70 TABLE V- 96 W= 40

SHELL DATA:

L = 70.0 FT.
W = 40.0 FT.
R = 32.5 FT.
T = 2.5 IN.
PHI= 37.1 DEG.
G = 50.0 LB/FT2

BEAM DATA:

B = 0.75 FT.
D1 = 4.25 FT.
D2 = 3.50 FT.

EXTERNAL BEAM RESULTS:
F = 155290 LB.
M1 = 305857 LB.FT.
M2 = 4680 LB.FT.
S1 = -23304 LB.
S2 = -2407 LB.

INTERNAL BEAM RESULTS:
F = 232707 LB.
M1 = 113522 LB.FT.
S1 = -19443 LB.

EXTERNAL-HALF SHELL RESULTS:

POINT	T1	S	M2	T2
0	-5664	-6969	346	-196
1	-11656	-4711	336	-1424
2	-9180	-2134	-20	-2089
3	-4041	-579	-297	-2305
4	-1620	0	-390	-2335

INTERNAL-HALF SHELL RESULTS:

POINT	T1	S	M2	T2
0	8431	-5221	-799	698
1	-2519	-5827	288	-479
2	-8255	-4448	244	-1588
3	-9640	-2274	-117	-2308
4	-9579	0	-293	-2549

L= 72 TABLE V- 97 W= 32

****

SHELL DATA:

L = 72.0 FT.
W = 32.0 FT.
R = 25.0 FT.
T = 2.5 IN.
PHI= 38.7 DEG.
G = 50.0 LB/FT2

BEAM DATA:

B = 0.75 FT.
D1 = 4.25 FT.
D2 = 3.50 FT.

EXTERNAL BEAM RESULTS:
F = 151241 LB.
M1 = 290065 LB.FT.
M2 = 4382 LB.FT.
S1 = -21722 LB.
S2 = -2289 LB.

INTERNAL BEAM RESULTS:
F = 242808 LB.
M1 = 127612 LB.FT.
S1 = -20117 LB.

EXTERNAL-HALF SHELL RESULTS:

POINT	T1	S	M2	T2
0	-5141	-6596	704	-130
1	-11342	-4946	444	-1061
2	-10648	-2850	-38	-1667
3	-7462	-1180	-421	-1962
4	-5860	0	-560	-2044

INTERNAL-HALF SHELL RESULTS:

POINT	T1	S	M2	T2
0	8250	-5297	-528	578
1	-2564	-5763	202	-321
2	-9487	-4593	181	-1184
3	-12881	-2488	-68	-1778
4	-13819	0	-192	-1987

L= 72 TABLE V- 98 W= 34

****

SHELL DATA:

L = 72.0 FT.
W = 34.0 FT.
R = 27.5 FT.
T = 2.5 IN.
PHI= 37.2 DEG.
G = 50.0 LB/FT2

BEAM DATA:

B = 0.75 FT.
D1 = 4.25 FT.
D2 = 3.50 FT.

EXTERNAL BEAM RESULTS:
F = 155586 LB.
M1 = 299820 LB.FT.
M2 = 4392 LB.FT.
S1 = -22409 LB.
S2 = -2359 LB.

INTERNAL BEAM RESULTS:
F = 247752 LB.
M1 = 128672 LB.FT.
S1 = -20462 LB.

EXTERNAL-HALF SHELL RESULTS:

POINT	T1	S	M2	T2
0	-5435	-6788	611	-151
1	-11503	-4994	423	-1164
2	-10412	-2778	-30	-1805
3	-6814	-1100	-396	-2103
4	-5040	0	-530	-2182

INTERNAL-HALF SHELL RESULTS:

POINT	T1	S	M2	T2
0	8510	-5405	-594	617
1	-2493	-5926	222	-367
2	-9330	-4706	196	-1312
3	-12469	-2534	-81	-1959
4	-13272	0	-218	-2185

L= 72 TABLE V- 99 W= 36

****

SHELL DATA:

L = 72.0 FT.
W = 36.0 FT.
R = 30.0 FT.
T = 2.5 IN.
PHI= 36.0 DEG.
G = 50.0 LB/FT2

BEAM DATA:

B = 0.75 FT.
D1 = 4.25 FT.
D2 = 3.50 FT.

EXTERNAL BEAM RESULTS:
F = 159953 LB.
M1 = 308069 LB.FT.
M2 = 4388 LB.FT.
S1 = -23032 LB.
S2 = -2430 LB.

INTERNAL BEAM RESULTS:
F = 252166 LB.
M1 = 129537 LB.FT.
S1 = -20766 LB.

EXTERNAL-HALF SHELL RESULTS:

POINT	T1	S	M2	T2
0	-5644	-6979	519	-168
1	-11644	-5043	404	-1268
2	-10210	-2708	-21	-1945
3	-6234	-1020	-372	-2244
4	-4299	0	-500	-2319

INTERNAL-HALF SHELL RESULTS:

POINT	T1	S	M2	T2
0	8747	-5501	-662	655
1	-2464	-6073	243	-416
2	-9197	-4798	213	-1442
3	-12035	-2563	-93	-2139
4	-12678	0	-245	-2381

L= 72 TABLE V-100 W= 38

SHELL DATA:

L = 72.0 FT.
W = 38.0 FT.
R = 32.5 FT.
T = 2.5 IN.
PHI= 35.0 DEG.
G = 50.0 LB/FT2

BEAM DATA:

B = 0.75 FT.
D1 = 4.25 FT.
D2 = 3.50 FT.

EXTERNAL BEAM RESULTS:
F = 164331 LB.
M1 = 314988 LB.FT.
M2 = 4371 LB.FT.
S1 = -23597 LB.
S2 = -2502 LB.

INTERNAL BEAM RESULTS:
F = 256253 LB.
M1 = 130393 LB.FT.
S1 = -21050 LB.

EXTERNAL-HALF SHELL RESULTS:

POINT	T1	S	M2	T2
0	-5780	-7170	430	-184
1	-11767	-5098	387	-1374
2	-10040	-2642	-12	-2087
3	-5715	-940	-350	-2386
4	-3629	0	-472	-2455

INTERNAL-HALF SHELL RESULTS:

POINT	T1	S	M2	T2
0	8963	-5590	-731	692
1	-2481	-6206	266	-468
2	-9095	-4866	230	-1575
3	-11586	-2575	-105	-2318
4	-12043	0	-271	-2574

L= 72 TABLE V-101 W= 40

SHELL DATA:

L = 72.0 FT.
W = 40.0 FT.
R = 32.5 FT.
T = 2.5 IN.
PHI= 37.1 DEG.
G = 50.0 LB/FT2

BEAM DATA:

B = 0.75 FT.
D1 = 4.25 FT.
D2 = 3.50 FT.

EXTERNAL BEAM RESULTS:
F = 164179 LB.
M1 = 319746 LB.FT.
M2 = 4891 LB.FT.
S1 = -23794 LB.
S2 = -2477 LB.

INTERNAL BEAM RESULTS:
F = 244865 LB.
M1 = 117462 LB.FT.
S1 = -19803 LB.

EXTERNAL-HALF SHELL RESULTS:

POINT	T1	S	M2	T2
0	-5800	-7163	407	-191
1	-12199	-4880	364	-1423
2	-9708	-2245	-25	-2099
3	-4398	-632	-326	-2325
4	-1889	0	-428	-2359

INTERNAL-HALF SHELL RESULTS:

POINT	T1	S	M2	T2
0	8991	-5342	-809	710
1	-2461	-6000	292	-463
2	-8618	-4622	249	-1579
3	-10321	-2386	-118	-2311
4	-10382	0	-298	-2558

L= 72 TABLE V-102 W= 42

SHELL DATA:

L = 72.0 FT.
W = 42.0 FT.
R = 35.0 FT.
T = 2.5 IN.
PHI= 36.1 DEG.
G = 50.0 LB/FT2

BEAM DATA:

B = 0.75 FT.
D1 = 4.25 FT.
D2 = 3.50 FT.

EXTERNAL BEAM RESULTS:
F = 168813 LB.
M1 = 324483 LB.FT.
M2 = 4802 LB.FT.
S1 = -24280 LB.
S2 = -2556 LB.

INTERNAL BEAM RESULTS:
F = 249854 LB.
M1 = 119819 LB.FT.
S1 = -20206 LB.

EXTERNAL-HALF SHELL RESULTS:

POINT	T1	S	M2	T2
0	-5828	-7365	325	-205
1	-12271	-4958	350	-1535
2	-9606	-2200	-18	-2248
3	-4064	-565	-305	-2472
4	-1453	0	-401	-2499

INTERNAL-HALF SHELL RESULTS:

POINT	T1	S	M2	T2
0	9176	-5450	-877	739
1	-2622	-6135	315	-527
2	-8654	-4659	265	-1720
3	-9912	-2361	-129	-2488
4	-9732	0	-320	-2743

L= 74 TABLE V-103 W= 32

SHELL DATA:

```
L   = 74.0 FT.
W   = 32.0 FT.
R   = 25.0 FT.
T   =  2.5 IN.
PHI= 38.7 DEG.
G   = 50.0 LB/FT2
```

BEAM DATA:

```
B  = 0.75 FT.
D1 = 4.50 FT.
D2 = 3.75 FT.
```

EXTERNAL BEAM RESULTS:

```
F  = 156688 LB.
M1 = 316310 LB.FT.
M2 =   4212 LB.FT.
S1 = -23398 LB.
S2 =  -2321 LB.
```

INTERNAL BEAM RESULTS:

```
F  = 256033 LB.
M1 = 148707 LB.FT.
S1 = -22599 LB.
```

EXTERNAL-HALF SHELL RESULTS:

POINT	T1	S	M2	T2
0	-5603	-6652	659	-135
1	-11393	-5012	431	-1046
2	-10855	-2952	-31	-1648
3	-7992	-1261	-406	-1950
4	-6540	0	-544	-2037

INTERNAL-HALF SHELL RESULTS:

POINT	T1	S	M2	T2
0	7644	-5434	-511	530
1	-2967	-5799	197	-355
2	-9821	-4597	176	-1194
3	-13238	-2487	-65	-1771
4	-14201	0	-185	-1974

L= 74 TABLE V-104 W= 34

SHELL DATA:

```
L   = 74.0 FT.
W   = 34.0 FT.
R   = 27.5 FT.
T   =  2.5 IN.
PHI= 37.2 DEG.
G   = 50.0 LB/FT2
```

BEAM DATA:

```
B  = 0.75 FT.
D1 = 4.50 FT.
D2 = 3.75 FT.
```

EXTERNAL BEAM RESULTS:

```
F  = 161033 LB.
M1 = 326255 LB.FT.
M2 =   4216 LB.FT.
S1 = -24098 LB.
S2 =  -2389 LB.
```

INTERNAL BEAM RESULTS:

```
F  = 261337 LB.
M1 = 149804 LB.FT.
S1 = -22985 LB.
```

EXTERNAL-HALF SHELL RESULTS:

POINT	T1	S	M2	T2
0	-5878	-6836	566	-155
1	-11518	-5059	412	-1146
2	-10597	-2890	-21	-1784
3	-7340	-1190	-382	-2093
4	-5724	0	-516	-2178

INTERNAL-HALF SHELL RESULTS:

POINT	T1	S	M2	T2
0	7912	-5547	-576	566
1	-2875	-5963	216	-403
2	-9646	-4713	192	-1323
3	-12824	-2537	-77	-1951
4	-13660	0	-211	-2171

L= 74 TABLE V-105 W= 36

SHELL DATA:

```
L   = 74.0 FT.
W   = 36.0 FT.
R   = 30.0 FT.
T   =  2.5 IN.
PHI= 36.0 DEG.
G   = 50.0 LB/FT2
```

BEAM DATA:

```
B  = 0.75 FT.
D1 = 4.50 FT.
D2 = 3.75 FT.
```

EXTERNAL BEAM RESULTS:

```
F  = 165400 LB.
M1 = 334652 LB.FT.
M2 =   4207 LB.FT.
S1 = -24733 LB.
S2 =  -2459 LB.
```

INTERNAL BEAM RESULTS:

```
F  = 266074 LB.
M1 = 150666 LB.FT.
S1 = -23325 LB.
```

EXTERNAL-HALF SHELL RESULTS:

POINT	T1	S	M2	T2
0	-6074	-7021	474	-173
1	-11628	-5111	395	-1247
2	-10376	-2831	-12	-1922
3	-6756	-1118	-360	-2235
4	-4984	0	-488	-2318

INTERNAL-HALF SHELL RESULTS:

POINT	T1	S	M2	T2
0	8157	-5647	-643	601
1	-2821	-6113	237	-454
2	-9493	-4809	208	-1453
3	-12390	-2571	-90	-2131
4	-13076	0	-237	-2367

L= 74 TABLE V-106 W= 38

SHELL DATA:

L = 74.0 FT.
W = 38.0 FT.
R = 32.5 FT.
T = 2.5 IN.
PHI= 35.0 DEG.
G = 50.0 LB/FT2

BEAM DATA:

B = 0.75 FT.
D1 = 4.50 FT.
D2 = 3.75 FT.

EXTERNAL BEAM RESULTS:
F = 169779 LB.
M1 = 341685 LB.FT.
M2 = 4187 LB.FT.
S1 = -25311 LB.
S2 = -2529 LB.

INTERNAL BEAM RESULTS:
F = 270443 LB.
M1 = 151507 LB.FT.
S1 = -23640 LB.

EXTERNAL-HALF SHELL RESULTS:

POINT	T1	S	M2	T2
0	-6200	-7207	385	-188
1	-11724	-5167	379	-1350
2	-10188	-2773	-2	-2063
3	-6232	-1047	-338	-2378
4	-4314	0	-462	-2457

INTERNAL-HALF SHELL RESULTS:

POINT	T1	S	M2	T2
0	8380	-5740	-712	635
1	-2812	-6249	259	-507
2	-9371	-4885	225	-1585
3	-11941	-2588	-102	-2310
4	-12453	0	-263	-2560

L= 74 TABLE V-107 W= 40

SHELL DATA:

L = 74.0 FT.
W = 40.0 FT.
R = 32.5 FT.
T = 2.5 IN.
PHI= 37.1 DEG.
G = 50.0 LB/FT2

BEAM DATA:

B = 0.75 FT.
D1 = 4.50 FT.
D2 = 3.75 FT.

EXTERNAL BEAM RESULTS:
F = 169357 LB.
M1 = 347268 LB.FT.
M2 = 4707 LB.FT.
S1 = -25522 LB.
S2 = -2500 LB.

INTERNAL BEAM RESULTS:
F = 258817 LB.
M1 = 136578 LB.FT.
S1 = -22267 LB.

EXTERNAL-HALF SHELL RESULTS:

POINT	T1	S	M2	T2
0	-6275	-7189	363	-195
1	-12124	-4942	359	-1396
2	-9814	-2383	-13	-2073
3	-4877	-747	-318	-2322
4	-2537	0	-424	-2368

INTERNAL-HALF SHELL RESULTS:

POINT	T1	S	M2	T2
0	8482	-5493	-791	661
1	-2726	-6051	286	-496
2	-8859	-4658	245	-1587
3	-10689	-2415	-115	-2304
4	-10827	0	-292	-2548

L= 74 TABLE V-108 W= 42

SHELL DATA:

L = 74.0 FT.
W = 42.0 FT.
R = 35.0 FT.
T = 2.5 IN.
PHI= 36.1 DEG.
G = 50.0 LB/FT2

BEAM DATA:

B = 0.75 FT.
D1 = 4.50 FT.
D2 = 3.75 FT.

EXTERNAL BEAM RESULTS:
F = 174026 LB.
M1 = 352064 LB.FT.
M2 = 4615 LB.FT.
S1 = -26023 LB.
S2 = -2578 LB.

INTERNAL BEAM RESULTS:
F = 264008 LB.
M1 = 139108 LB.FT.
S1 = -22706 LB.

EXTERNAL-HALF SHELL RESULTS:

POINT	T1	S	M2	T2
0	-6296	-7388	279	-209
1	-12181	-5021	346	-1504
2	-9704	-2346	-5	-2220
3	-4538	-686	-298	-2469
4	-2094	0	-399	-2511

INTERNAL-HALF SHELL RESULTS:

POINT	T1	S	M2	T2
0	8663	-5604	-859	689
1	-2865	-6185	308	-559
2	-8874	-4700	261	-1726
3	-10279	-2397	-126	-2481
4	-10188	0	-315	-2733

L= 76 TABLE V-109 W= 34

SHELL DATA:

L = 76.0 FT.
W = 34.0 FT.
R = 27.5 FT.
T = 2.5 IN.
PHI= 37.2 DEG.
G = 50.0 LB/FT2

BEAM DATA:

B = 0.75 FT.
D1 = 4.50 FT.
D2 = 3.75 FT.

EXTERNAL BEAM RESULTS:
F = 170067 LB.
M1 = 338146 LB.FT.
M2 = 4342 LB.FT.
S1 = -24515 LB.
S2 = -2462 LB.

INTERNAL BEAM RESULTS:
F = 275140 LB.
M1 = 156075 LB.FT.
S1 = -23494 LB.

EXTERNAL-HALF SHELL RESULTS:

POINT	T1	S	M2	T2
0	-5911	-7030	632	-150
1	-11967	-5250	442	-1147
2	-11196	-3036	-26	-1797
3	-7952	-1270	-413	-2117
4	-6325	0	-556	-2206

INTERNAL-HALF SHELL RESULTS:

POINT	T1	S	M2	T2
0	8420	-5686	-579	572
1	-2923	-6132	219	-397
2	-10127	-4864	195	-1320
3	-13601	-2626	-76	-1952
4	-14545	0	-212	-2174

L= 76 TABLE V-110 W= 36

SHELL DATA:

L = 76.0 FT.
W = 36.0 FT.
R = 30.0 FT.
T = 2.5 IN.
PHI= 36.0 DEG.
G = 50.0 LB/FT2

BEAM DATA:

B = 0.75 FT.
D1 = 4.50 FT.
D2 = 3.75 FT.

EXTERNAL BEAM RESULTS:
F = 174581 LB.
M1 = 347591 LB.FT.
M2 = 4352 LB.FT.
S1 = -25186 LB.
S2 = -2531 LB.

INTERNAL BEAM RESULTS:
F = 279991 LB.
M1 = 156730 LB.FT.
S1 = -23823 LB.

EXTERNAL-HALF SHELL RESULTS:

POINT	T1	S	M2	T2
0	-6146	-7216	538	-168
1	-12099	-5297	424	-1248
2	-10954	-2971	-16	-1934
3	-7308	-1195	-390	-2258
4	-5511	0	-528	-2345

INTERNAL-HALF SHELL RESULTS:

POINT	T1	S	M2	T2
0	8683	-5786	-647	608
1	-2847	-6286	240	-446
2	-9955	-4967	211	-1449
3	-13154	-2666	-89	-2132
4	-13952	0	-239	-2371

L= 76 TABLE V-111 W= 38

SHELL DATA:

L = 76.0 FT.
W = 38.0 FT.
R = 32.5 FT.
T = 2.5 IN.
PHI= 35.0 DEG.
G = 50.0 LB/FT2

BEAM DATA:

B = 0.75 FT.
D1 = 4.50 FT.
D2 = 3.75 FT.

EXTERNAL BEAM RESULTS:
F = 179119 LB.
M1 = 355583 LB.FT.
M2 = 4347 LB.FT.
S1 = -25798 LB.
S2 = -2601 LB.

INTERNAL BEAM RESULTS:
F = 284426 LB.
M1 = 157340 LB.FT.
S1 = -24125 LB.

EXTERNAL-HALF SHELL RESULTS:

POINT	T1	S	M2	T2
0	-6307	-7404	447	-183
1	-12215	-5349	407	-1350
2	-10749	-2907	-7	-2074
3	-6732	-1116	-367	-2400
4	-4774	0	-500	-2484

INTERNAL-HALF SHELL RESULTS:

POINT	T1	S	M2	T2
0	8923	-5878	-717	644
1	-2817	-6427	262	-497
2	-9812	-5051	229	-1580
3	-12693	-2689	-102	-2312
4	-13320	0	-266	-2566

L= 76 TABLE V-112 W= 40

SHELL DATA:

L = 76.0 FT.
W = 40.0 FT.
R = 32.5 FT.
T = 2.5 IN.
PHI= 37.1 DEG.
G = 50.0 LB/FT2

BEAM DATA:

B = 0.75 FT.
D1 = 4.50 FT.
D2 = 3.75 FT.

EXTERNAL BEAM RESULTS:
F = 178561 LB.
M1 = 361974 LB.FT.
M2 = 4903 LB.FT.
S1 = -26028 LB.
S2 = -2569 LB.

INTERNAL BEAM RESULTS:
F = 271829 LB.
M1 = 141348 LB.FT.
S1 = -22684 LB.

EXTERNAL-HALF SHELL RESULTS:

POINT	T1	S	M2	T2
0	-6417	-7381	423	-191
1	-12650	-5110	387	-1395
2	-10348	-2496	-17	-2082
3	-5276	-804	-346	-2341
4	-2865	0	-462	-2392

INTERNAL-HALF SHELL RESULTS:

POINT	T1	S	M2	T2
0	9023	-5618	-799	671
1	-2692	-6221	290	-483
2	-9245	-4825	250	-1580
3	-11387	-2520	-116	-2307
4	-11645	0	-296	-2555

L= 76 TABLE V-113 W= 42

SHELL DATA:

L = 76.0 FT.
W = 42.0 FT.
R = 35.0 FT.
T = 2.5 IN.
PHI= 36.1 DEG.
G = 50.0 LB/FT2

BEAM DATA:

B = 0.75 FT.
D1 = 4.50 FT.
D2 = 3.75 FT.

EXTERNAL BEAM RESULTS:
F = 183441 LB.
M1 = 367543 LB.FT.
M2 = 4820 LB.FT.
S1 = -26561 LB.
S2 = -2648 LB.

INTERNAL BEAM RESULTS:
F = 277072 LB.
M1 = 143710 LB.FT.
S1 = -23107 LB.

EXTERNAL-HALF SHELL RESULTS:

POINT	T1	S	M2	T2
0	-6462	-7582	336	-205
1	-12725	-5188	372	-1504
2	-10228	-2456	-9	-2229
3	-4901	-739	-325	-2488
4	-2376	0	-435	-2533

INTERNAL-HALF SHELL RESULTS:

POINT	T1	S	M2	T2
0	9217	-5726	-869	701
1	-2814	-6359	313	-544
2	-9245	-4875	266	-1717
3	-10968	-2509	-128	-2484
4	-11000	0	-320	-2743

L= 76 TABLE V-114 W= 44

SHELL DATA:

L = 76.0 FT.
W = 44.0 FT.
R = 37.5 FT.
T = 2.5 IN.
PHI= 35.2 DEG.
G = 50.0 LB/FT2

BEAM DATA:

B = 0.75 FT.
D1 = 4.50 FT.
D2 = 3.75 FT.

EXTERNAL BEAM RESULTS:
F = 188252 LB.
M1 = 372072 LB.FT.
M2 = 4736 LB.FT.
S1 = -27047 LB.
S2 = -2726 LB.

INTERNAL BEAM RESULTS:
F = 282217 LB.
M1 = 146153 LB.FT.
S1 = -23528 LB.

EXTERNAL-HALF SHELL RESULTS:

POINT	T1	S	M2	T2
0	-6455	-7781	254	-218
1	-12788	-5267	360	-1614
2	-10126	-2413	-2	-2378
3	-4556	-673	-306	-2635
4	-1920	0	-410	-2675

INTERNAL-HALF SHELL RESULTS:

POINT	T1	S	M2	T2
0	9401	-5832	-939	728
1	-2979	-6489	336	-610
2	-9274	-4905	282	-1859
3	-10532	-2479	-139	-2660
4	-10311	0	-343	-2926

```
L= 78        TABLE V-115        W= 34
**************************************

SHELL DATA:

     L   = 78.0 FT.
     W   = 34.0 FT.
     R   = 27.5 FT.
     T   =  2.5 IN.
     PHI= 37.2 DEG.
     G   = 50.0 LB/FT2

BEAM DATA:

     B  = 0.75 FT.
     D1 = 4.75 FT.
     D2 = 4.00 FT.

EXTERNAL BEAM RESULTS:
     F  = 175805 LB.
     M1 = 366784 LB.FT.
     M2 =   4171 LB.FT.
     S1 = -26272 LB.
     S2 =  -2492 LB.

INTERNAL BEAM RESULTS:
     F  = 289022 LB.
     M1 = 180119 LB.FT.
     S1 = -26153 LB.

EXTERNAL-HALF SHELL RESULTS:

POINT      T1       S      M2      T2

  0     -6305   -7080     588    -154
  1    -12006   -5318     430   -1131
  2    -11413   -3145     -18   -1778
  3     -8496   -1353    -398   -2106
  4     -7020       0    -540   -2200

INTERNAL-HALF SHELL RESULTS:

POINT      T1       S      M2      T2

  0      7798   -5820    -561     521
  1     -3327   -6164     213    -432
  2    -10454   -4865     190   -1331
  3    -13951   -2625     -73   -1945
  4    -14918       0    -204   -2160
```

```
L= 78        TABLE V-116        W= 36
**************************************

SHELL DATA:

     L   = 78.0 FT.
     W   = 36.0 FT.
     R   = 30.0 FT.
     T   =  2.5 IN.
     PHI= 36.0 DEG.
     G   = 50.0 LB/FT2

BEAM DATA:

     B  = 0.75 FT.
     D1 = 4.75 FT.
     D2 = 4.00 FT.

EXTERNAL BEAM RESULTS:
     F  = 180320 LB.
     M1 = 376372 LB.FT.
     M2 =   4177 LB.FT.
     S1 = -26954 LB.
     S2 =  -2559 LB.

INTERNAL BEAM RESULTS:
     F  = 294229 LB.
     M1 = 180739 LB.FT.
     S1 = -26520 LB.

EXTERNAL-HALF SHELL RESULTS:

POINT      T1       S      M2      T2

  0     -6523   -7262     495    -171
  1    -12105   -5367     413   -1230
  2    -11152   -3088      -8   -1914
  3     -7849   -1285    -376   -2249
  4     -6208       0    -514   -2342

INTERNAL-HALF SHELL RESULTS:

POINT      T1       S      M2      T2

  0      8071   -5925    -628     555
  1     -3228   -6320     234    -483
  2    -10264   -4972     206   -1460
  3    -13503   -2668     -86   -2125
  4    -14334       0    -231   -2357
```

```
L= 78        TABLE V-117        W= 38
**************************************

SHELL DATA:

     L   = 78.0 FT.
     W   = 38.0 FT.
     R   = 32.5 FT.
     T   =  2.5 IN.
     PHI= 35.0 DEG.
     G   = 50.0 LB/FT2

BEAM DATA:

     B  = 0.75 FT.
     D1 = 4.75 FT.
     D2 = 4.00 FT.

EXTERNAL BEAM RESULTS:
     F  = 184860 LB.
     M1 = 384477 LB.FT.
     M2 =   4169 LB.FT.
     S1 = -27578 LB.
     S2 =  -2628 LB.

INTERNAL BEAM RESULTS:
     F  = 298981 LB.
     M1 = 181295 LB.FT.
     S1 = -26855 LB.

EXTERNAL-HALF SHELL RESULTS:

POINT      T1       S      M2      T2

  0     -6671   -7445     404    -187
  1    -12194   -5422     398   -1330
  2    -10928   -3032       2   -2052
  3     -7267   -1216    -355   -2392
  4     -5471       0    -489   -2484

INTERNAL-HALF SHELL RESULTS:

POINT      T1       S      M2      T2

  0      8321   -6021    -698     588
  1     -3173   -6462     256    -535
  2    -10102   -5060     224   -1591
  3    -13042   -2696     -98   -2304
  4    -13714       0    -258   -2551
```

```
L= 78        TABLE V-118         W= 40
**************************************

SHELL DATA:

    L   = 78.0 FT.
    W   = 40.0 FT.
    R   = 32.5 FT.
    T   =  2.5 IN.
    PHI= 37.1 DEG.
    G   = 50.0 LB/FT2

BEAM DATA:

    B   = 0.75 FT.
    D1  = 4.75 FT.
    D2  = 4.00 FT.

EXTERNAL BEAM RESULTS:
    F   = 184003 LB.
    M1  = 391811 LB.FT.
    M2  =   4727 LB.FT.
    S1  = -27818 LB.
    S2  =  -2593 LB.

INTERNAL BEAM RESULTS:
    F   = 286242 LB.
    M1  = 163039 LB.FT.
    S1  = -25287 LB.

EXTERNAL-HALF SHELL RESULTS:

POINT      T1       S      M2      T2

  0      -6830   -7411    381    -195
  1     -12593   -5176    381   -1371
  2     -10481   -2631     -7   -2059
  3      -5774    -912   -337   -2337
  4      -3528       0   -456   -2398

INTERNAL-HALF SHELL RESULTS:

POINT      T1       S      M2      T2

  0       8498   -5764   -781     623
  1      -2985   -6268    284    -516
  2      -9505   -4853    245   -1588
  3     -11751   -2542   -113   -2300
  4     -12076       0   -289   -2544
```

```
L= 78        TABLE V-119         W= 42
**************************************

SHELL DATA:

    L   = 78.0 FT.
    W   = 42.0 FT.
    R   = 35.0 FT.
    T   =  2.5 IN.
    PHI= 36.1 DEG.
    G   = 50.0 LB/FT2

BEAM DATA:

    B   = 0.75 FT.
    D1  = 4.75 FT.
    D2  = 4.00 FT.

EXTERNAL BEAM RESULTS:
    F   = 188913 LB.
    M1  = 397423 LB.FT.
    M2  =   4641 LB.FT.
    S1  = -28365 LB.
    S2  =  -2670 LB.

INTERNAL BEAM RESULTS:
    F   = 291711 LB.
    M1  = 165545 LB.FT.
    S1  = -25747 LB.

EXTERNAL-HALF SHELL RESULTS:

POINT      T1       S      M2      T2

  0      -6867   -7608    293    -209
  1     -12650   -5255    368   -1477
  2     -10351   -2596      1   -2204
  3      -5393    -853   -317   -2485
  4      -3033       0   -431   -2542

INTERNAL-HALF SHELL RESULTS:

POINT      T1       S      M2      T2

  0       8692   -5874   -851     650
  1      -3083   -6407    307    -577
  2      -9483   -4908    262   -1725
  3     -11333   -2537   -125   -2477
  4     -11442       0   -313   -2732
```

```
L= 78        TABLE V-120         W= 44
**************************************

SHELL DATA:

    L   = 78.0 FT.
    W   = 44.0 FT.
    R   = 37.5 FT.
    T   =  2.5 IN.
    PHI= 35.2 DEG.
    G   = 50.0 LB/FT2

BEAM DATA:

    B   = 0.75 FT.
    D1  = 4.75 FT.
    D2  = 4.00 FT.

EXTERNAL BEAM RESULTS:
    F   = 193758 LB.
    M1  = 401963 LB.FT.
    M2  =   4554 LB.FT.
    S1  = -28866 LB.
    S2  =  -2746 LB.

INTERNAL BEAM RESULTS:
    F   = 297052 LB.
    M1  = 168137 LB.FT.
    S1  = -26202 LB.

EXTERNAL-HALF SHELL RESULTS:

POINT      T1       S      M2      T2

  0      -6854   -7803    210    -221
  1     -12700   -5336    358   -1584
  2     -10241   -2560     10   -2352
  3      -5042    -793   -299   -2633
  4      -2570       0   -409   -2686

INTERNAL-HALF SHELL RESULTS:

POINT      T1       S      M2      T2

  0       8873   -5982   -920     675
  1      -3224   -6536    330    -643
  2      -9492   -4943    278   -1866
  3     -10896   -2514   -136   -2653
  4     -10765       0   -337   -2916
```

L= 80 TABLE V-121 W= 34

SHELL DATA:

L = 80.0 FT.
W = 34.0 FT.
R = 27.5 FT.
T = 2.5 IN.
PHI= 37.2 DEG.
G = 50.0 LB/FT2

BEAM DATA:

B = 0.75 FT.
D1 = 4.75 FT.
D2 = 4.00 FT.

EXTERNAL BEAM RESULTS:
F = 185230 LB.
M1 = 379319 LB.FT.
M2 = 4281 LB.FT.
S1 = -26708 LB.
S2 = -2565 LB.

INTERNAL BEAM RESULTS:
F = 303604 LB.
M1 = 187500 LB.FT.
S1 = -26718 LB.

EXTERNAL-HALF SHELL RESULTS:

POINT	T1	S	M2	T2
0	-6336	-7273	652	-149
1	-12450	-5509	460	-1132
2	-12030	-3295	-23	-1790
3	-9155	-1435	-428	-2129
4	-7680	0	-579	-2228

INTERNAL-HALF SHELL RESULTS:

POINT	T1	S	M2	T2
0	8277	-5961	-563	527
1	-3403	-6330	216	-427
2	-10959	-5010	193	-1328
3	-14744	-2710	-72	-1946
4	-15815	0	-205	-2163

L= 80 TABLE V-122 W= 36

SHELL DATA:

L = 80.0 FT.
W = 36.0 FT.
R = 30.0 FT.
T = 2.5 IN.
PHI= 36.0 DEG.
G = 50.0 LB/FT2

BEAM DATA:

B = 0.75 FT.
D1 = 4.75 FT.
D2 = 4.00 FT.

EXTERNAL BEAM RESULTS:
F = 189879 LB.
M1 = 390006 LB.FT.
M2 = 4306 LB.FT.
S1 = -27425 LB.
S2 = -2632 LB.

INTERNAL BEAM RESULTS:
F = 308953 LB.
M1 = 187894 LB.FT.
S1 = -27076 LB.

EXTERNAL-HALF SHELL RESULTS:

POINT	T1	S	M2	T2
0	-6592	-7456	557	-167
1	-12568	-5555	442	-1230
2	-11746	-3231	-12	-1926
3	-8447	-1362	-405	-2271
4	-6794	0	-552	-2369

INTERNAL-HALF SHELL RESULTS:

POINT	T1	S	M2	T2
0	8570	-6066	-631	562
1	-3281	-6490	236	-476
2	-10748	-5124	209	-1457
3	-14283	-2759	-85	-2126
4	-15222	0	-232	-2360

L= 80 TABLE V-123 W= 38

SHELL DATA:

L = 80.0 FT.
W = 38.0 FT.
R = 32.5 FT.
T = 2.5 IN.
PHI= 35.0 DEG.
G = 50.0 LB/FT2

BEAM DATA:

B = 0.75 FT.
D1 = 4.75 FT.
D2 = 4.00 FT.

EXTERNAL BEAM RESULTS:
F = 194564 LB.
M1 = 399132 LB.FT.
M2 = 4314 LB.FT.
S1 = -28083 LB.
S2 = -2700 LB.

INTERNAL BEAM RESULTS:
F = 313800 LB.
M1 = 188198 LB.FT.
S1 = -27399 LB.

EXTERNAL-HALF SHELL RESULTS:

POINT	T1	S	M2	T2
0	-6776	-7640	464	-183
1	-12674	-5604	426	-1330
2	-11503	-3169	-2	-2063
3	-7811	-1288	-383	-2413
4	-5988	0	-526	-2509

INTERNAL-HALF SHELL RESULTS:

POINT	T1	S	M2	T2
0	8838	-6161	-702	596
1	-3204	-6637	258	-527
2	-10567	-5220	227	-1587
3	-13809	-2792	-98	-2305
4	-14593	0	-260	-2556

```
L= 80        TABLE V-124          W= 40
**************************************

SHELL DATA:

    L   = 80.0 FT.
    W   = 40.0 FT.
    R   = 32.5 FT.
    T   =  2.5 IN.
    PHI= 37.1 DEG.
    G   = 50.0 LB/FT2

BEAM DATA:

    B   = 0.75 FT.
    D1  = 4.75 FT.
    D2  = 4.00 FT.

EXTERNAL BEAM RESULTS:
    F   = 193533 LB.
    M1  = 407360 LB.FT.
    M2  =   4908 LB.FT.
    S1  = -28340 LB.
    S2  =  -2661 LB.

INTERNAL BEAM RESULTS:
    F   = 300091 LB.
    M1  = 168733 LB.FT.
    S1  = -25761 LB.

EXTERNAL-HALF SHELL RESULTS:

POINT      T1       S      M2     T2

  0     -6974   -7600    440   -190
  1    -13105   -5346    409  -1371
  2    -11023   -2751    -11  -2069
  3     -6215    -973   -366  -2356
  4     -3912       0   -493  -2421

INTERNAL-HALF SHELL RESULTS:

POINT      T1       S      M2     T2

  0      9021   -5892   -788    632
  1     -2975   -6437    287   -505
  2     -9913   -5014    249  -1582
  3    -12466   -2642   -113  -2302
  4    -12906       0   -292  -2550
```

```
L= 80        TABLE V-125          W= 42
**************************************

SHELL DATA:

    L   = 80.0 FT.
    W   = 42.0 FT.
    R   = 35.0 FT.
    T   =  2.5 IN.
    PHI= 36.1 DEG.
    G   = 50.0 LB/FT2

BEAM DATA:

    B   = 0.75 FT.
    D1  = 4.75 FT.
    D2  = 4.00 FT.

EXTERNAL BEAM RESULTS:
    F   = 198643 LB.
    M1  = 413822 LB.FT.
    M2  =   4832 LB.FT.
    S1  = -28921 LB.
    S2  =  -2739 LB.

INTERNAL BEAM RESULTS:
    F   = 305630 LB.
    M1  = 171049 LB.FT.
    S1  = -26207 LB.

EXTERNAL-HALF SHELL RESULTS:

POINT      T1       S      M2     T2

  0     -7037   -7800    350   -205
  1    -13179   -5422    394  -1476
  2    -10882   -2710     -2  -2213
  3     -5796    -910   -344  -2503
  4     -3369       0   -467  -2564

INTERNAL-HALF SHELL RESULTS:

POINT      T1       S      M2     T2

  0      9228   -6001   -859    661
  1     -3055   -6579    311   -564
  2     -9876   -5077    267  -1718
  3    -12038   -2643   -126  -2480
  4    -12268       0   -318  -2740
```

```
L= 80        TABLE V-126          W= 44
**************************************

SHELL DATA:

    L   = 80.0 FT.
    W   = 44.0 FT.
    R   = 37.5 FT.
    T   =  2.5 IN.
    PHI= 35.2 DEG.
    G   = 50.0 LB/FT2

BEAM DATA:

    B   = 0.75 FT.
    D1  = 4.75 FT.
    D2  = 4.00 FT.

EXTERNAL BEAM RESULTS:
    F   = 203694 LB.
    M1  = 419111 LB.FT.
    M2  =   4754 LB.FT.
    S1  = -29451 LB.
    S2  =  -2816 LB.

INTERNAL BEAM RESULTS:
    F   = 311008 LB.
    M1  = 173440 LB.FT.
    S1  = -26645 LB.

EXTERNAL-HALF SHELL RESULTS:

POINT      T1       S      M2     T2

  0     -7046   -7999    263   -218
  1    -13242   -5500    382  -1583
  2    -10761   -2669      6  -2359
  3     -5411    -846   -324  -2650
  4     -2863       0   -442  -2707

INTERNAL-HALF SHELL RESULTS:

POINT      T1       S      M2     T2

  0      9421   -6106   -931    688
  1     -3179   -6712    334   -627
  2     -9869   -5120    284  -1857
  3    -11593   -2627   -137  -2657
  4    -11585       0   -343  -2926
```

L= 82 TABLE V-127 W= 34

SHELL DATA:

L = 82.0 FT.
W = 34.0 FT.
R = 27.5 FT.
T = 2.5 IN.
PHI= 37.2 DEG.
G = 50.0 LB/FT2

BEAM DATA:

B = 0.75 FT.
D1 = 5.00 FT.
D2 = 4.00 FT.

EXTERNAL BEAM RESULTS:
F = 190736 LB.
M1 = 450962 LB.FT.
M2 = 4462 LB.FT.
S1 = -28230 LB.
S2 = -2575 LB.

INTERNAL BEAM RESULTS:
F = 323486 LB.
M1 = 229749 LB.FT.
S1 = -27372 LB.

EXTERNAL-HALF SHELL RESULTS:

POINT	T1	S	M2	T2
0	-5458	-7307	681	-145
1	-12335	-5668	479	-1118
2	-12483	-3476	-21	-1788
3	-9958	-1551	-444	-2144
4	-8602	0	-603	-2250

INTERNAL-HALF SHELL RESULTS:

POINT	T1	S	M2	T2
0	8781	-6196	-574	565
1	-3577	-6582	222	-404
2	-11639	-521[illegible]	199	-1321
3	-15746	-2828	-72	-1950
4	-16931	0	-208	-2172

L= 82 TABLE V-128 W= 36

SHELL DATA:

L = 82.0 FT.
W = 36.0 FT.
R = 30.0 FT.
T = 2.5 IN.
PHI= 36.0 DEG.
G = 50.0 LB/FT2

BEAM DATA:

B = 0.75 FT.
D1 = 5.00 FT.
D2 = 4.00 FT.

EXTERNAL BEAM RESULTS:
F = 195206 LB.
M1 = 463785 LB.FT.
M2 = 4499 LB.FT.
S1 = -28979 LB.
S2 = -2637 LB.

INTERNAL BEAM RESULTS:
F = 329073 LB.
M1 = 229844 LB.FT.
S1 = -27698 LB.

EXTERNAL-HALF SHELL RESULTS:

POINT	T1	S	M2	T2
0	-5706	-7478	584	-163
1	-12445	-5711	461	-1215
2	-12175	-3414	-10	-1923
3	-9208	-1479	-422	-2286
4	-7665	0	-578	-2392

INTERNAL-HALF SHELL RESULTS:

POINT	T1	S	M2	T2
0	9088	-6303	-644	603
1	-3435	-6747	242	-451
2	-11403	-5340	215	-1448
3	-15261	-2881	-86	-2131
4	-16313	0	-236	-2370

L= 82 TABLE V-129 W= 38

SHELL DATA:

L = 82.0 FT.
W = 38.0 FT.
R = 32.5 FT.
T = 2.5 IN.
PHI= 35.0 DEG.
G = 50.0 LB/FT2

BEAM DATA:

B = 0.75 FT.
D1 = 5.00 FT.
D2 = 4.00 FT.

EXTERNAL BEAM RESULTS:
F = 199760 LB.
M1 = 474895 LB.FT.
M2 = 4520 LB.FT.
S1 = -29667 LB.
S2 = -2701 LB.

INTERNAL BEAM RESULTS:
F = 334092 LB.
M1 = 229807 LB.FT.
S1 = -27987 LB.

EXTERNAL-HALF SHELL RESULTS:

POINT	T1	S	M2	T2
0	-5888	-7653	489	-179
1	-12550	-5759	445	-1313
2	-11913	-3352	0	-2059
3	-8532	-1406	-401	-2429
4	-6810	0	-552	-2533

INTERNAL-HALF SHELL RESULTS:

POINT	T1	S	M2	T2
0	9368	-6399	-716	641
1	-3336	-6899	265	-499
2	-11196	-5441	233	-1578
3	-14762	-2920	-100	-2311
4	-15661	0	-265	-2567

L= 82 TABLE V-130 W= 40

****

SHELL DATA:

L = 82.0 FT.
W = 40.0 FT.
R = 32.5 FT.
T = 2.5 IN.
PHI= 37.1 DEG.
G = 50.0 LB/FT2

BEAM DATA:

B = 0.75 FT.
D1 = 5.00 FT.
D2 = 4.00 FT.

EXTERNAL BEAM RESULTS:
F = 198132 LB.
M1 = 484661 LB.FT.
M2 = 5164 LB.FT.
S1 = -29947 LB.
S2 = -2653 LB.

INTERNAL BEAM RESULTS:
F = 319416 LB.
M1 = 205289 LB.FT.
S1 = -26206 LB.

EXTERNAL-HALF SHELL RESULTS:

POINT	T1	S	M2	T2
0	-6083	-7590	464	-187
1	-12979	-5490	430	-1350
2	-11397	-2926	-7	-2062
3	-6852	-1088	-384	-2371
4	-4629	0	-522	-2446

INTERNAL-HALF SHELL RESULTS:

POINT	T1	S	M2	T2
0	9533	-6118	-805	679
1	-3072	-6689	295	-476
2	-10478	-5232	257	-1572
3	-13343	-2770	-115	-2308
4	-13895	0	-299	-2562

L= 82 TABLE V-131 W= 42

****

SHELL DATA:

L = 82.0 FT.
W = 42.0 FT.
R = 35.0 FT.
T = 2.5 IN.
PHI= 36.1 DEG.
G = 50.0 LB/FT2

BEAM DATA:

B = 0.75 FT.
D1 = 5.00 FT.
D2 = 4.00 FT.

EXTERNAL BEAM RESULTS:
F = 203243 LB.
M1 = 492871 LB.FT.
M2 = 5097 LB.FT.
S1 = -30556 LB.
S2 = -2729 LB.

INTERNAL BEAM RESULTS:
F = 325072 LB.
M1 = 207722 LB.FT.
S1 = -26625 LB.

EXTERNAL-HALF SHELL RESULTS:

POINT	T1	S	M2	T2
0	-6158	-7786	371	-202
1	-13067	-5565	415	-1454
2	-11249	-2888	2	-2205
3	-6401	-1025	-363	-2518
4	-4041	0	-495	-2590

INTERNAL-HALF SHELL RESULTS:

POINT	T1	S	M2	T2
0	9750	-6227	-879	711
1	-3135	-6835	319	-532
2	-10420	-5302	275	-1706
3	-12896	-2776	-128	-2486
4	-13239	0	-326	-2753

L= 82 TABLE V-132 W= 44

****

SHELL DATA:

L = 82.0 FT.
W = 44.0 FT.
R = 37.5 FT.
T = 2.5 IN.
PHI= 35.2 DEG.
G = 50.0 LB/FT2

BEAM DATA:

B = 0.75 FT.
D1 = 5.00 FT.
D2 = 4.00 FT.

EXTERNAL BEAM RESULTS:
F = 208335 LB.
M1 = 499749 LB.FT.
M2 = 5027 LB.FT.
S1 = -31113 LB.
S2 = -2804 LB.

INTERNAL BEAM RESULTS:
F = 330523 LB.
M1 = 210234 LB.FT.
S1 = -27035 LB.

EXTERNAL-HALF SHELL RESULTS:

POINT	T1	S	M2	T2
0	-6181	-7981	282	-215
1	-13149	-5643	403	-1559
2	-11125	-2846	11	-2350
3	-5985	-960	-343	-2666
4	-3492	0	-471	-2733

INTERNAL-HALF SHELL RESULTS:

POINT	T1	S	M2	T2
0	9953	-6331	-953	741
1	-3243	-6971	343	-592
2	-10393	-5351	293	-1844
3	-12432	-2765	-141	-2664
4	-12538	0	-352	-2941

```
L= 84       TABLE V-133        W= 34
************************************

SHELL DATA:

    L   = 84.0 FT.
    W   = 34.0 FT.
    R   = 27.5 FT.
    T   =  2.5 IN.
    PHI= 37.2 DEG.
    G   = 50.0 LB/FT2

BEAM DATA:

    B   = 0.75 FT.
    D1  = 5.25 FT.
    D2  = 4.25 FT.

EXTERNAL BEAM RESULTS:
    F   = 197180 LB.
    M1  = 484752 LB.FT.
    M2  =   4282 LB.FT.
    S1  = -30105 LB.
    S2  =  -2611 LB.

INTERNAL BEAM RESULTS:
    F   = 337840 LB.
    M1  = 260993 LB.FT.
    S1  = -30272 LB.

EXTERNAL-HALF SHELL RESULTS:

POINT      T1       S      M2     T2

  0     -5900   -7374    636   -149
  1    -12444   -5737    466  -1106
  2    -12746   -3571    -14  -1772
  3    -10516   -1622   -428  -2131
  4     -9296       0   -585  -2241

INTERNAL-HALF SHELL RESULTS:

POINT      T1       S      M2     T2

  0      8105   -6317   -556    513
  1     -4018   -6602    216   -440
  2    -11979   -5208    194  -1332
  3    -16084   -2819    -69  -1943
  4    -17282       0   -200  -2158
```

```
L= 84       TABLE V-134        W= 36
************************************

SHELL DATA:

    L   = 84.0 FT.
    W   = 36.0 FT.
    R   = 30.0 FT.
    T   =  2.5 IN.
    PHI= 36.0 DEG.
    G   = 50.0 LB/FT2

BEAM DATA:

    B   = 0.75 FT.
    D1  = 5.25 FT.
    D2  = 4.25 FT.

EXTERNAL BEAM RESULTS:
    F   = 201687 LB.
    M1  = 497659 LB.FT.
    M2  =   4315 LB.FT.
    S1  = -30862 LB.
    S2  =  -2672 LB.

INTERNAL BEAM RESULTS:
    F   = 343845 LB.
    M1  = 260978 LB.FT.
    S1  = -30639 LB.

EXTERNAL-HALF SHELL RESULTS:

POINT      T1       S      M2     T2

  0     -6130   -7543    541   -167
  1    -12521   -5783    449  -1201
  2    -12417   -3517     -3  -1905
  3     -9763   -1558   -407  -2275
  4     -8364       0   -561  -2385

INTERNAL-HALF SHELL RESULTS:

POINT      T1       S      M2     T2

  0      8423   -6429   -625    548
  1     -3854   -6769    236   -488
  2    -11727   -5332    210  -1460
  3    -15599   -2875    -82  -2123
  4    -16674       0   -228  -2356
```

```
L= 84       TABLE V-135        W= 38
************************************

SHELL DATA:

    L   = 84.0 FT.
    W   = 38.0 FT.
    R   = 32.5 FT.
    T   =  2.5 IN.
    PHI= 35.0 DEG.
    G   = 50.0 LB/FT2

BEAM DATA:

    B   = 0.75 FT.
    D1  = 5.25 FT.
    D2  = 4.25 FT.

EXTERNAL BEAM RESULTS:
    F   = 206270 LB.
    M1  = 508840 LB.FT.
    M2  =   4333 LB.FT.
    S1  = -31560 LB.
    S2  =  -2735 LB.

INTERNAL BEAM RESULTS:
    F   = 349245 LB.
    M1  = 260802 LB.FT.
    S1  = -30962 LB.

EXTERNAL-HALF SHELL RESULTS:

POINT      T1       S      M2     T2

  0     -6299   -7714    446   -183
  1    -12596   -5834    435  -1297
  2    -12135   -3465      8  -2040
  3     -9082   -1493   -387  -2419
  4     -7511       0   -538  -2529

INTERNAL-HALF SHELL RESULTS:

POINT      T1       S      M2     T2

  0      8716   -6530   -696    583
  1     -3733   -6923    258   -538
  2    -11504   -5437    228  -1589
  3    -15102   -2917    -96  -2303
  4    -16032       0   -257  -2552
```

L= 84 TABLE V-136 W= 40
****

SHELL DATA:

L = 84.0 FT.
W = 40.0 FT.
R = 32.5 FT.
T = 2.5 IN.
PHI= 37.1 DEG.
G = 50.0 LB/FT2

BEAM DATA:

B = 0.75 FT.
D1 = 5.25 FT.
D2 = 4.25 FT.

EXTERNAL BEAM RESULTS:
F = 204316 LB.
M1 = 519646 LB.FT.
M2 = 4978 LB.FT.
S1 = -31845 LB.
S2 = -2684 LB.

INTERNAL BEAM RESULTS:
F = 334592 LB.
M1 = 233352 LB.FT.
S1 = -29046 LB.

EXTERNAL-HALF SHELL RESULTS:

POINT	T1	S	M2	T2
0	-6534	-7641	424	-191
1	-12981	-5550	422	-1332
2	-11568	-3040	0	-2043
3	-7366	-1184	-374	-2366
4	-5300	0	-513	-2448

INTERNAL-HALF SHELL RESULTS:

POINT	T1	S	M2	T2
0	8968	-6256	-786	628
1	-3405	-6728	288	-510
2	-10760	-5248	252	-1581
3	-13704	-2782	-112	-2301
4	-14310	0	-292	-2550

L= 84 TABLE V-137 W= 42
****

SHELL DATA:

L = 84.0 FT.
W = 42.0 FT.
R = 35.0 FT.
T = 2.5 IN.
PHI= 36.1 DEG.
G = 50.0 LB/FT2

BEAM DATA:

B = 0.75 FT.
D1 = 5.25 FT.
D2 = 4.25 FT.

EXTERNAL BEAM RESULTS:
F = 209466 LB.
M1 = 527897 LB.FT.
M2 = 4908 LB.FT.
S1 = -32467 LB.
S2 = -2758 LB.

INTERNAL BEAM RESULTS:
F = 340521 LB.
M1 = 235870 LB.FT.
S1 = -29502 LB.

EXTERNAL-HALF SHELL RESULTS:

POINT	T1	S	M2	T2
0	-6600	-7834	330	-205
1	-13049	-5636	409	-1433
2	-11407	-3015	11	-2185
3	-6909	-1127	-353	-2513
4	-4710	0	-489	-2594

INTERNAL-HALF SHELL RESULTS:

POINT	T1	S	M2	T2
0	9187	-6367	-859	659
1	-3445	-6874	312	-567
2	-10683	-5322	270	-1716
3	-13258	-2793	-125	-2479
4	-13665	0	-319	-2741

L= 84 TABLE V-138 W= 44
****

SHELL DATA:

L = 84.0 FT.
W = 44.0 FT.
R = 37.5 FT.
T = 2.5 IN.
PHI= 35.2 DEG.
G = 50.0 LB/FT2

BEAM DATA:

B = 0.75 FT.
D1 = 5.25 FT.
D2 = 4.25 FT.

EXTERNAL BEAM RESULTS:
F = 214597 LB.
M1 = 534789 LB.FT.
M2 = 4835 LB.FT.
S1 = -33037 LB.
S2 = -2832 LB.

INTERNAL BEAM RESULTS:
F = 346211 LB.
M1 = 238463 LB.FT.
S1 = -29946 LB.

EXTERNAL-HALF SHELL RESULTS:

POINT	T1	S	M2	T2
0	-6617	-8025	240	-218
1	-13113	-5716	398	-1536
2	-11272	-2980	21	-2328
3	-6488	-1069	-334	-2662
4	-4157	0	-466	-2740

INTERNAL-HALF SHELL RESULTS:

POINT	T1	S	M2	T2
0	9391	-6474	-933	687
1	-3529	-7010	336	-627
2	-10636	-5376	288	-1852
3	-12794	-2788	-137	-2656
4	-12977	0	-345	-2929

L= 84 TABLE V-139 W= 46

SHELL DATA:

L = 84.0 FT.
W = 46.0 FT.
R = 37.5 FT.
T = 2.5 IN.
PHI= 37.1 DEG.
G = 50.0 LB/FT2

BEAM DATA:

B = 0.75 FT.
D1 = 5.25 FT.
D2 = 4.25 FT.

EXTERNAL BEAM RESULTS:
F = 214434 LB.
M1 = 539794 LB.FT.
M2 = 5312 LB.FT.
S1 = -33214 LB.
S2 = -2813 LB.

INTERNAL BEAM RESULTS:
F = 334789 LB.
M1 = 220118 LB.FT.
S1 = -28566 LB.

EXTERNAL-HALF SHELL RESULTS:

POINT	T1	S	M2	T2
0	-6611	-8019	222	-226
1	-13479	-5518	378	-1585
2	-10944	-2619	8	-2345
3	-5250	-786	-314	-2614
4	-2527	0	-426	-2660

INTERNAL-HALF SHELL RESULTS:

POINT	T1	S	M2	T2
0	9474	-6260	-1013	705
1	-3593	-6821	364	-628
2	-10251	-5122	306	-1861
3	-11523	-2581	-150	-2648
4	-11252	0	-371	-2909

L= 84 TABLE V-140 W= 48

SHELL DATA:

L = 84.0 FT.
W = 48.0 FT.
R = 40.0 FT.
T = 2.5 IN.
PHI= 36.2 DEG.
G = 50.0 LB/FT2

BEAM DATA:

B = 0.75 FT.
D1 = 5.25 FT.
D2 = 4.25 FT.

EXTERNAL BEAM RESULTS:
F = 219765 LB.
M1 = 544512 LB.FT.
M2 = 5187 LB.FT.
S1 = -33715 LB.
S2 = -2892 LB.

INTERNAL BEAM RESULTS:
F = 341465 LB.
M1 = 224768 LB.FT.
S1 = -29146 LB.

EXTERNAL-HALF SHELL RESULTS:

POINT	T1	S	M2	T2
0	-6558	-8219	139	-237
1	-13521	-5613	370	-1693
2	-10867	-2597	16	-2495
3	-4959	-734	-298	-2767
4	-2134	0	-406	-2808

INTERNAL-HALF SHELL RESULTS:

POINT	T1	S	M2	T2
0	9653	-6385	-1083	727
1	-3808	-6956	387	-701
2	-10326	-5143	321	-2006
3	-11087	-2540	-161	-2823
4	-10537	0	-393	-3088

L= 84 TABLE V-141 W= 50

SHELL DATA:

L = 84.0 FT.
W = 50.0 FT.
R = 40.0 FT.
T = 2.5 IN.
PHI= 38.0 DEG.
G = 50.0 LB/FT2

BEAM DATA:

B = 0.75 FT.
D1 = 5.25 FT.
D2 = 4.25 FT.

EXTERNAL BEAM RESULTS:
F = 220304 LB.
M1 = 545738 LB.FT.
M2 = 5592 LB.FT.
S1 = -33794 LB.
S2 = -2884 LB.

INTERNAL BEAM RESULTS:
F = 332703 LB.
M1 = 211374 LB.FT.
S1 = -28113 LB.

EXTERNAL-HALF SHELL RESULTS:

POINT	T1	S	M2	T2
0	-6405	-8239	133	-243
1	-13825	-5457	346	-1748
2	-10611	-2273	-3	-2520
3	-3972	-476	-275	-2722
4	-845	0	-358	-2730

INTERNAL-HALF SHELL RESULTS:

POINT	T1	S	M2	T2
0	9690	-6221	-1146	725
1	-4090	-6781	409	-727
2	-10151	-4845	329	-2029
3	-9886	-2282	-171	-2808
4	-8804	0	-407	-3050

L= 86 TABLE V-142 W= 34

SHELL DATA:

L = 86.0 FT.
W = 34.0 FT.
R = 27.5 FT.
T = 2.5 IN.
PHI= 37.2 DEG.
G = 50.0 LB/FT2

BEAM DATA:

B = 0.75 FT.
D1 = 5.25 FT.
D2 = 4.25 FT.

EXTERNAL BEAM RESULTS:
F = 207263 LB.
M1 = 500304 LB.FT.
M2 = 4380 LB.FT.
S1 = -30570 LB.
S2 = -2685 LB.

INTERNAL BEAM RESULTS:
F = 353829 LB.
M1 = 271286 LB.FT.
S1 = -30892 LB.

EXTERNAL-HALF SHELL RESULTS:

POINT	T1	S	M2	T2
0	-5907	-7571	700	-144
1	-12885	-5932	494	-1108
2	-13402	-3724	-18	-1784
3	-11252	-1706	-457	-2153
4	-10050	0	-622	-2267

INTERNAL-HALF SHELL RESULTS:

POINT	T1	S	M2	T2
0	8566	-6462	-557	518
1	-4129	-6767	218	-435
2	-12528	-5350	196	-1330
3	-16922	-2901	-67	-1944
4	-18223	0	-200	-2160

L= 86 TABLE V-143 W= 36

SHELL DATA:

L = 86.0 FT.
W = 36.0 FT.
R = 30.0 FT.
T = 2.5 IN.
PHI= 36.0 DEG.
G = 50.0 LB/FT2

BEAM DATA:

B = 0.75 FT.
D1 = 5.25 FT.
D2 = 4.25 FT.

EXTERNAL BEAM RESULTS:
F = 211862 LB.
M1 = 514473 LB.FT.
M2 = 4433 LB.FT.
S1 = -31362 LB.
S2 = -2745 LB.

INTERNAL BEAM RESULTS:
F = 360020 LB.
M1 = 270976 LB.FT.
S1 = -31250 LB.

EXTERNAL-HALF SHELL RESULTS:

POINT	T1	S	M2	T2
0	-6172	-7739	603	-162
1	-12975	-5974	477	-1202
2	-13046	-3665	-7	-1917
3	-10436	-1638	-436	-2297
4	-9040	0	-599	-2411

INTERNAL-HALF SHELL RESULTS:

POINT	T1	S	M2	T2
0	8905	-6575	-627	554
1	-3942	-6939	239	-483
2	-12256	-5480	213	-1457
3	-16422	-2962	-82	-2124
4	-17604	0	-229	-2358

L= 86 TABLE V-144 W= 38

SHELL DATA:

L = 86.0 FT.
W = 38.0 FT.
R = 32.5 FT.
T = 2.5 IN.
PHI= 35.0 DEG.
G = 50.0 LB/FT2

BEAM DATA:

B = 0.75 FT.
D1 = 5.25 FT.
D2 = 4.25 FT.

EXTERNAL BEAM RESULTS:
F = 216549 LB.
M1 = 526852 LB.FT.
M2 = 4467 LB.FT.
S1 = -32093 LB.
S2 = -2808 LB.

INTERNAL BEAM RESULTS:
F = 365555 LB.
M1 = 270470 LB.FT.
S1 = -31562 LB.

EXTERNAL-HALF SHELL RESULTS:

POINT	T1	S	M2	T2
0	-6372	-7910	507	-179
1	-13064	-6020	462	-1298
2	-12740	-3606	3	-2051
3	-9698	-1569	-415	-2440
4	-8117	0	-575	-2554

INTERNAL-HALF SHELL RESULTS:

POINT	T1	S	M2	T2
0	9217	-6676	-699	590
1	-3798	-7097	261	-531
2	-12011	-5593	231	-1586
3	-15912	-3009	-95	-2304
4	-16952	0	-258	-2555

L= 86 TABLE V-145 W= 40

SHELL DATA:

L = 86.0 FT.
W = 40.0 FT.
R = 32.5 FT.
T = 2.5 IN.
PHI= 37.1 DEG.
G = 50.0 LB/FT2

BEAM DATA:

B = 0.75 FT.
D1 = 5.25 FT.
D2 = 4.25 FT.

EXTERNAL BEAM RESULTS:
F = 214332 LB.
M1 = 538703 LB.FT.
M2 = 5150 LB.FT.
S1 = -32395 LB.
S2 = -2753 LB.

INTERNAL BEAM RESULTS:
F = 349921 LB.
M1 = 241390 LB.FT.
S1 = -29573 LB.

EXTERNAL-HALF SHELL RESULTS:

POINT	T1	S	M2	T2
0	-6644	-7828	484	-186
1	-13472	-5732	450	-1332
2	-12131	-3175	-3	-2053
3	-7874	-1250	-402	-2384
4	-5771	0	-550	-2471

INTERNAL-HALF SHELL RESULTS:

POINT	T1	S	M2	T2
0	9478	-6391	-791	636
1	-3427	-6896	291	-501
2	-11210	-5404	255	-1577
3	-14458	-2877	-112	-2303
4	-15179	0	-294	-2555

L= 86 TABLE V-146 W= 42

SHELL DATA:

L = 86.0 FT.
W = 42.0 FT.
R = 35.0 FT.
T = 2.5 IN.
PHI= 36.1 DEG.
G = 50.0 LB/FT2

BEAM DATA:

B = 0.75 FT.
D1 = 5.25 FT.
D2 = 4.25 FT.

EXTERNAL BEAM RESULTS:
F = 219649 LB.
M1 = 548010 LB.FT.
M2 = 5091 LB.FT.
S1 = -33051 LB.
S2 = -2827 LB.

INTERNAL BEAM RESULTS:
F = 355950 LB.
M1 = 243653 LB.FT.
S1 = -30015 LB.

EXTERNAL-HALF SHELL RESULTS:

POINT	T1	S	M2	T2
0	-6736	-8023	388	-201
1	-13553	-5806	435	-1433
2	-11956	-3136	7	-2194
3	-7377	-1189	-380	-2531
4	-5130	0	-524	-2616

INTERNAL-HALF SHELL RESULTS:

POINT	T1	S	M2	T2
0	9712	-6501	-865	668
1	-3448	-7047	316	-556
2	-11116	-5485	274	-1710
3	-14003	-2894	-125	-2481
4	-14529	0	-322	-2747

L= 86 TABLE V-147 W= 44

SHELL DATA:

L = 86.0 FT.
W = 44.0 FT.
R = 37.5 FT.
T = 2.5 IN.
PHI= 35.2 DEG.
G = 50.0 LB/FT2

BEAM DATA:

B = 0.75 FT.
D1 = 5.25 FT.
D2 = 4.25 FT.

EXTERNAL BEAM RESULTS:
F = 224954 LB.
M1 = 555848 LB.FT.
M2 = 5029 LB.FT.
S1 = -33652 LB.
S2 = -2902 LB.

INTERNAL BEAM RESULTS:
F = 361701 LB.
M1 = 245972 LB.FT.
S1 = -30442 LB.

EXTERNAL-HALF SHELL RESULTS:

POINT	T1	S	M2	T2
0	-6775	-8217	295	-215
1	-13629	-5882	423	-1536
2	-11807	-3096	18	-2337
3	-6919	-1127	-360	-2679
4	-4531	0	-500	-2761

INTERNAL-HALF SHELL RESULTS:

POINT	T1	S	M2	T2
0	9930	-6606	-941	698
1	-3514	-7186	340	-615
2	-11052	-5547	293	-1845
3	-13530	-2895	-138	-2659
4	-13835	0	-349	-2937

L= 86 TABLE V-148 W= 46

SHELL DATA:

L = 86.0 FT.
W = 46.0 FT.
R = 37.5 FT.
T = 2.5 IN.
PHI= 37.1 DEG.
G = 50.0 LB/FT2

BEAM DATA:

B = 0.75 FT.
D1 = 5.25 FT.
D2 = 4.25 FT.

EXTERNAL BEAM RESULTS:
F = 224697 LB.
M1 = 561607 LB.FT.
M2 = 5534 LB.FT.
S1 = -33848 LB.
S2 = -2880 LB.

INTERNAL BEAM RESULTS:
F = 349332 LB.
M1 = 226386 LB.FT.
S1 = -28993 LB.

EXTERNAL-HALF SHELL RESULTS:

POINT	T1	S	M2	T2
0	-6791	-8208	274	-222
1	-14023	-5675	402	-1584
2	-11460	-2721	4	-2352
3	-5603	-834	-339	-2629
4	-2798	0	-460	-2679

INTERNAL-HALF SHELL RESULTS:

POINT	T1	S	M2	T2
0	10008	-6380	-1024	718
1	-3546	-6991	369	-613
2	-10621	-5294	312	-1852
3	-12214	-2693	-152	-2651
4	-12066	0	-377	-2518

L= 86 TABLE V-149 W= 48

SHELL DATA:

L = 86.0 FT.
W = 48.0 FT.
R = 40.0 FT.
T = 2.5 IN.
PHI= 36.2 DEG.
G = 50.0 LB/FT2

BEAM DATA:

B = 0.75 FT.
D1 = 5.25 FT.
D2 = 4.25 FT.

EXTERNAL BEAM RESULTS:
F = 230236 LB.
M1 = 567073 LB.FT.
M2 = 5415 LB.FT.
S1 = -34376 LB.
S2 = -2960 LB.

INTERNAL BEAM RESULTS:
F = 356067 LB.
M1 = 230842 LB.FT.
S1 = -29556 LB.

EXTERNAL-HALF SHELL RESULTS:

POINT	T1	S	M2	T2
0	-6753	-8410	188	-234
1	-14074	-5769	393	-1692
2	-11374	-2695	13	-2501
3	-5287	-779	-322	-2781
4	-2373	0	-438	-2826

INTERNAL-HALF SHELL RESULTS:

POINT	T1	S	M2	T2
0	10198	-6503	-1097	741
1	-3747	-7129	393	-682
2	-10683	-5322	328	-1995
3	-11769	-2658	-164	-2826
4	-11346	0	-401	-3099

L= 86 TABLE V-150 W= 50

SHELL DATA:

L = 86.0 FT.
W = 50.0 FT.
R = 40.0 FT.
T = 2.5 IN.
PHI= 38.0 DEG.
G = 50.0 LB/FT2

BEAM DATA:

B = 0.75 FT.
D1 = 5.25 FT.
D2 = 4.25 FT.

EXTERNAL BEAM RESULTS:
F = 230770 LB.
M1 = 568748 LB.FT.
M2 = 5842 LB.FT.
S1 = -34469 LB.
S2 = -2952 LB.

INTERNAL BEAM RESULTS:
F = 346440 LB.
M1 = 216502 LB.FT.
S1 = -28462 LB.

EXTERNAL-HALF SHELL RESULTS:

POINT	T1	S	M2	T2
0	-6606	-8430	180	-240
1	-14406	-5606	368	-1747
2	-11109	-2358	-6	-2525
3	-4237	-511	-298	-2735
4	-996	0	-388	-2746

INTERNAL-HALF SHELL RESULTS:

POINT	T1	S	M2	T2
0	10225	-6327	-1164	741
1	-4010	-6947	417	-705
2	-10472	-5026	339	-2017
3	-10525	-2404	-174	-2812
4	-9566	0	-417	-3064

L= 88 TABLE V-151 W= 34

SHELL DATA:

L = 88.0 FT.
W = 34.0 FT.
R = 27.5 FT.
T = 2.5 IN.
PHI= 37.2 DEG.
G = 50.0 LB/FT2

BEAM DATA:

B = 0.75 FT.
D1 = 5.50 FT.
D2 = 4.50 FT.

EXTERNAL BEAM RESULTS:
F = 213967 LB.
M1 = 536691 LB.FT.
M2 = 4204 LB.FT.
S1 = -32518 LB.
S2 = -2720 LB.

INTERNAL BEAM RESULTS:
F = 368352 LB.
M1 = 306151 LB.FT.
S1 = -33920 LB.

EXTERNAL-HALF SHELL RESULTS:

POINT	T1	S	M2	T2
0	-6312	-7638	655	-148
1	-13014	-6001	480	-1097
2	-13693	-3814	-11	-1768
3	-11827	-1773	-440	-2141
4	-10755	0	-603	-2257

INTERNAL-HALF SHELL RESULTS:

POINT	T1	S	M2	T2
0	7869	-6575	-539	467
1	-4588	-6782	213	-470
2	-12874	-5335	192	-1340
3	-17253	-2888	-64	-1937
4	-18561	0	-192	-2146

L= 88 TABLE V-152 W= 36

SHELL DATA:

L = 88.0 FT.
W = 36.0 FT.
R = 30.0 FT.
T = 2.5 IN.
PHI= 36.0 DEG.
G = 50.0 LB/FT2

BEAM DATA:

B = 0.75 FT.
D1 = 5.50 FT.
D2 = 4.50 FT.

EXTERNAL BEAM RESULTS:
F = 218612 LB.
M1 = 550901 LB.FT.
M2 = 4252 LB.FT.
S1 = -33317 LB.
S2 = -2780 LB.

INTERNAL BEAM RESULTS:
F = 374992 LB.
M1 = 305696 LB.FT.
S1 = -34321 LB.

EXTERNAL-HALF SHELL RESULTS:

POINT	T1	S	M2	T2
0	-6555	-7804	560	-166
1	-13070	-6047	464	-1190
2	-13315	-3764	0	-1900
3	-11008	-1712	-420	-2285
4	-9750	0	-581	-2403

INTERNAL-HALF SHELL RESULTS:

POINT	T1	S	M2	T2
0	8220	-6693	-608	500
1	-4380	-6956	233	-519
2	-12588	-5468	208	-1469
3	-16753	-2952	-78	-2116
4	-17949	0	-221	-2344

L= 88 TABLE V-153 W= 38

SHELL DATA:

L = 88.0 FT.
W = 38.0 FT.
R = 32.5 FT.
T = 2.5 IN.
PHI= 35.0 DEG.
G = 50.0 LB/FT2

BEAM DATA:

B = 0.75 FT.
D1 = 5.50 FT.
D2 = 4.50 FT.

EXTERNAL BEAM RESULTS:
F = 223334 LB.
M1 = 563319 LB.FT.
M2 = 4284 LB.FT.
S1 = -34056 LB.
S2 = -2841 LB.

INTERNAL BEAM RESULTS:
F = 380942 LB.
M1 = 305010 LB.FT.
S1 = -34670 LB.

EXTERNAL-HALF SHELL RESULTS:

POINT	T1	S	M2	T2
0	-6740	-7973	465	-182
1	-13127	-6097	451	-1283
2	-12989	-3715	11	-2034
3	-10266	-1651	-400	-2430
4	-8831	0	-559	-2548

INTERNAL-HALF SHELL RESULTS:

POINT	T1	S	M2	T2
0	8545	-6799	-679	534
1	-4214	-7117	254	-570
2	-12328	-5583	225	-1598
3	-16244	-3002	-92	-2296
4	-17308	0	-249	-2541

```
L= 88         TABLE V-154          W= 40
****************************************

SHELL DATA:

    L   = 88.0 FT.
    W   = 40.0 FT.
    R   = 32.5 FT.
    T   =  2.5 IN.
    PHI= 37.1 DEG.
    G   = 50.0 LB/FT2

BEAM DATA:

    B  = 0.75 FT.
    D1 = 5.50 FT.
    D2 = 4.50 FT.

EXTERNAL BEAM RESULTS:
    F  = 220782 LB.
    M1 = 576256 LB.FT.
    M2 =   4969 LB.FT.
    S1 = -34358 LB.
    S2 =  -2783 LB.

INTERNAL BEAM RESULTS:
    F  = 365440 LB.
    M1 = 272718 LB.FT.
    S1 = -32550 LB.
```

EXTERNAL-HALF SHELL RESULTS:

POINT	T1	S	M2	T2
0	-7045	-7881	445	-190
1	-13489	-5804	442	-1316
2	-12328	-3291	4	-2035
3	-8405	-1340	-391	-2378
4	-6456	0	-540	-2471

INTERNAL-HALF SHELL RESULTS:

POINT	T1	S	M2	T2
0	8894	-6523	-771	586
1	-3782	-6931	285	-535
2	-11505	-5414	250	-1586
3	-14814	-2884	-108	-2296
4	-15579	0	-286	-2542

```
L= 88         TABLE V-155          W= 42
****************************************

SHELL DATA:

    L   = 88.0 FT.
    W   = 42.0 FT.
    R   = 35.0 FT.
    T   =  2.5 IN.
    PHI= 36.1 DEG.
    G   = 50.0 LB/FT2

BEAM DATA:

    B  = 0.75 FT.
    D1 = 5.50 FT.
    D2 = 4.50 FT.

EXTERNAL BEAM RESULTS:
    F  = 226136 LB.
    M1 = 585603 LB.FT.
    M2 =   4908 LB.FT.
    S1 = -35027 LB.
    S2 =  -2856 LB.

INTERNAL BEAM RESULTS:
    F  = 371770 LB.
    M1 = 275029 LB.FT.
    S1 = -33029 LB.
```

EXTERNAL-HALF SHELL RESULTS:

POINT	T1	S	M2	T2
0	-7127	-8073	348	-204
1	-13550	-5880	428	-1415
2	-12139	-3259	16	-2175
3	-7903	-1286	-370	-2526
4	-5813	0	-516	-2618

INTERNAL-HALF SHELL RESULTS:

POINT	T1	S	M2	T2
0	9133	-6636	-845	616
1	-3781	-7082	309	-591
2	-11393	-5499	268	-1719
3	-14360	-2906	-121	-2474
4	-14940	0	-314	-2734

```
L= 88         TABLE V-156          W= 44
****************************************

SHELL DATA:

    L   = 88.0 FT.
    W   = 44.0 FT.
    R   = 37.5 FT.
    T   =  2.5 IN.
    PHI= 35.2 DEG.
    G   = 50.0 LB/FT2

BEAM DATA:

    B  = 0.75 FT.
    D1 = 5.50 FT.
    D2 = 4.50 FT.

EXTERNAL BEAM RESULTS:
    F  = 231477 LB.
    M1 = 593456 LB.FT.
    M2 =   4843 LB.FT.
    S1 = -35642 LB.
    S2 =  -2930 LB.

INTERNAL BEAM RESULTS:
    F  = 377788 LB.
    M1 = 277390 LB.FT.
    S1 = -33490 LB.
```

EXTERNAL-HALF SHELL RESULTS:

POINT	T1	S	M2	T2
0	-7159	-8263	256	-218
1	-13608	-5958	417	-1515
2	-11977	-3226	27	-2317
3	-7439	-1230	-351	-2675
4	-5211	0	-494	-2765

INTERNAL-HALF SHELL RESULTS:

POINT	T1	S	M2	T2
0	9354	-6743	-920	644
1	-3824	-7222	333	-650
2	-11310	-5565	287	-1855
3	-13888	-2911	-134	-2652
4	-14259	0	-341	-2924

L= 88 TABLE V-157 W= 46

SHELL DATA:

L = 88.0 FT.
W = 46.0 FT.
R = 37.5 FT.
T = 2.5 IN.
PHI= 37.1 DEG.
G = 50.0 LB/FT2

BEAM DATA:

B = 0.75 FT.
D1 = 5.50 FT.
D2 = 4.50 FT.

EXTERNAL BEAM RESULTS:
F = 230954 LB.
M1 = 600224 LB.FT.
M2 = 5350 LB.FT.
S1 = -35851 LB.
S2 = -2905 LB.

INTERNAL BEAM RESULTS:
F = 365219 LB.
M1 = 255374 LB.FT.
S1 = -31920 LB.

EXTERNAL-HALF SHELL RESULTS:

POINT	T1	S	M2	T2
0	-7218	-8245	235	-225
1	-13978	-5744	400	-1561
2	-11595	-2855	15	-2331
3	-6087	-943	-333	-2628
4	-3439	0	-458	-2688

INTERNAL-HALF SHELL RESULTS:

POINT	T1	S	M2	T2
0	9491	-6519	-1005	670
1	-3800	-7036	362	-643
2	-10848	-5330	307	-1858
3	-12583	-2724	-149	-2645
4	-12523	0	-371	-2909

L= 88 TABLE V-158 W= 48

SHELL DATA:

L = 88.0 FT.
W = 48.0 FT.
R = 40.0 FT.
T = 2.5 IN.
PHI= 36.2 DEG.
G = 50.0 LB/FT2

BEAM DATA:

B = 0.75 FT.
D1 = 5.50 FT.
D2 = 4.50 FT.

EXTERNAL BEAM RESULTS:
F = 236556 LB.
M1 = 605628 LB.FT.
M2 = 5228 LB.FT.
S1 = -36394 LB.
S2 = -2984 LB.

INTERNAL BEAM RESULTS:
F = 372150 LB.
M1 = 260093 LB.FT.
S1 = -32522 LB.

EXTERNAL-HALF SHELL RESULTS:

POINT	T1	S	M2	T2
0	-7174	-8445	148	-237
1	-14017	-5839	391	-1666
2	-11504	-2835	25	-2479
3	-5766	-892	-316	-2780
4	-3007	0	-438	-2837

INTERNAL-HALF SHELL RESULTS:

POINT	T1	S	M2	T2
0	9676	-6642	-1078	692
1	-3980	-7174	387	-712
2	-10891	-5363	324	-2001
3	-12139	-2695	-160	-2820
4	-11814	0	-395	-3091

L= 88 TABLE V-159 W= 50

SHELL DATA:

L = 88.0 FT.
W = 50.0 FT.
R = 40.0 FT.
T = 2.5 IN.
PHI= 38.0 DEG.
G = 50.0 LB/FT2

BEAM DATA:

B = 0.75 FT.
D1 = 5.50 FT.
D2 = 4.50 FT.

EXTERNAL BEAM RESULTS:
F = 236915 LB.
M1 = 608110 LB.FT.
M2 = 5655 LB.FT.
S1 = -36505 LB.
S2 = -2973 LB.

INTERNAL BEAM RESULTS:
F = 362185 LB.
M1 = 243865 LB.FT.
S1 = -31321 LB.

EXTERNAL-HALF SHELL RESULTS:

POINT	T1	S	M2	T2
0	-7070	-8457	138	-244
1	-14340	-5670	369	-1719
2	-11219	-2499	6	-2501
3	-4680	-627	-294	-2736
4	-1584	0	-392	-2761

INTERNAL-HALF SHELL RESULTS:

POINT	T1	S	M2	T2
0	9741	-6464	-1148	698
1	-4195	-6996	411	-729
2	-10649	-5081	336	-2019
3	-10898	-2455	-171	-2808
4	-10055	0	-413	-3059

```
L= 90        TABLE V-160        W= 34
***********************************

SHELL DATA:

    L   = 90.0 FT.
    W   = 34.0 FT.
    R   = 27.5 FT.
    T   =  2.5 IN.
    PHI= 37.2 DEG.
    G   = 50.0 LB/FT2

BEAM DATA:

    B  = 0.75 FT.
    D1 = 5.50 FT.
    D2 = 4.50 FT.

EXTERNAL BEAM RESULTS:
    F  = 224462 LB.
    M1 = 553046 LB.FT.
    M2 =   4288 LB.FT.
    S1 = -33007 LB.
    S2 =  -2795 LB.

INTERNAL BEAM RESULTS:
    F  = 385054 LB.
    M1 = 317917 LB.FT.
    S1 = -34595 LB.

EXTERNAL-HALF SHELL RESULTS:

POINT     T1      S      M2     T2

  0     -6321   -7835   716   -144
  1    -13458   -6196   507  -1098
  2    -14369   -3968   -15  -1780
  3    -12605   -1857  -467  -2162
  4    -11559       0  -638  -2281

INTERNAL-HALF SHELL RESULTS:

POINT     T1      S      M2     T2

  0      8300   -6720  -540    471
  1     -4724   -6944   215   -466
  2    -13443   -5473   194  -1339
  3    -18105   -2967   -62  -1937
  4    -19513       0  -191  -2148
```

```
L= 90        TABLE V-161        W= 36
***********************************

SHELL DATA:

    L   = 90.0 FT.
    W   = 36.0 FT.
    R   = 30.0 FT.
    T   =  2.5 IN.
    PHI= 36.0 DEG.
    G   = 50.0 LB/FT2

BEAM DATA:

    B  = 0.75 FT.
    D1 = 5.50 FT.
    D2 = 4.50 FT.

EXTERNAL BEAM RESULTS:
    F  = 229194 LB.
    M1 = 568534 LB.FT.
    M2 =   4356 LB.FT.
    S1 = -33838 LB.
    S2 =  -2853 LB.

INTERNAL BEAM RESULTS:
    F  = 391910 LB.
    M1 = 317148 LB.FT.
    S1 = -34990 LB.

EXTERNAL-HALF SHELL RESULTS:

POINT     T1      S      M2     T2

  0     -6597   -8000   620   -162
  1    -13524   -6238   491  -1191
  2    -13963   -3913    -4  -1912
  3    -11722   -1793  -447  -2306
  4    -10478       0  -617  -2427

INTERNAL-HALF SHELL RESULTS:

POINT     T1      S      M2     T2

  0      8673   -6840  -609    505
  1     -4494   -7123   235   -515
  2    -13136   -5611   210  -1466
  3    -17590   -3035   -77  -2117
  4    -18890       0  -221  -2346
```

```
L= 90        TABLE V-162        W= 38
***********************************

SHELL DATA:

    L   = 90.0 FT.
    W   = 38.0 FT.
    R   = 32.5 FT.
    T   =  2.5 IN.
    PHI= 35.0 DEG.
    G   = 50.0 LB/FT2

BEAM DATA:

    B  = 0.75 FT.
    D1 = 5.50 FT.
    D2 = 4.50 FT.

EXTERNAL BEAM RESULTS:
    F  = 234011 LB.
    M1 = 582184 LB.FT.
    M2 =   4405 LB.FT.
    S1 = -34609 LB.
    S2 =  -2914 LB.

INTERNAL BEAM RESULTS:
    F  = 398023 LB.
    M1 = 316108 LB.FT.
    S1 = -35329 LB.

EXTERNAL-HALF SHELL RESULTS:

POINT     T1      S      M2     T2

  0     -6813   -8168   524   -178
  1    -13592   -6283   477  -1285
  2    -13611   -3858     6  -2045
  3    -10922   -1728  -427  -2450
  4     -9488       0  -595  -2572

INTERNAL-HALF SHELL RESULTS:

POINT     T1      S      M2     T2

  0      9019   -6946  -682    540
  1     -4305   -7288   257   -564
  2    -12856   -5733   228  -1595
  3    -17067   -3090   -91  -2297
  4    -18237       0  -250  -2543
```

L= 90 TABLE V-163 W= 40

****

SHELL DATA:

L = 90.0 FT.
W = 40.0 FT.
R = 32.5 FT.
T = 2.5 IN.
PHI= 37.1 DEG.
G = 50.0 LB/FT2

BEAM DATA:

B = 0.75 FT.
D1 = 5.50 FT.
D2 = 4.50 FT.

EXTERNAL BEAM RESULTS:
F = 231166 LB.
M1 = 596205 LB.FT.
M2 = 5127 LB.FT.
S1 = -34925 LB.
S2 = -2852 LB.

INTERNAL BEAM RESULTS:
F = 381557 LB.
M1 = 282017 LB.FT.
S1 = -33134 LB.

EXTERNAL-HALF SHELL RESULTS:

POINT	T1	S	M2	T2
0	-7156	-8068	505	-186
1	-13973	-5978	469	-1316
2	-12906	-3420	0	-2045
3	-8953	-1408	-419	-2397
4	-6977	0	-577	-2493

INTERNAL-HALF SHELL RESULTS:

POINT	T1	S	M2	T2
0	9382	-6658	-775	593
1	-3828	-7097	288	-528
2	-11975	-5565	253	-1583
3	-15583	-2974	-108	-2297
4	-16458	0	-287	-2546

L= 90 TABLE V-164 W= 42

****

SHELL DATA:

L = 90.0 FT.
W = 42.0 FT.
R = 35.0 FT.
T = 2.5 IN.
PHI= 36.1 DEG.
G = 50.0 LB/FT2

BEAM DATA:

B = 0.75 FT.
D1 = 5.50 FT.
D2 = 4.50 FT.

EXTERNAL BEAM RESULTS:
F = 236676 LB.
M1 = 606678 LB.FT.
M2 = 5078 LB.FT.
S1 = -35628 LB.
S2 = -2925 LB.

INTERNAL BEAM RESULTS:
F = 388012 LB.
M1 = 284052 LB.FT.
S1 = -33601 LB.

EXTERNAL-HALF SHELL RESULTS:

POINT	T1	S	M2	T2
0	-7265	-8261	406	-201
1	-14046	-6050	454	-1415
2	-12700	-3382	12	-2184
3	-8408	-1350	-397	-2544
4	-6282	0	-551	-2639

INTERNAL-HALF SHELL RESULTS:

POINT	T1	S	M2	T2
0	9637	-6772	-851	624
1	-3807	-7252	312	-582
2	-11846	-5657	272	-1715
3	-15119	-3002	-122	-2476
4	-15814	0	-316	-2739

L= 90 TABLE V-165 W= 44

****

SHELL DATA:

L = 90.0 FT.
W = 44.0 FT.
R = 37.5 FT.
T = 2.5 IN.
PHI= 35.2 DEG.
G = 50.0 LB/FT2

BEAM DATA:

B = 0.75 FT.
D1 = 5.50 FT.
D2 = 4.50 FT.

EXTERNAL BEAM RESULTS:
F = 242182 LB.
M1 = 615549 LB.FT.
M2 = 5024 LB.FT.
S1 = -36274 LB.
S2 = -2999 LB.

INTERNAL BEAM RESULTS:
F = 394114 LB.
M1 = 286113 LB.FT.
S1 = -34047 LB.

EXTERNAL-HALF SHELL RESULTS:

POINT	T1	S	M2	T2
0	-7320	-8453	310	-214
1	-14115	-6125	442	-1515
2	-12523	-3345	23	-2325
3	-7906	-1290	-377	-2691
4	-5633	0	-528	-2785

INTERNAL-HALF SHELL RESULTS:

POINT	T1	S	M2	T2
0	9873	-6878	-927	654
1	-3831	-7396	337	-639
2	-11746	-5730	291	-1849
3	-14637	-3013	-135	-2654
4	-15127	0	-345	-2930

L= 90 TABLE V-166 W= 46

****

SHELL DATA:

L = 90.0 FT.
W = 46.0 FT.
R = 37.5 FT.
T = 2.5 IN.
PHI= 37.1 DEG.
G = 50.0 LB/FT2

BEAM DATA:

B = 0.75 FT.
D1 = 5.50 FT.
D2 = 4.50 FT.

EXTERNAL BEAM RESULTS:
F = 241534 LB.
M1 = 623174 LB.FT.
M2 = 5560 LB.FT.
S1 = -36501 LB.
S2 = -2972 LB.

INTERNAL BEAM RESULTS:
F = 380592 LB.
M1 = 262699 LB.FT.
S1 = -32405 LB.

EXTERNAL-HALF SHELL RESULTS:

POINT	T1	S	M2	T2
0	-7404	-8431	287	-222
1	-14511	-5901	424	-1560
2	-12118	-2960	12	-2338
3	-6474	-994	-358	-2643
4	-3756	0	-491	-2706

INTERNAL-HALF SHELL RESULTS:

POINT	T1	S	M2	T2
0	10011	-6642	-1015	632
1	-3773	-7204	367	-629
2	-11237	-5496	313	-1851
3	-13288	-2830	-150	-2648
4	-13350	0	-376	-2917

L= 90 TABLE V-167 W= 48

****

SHELL DATA:

L = 90.0 FT.
W = 48.0 FT.
R = 40.0 FT.
T = 2.5 IN.
PHI= 36.2 DEG.
G = 50.0 LB/FT2

BEAM DATA:

B = 0.75 FT.
D1 = 5.50 FT.
D2 = 4.50 FT.

EXTERNAL BEAM RESULTS:
F = 247338 LB.
M1 = 629396 LB.FT.
M2 = 5444 LB.FT.
S1 = -37073 LB.
S2 = -3051 LB.

INTERNAL BEAM RESULTS:
F = 387593 LB.
M1 = 267196 LB.FT.
S1 = -32990 LB.

EXTERNAL-HALF SHELL RESULTS:

POINT	T1	S	M2	T2
0	-7375	-8633	197	-234
1	-14559	-5995	414	-1665
2	-12016	-2936	22	-2485
3	-6126	-941	-340	-2794
4	-3291	0	-469	-2854

INTERNAL-HALF SHELL RESULTS:

POINT	T1	S	M2	T2
0	10207	-6764	-1090	705
1	-3938	-7345	392	-696
2	-11266	-5537	330	-1992
3	-12836	-2807	-162	-2824
4	-12636	0	-401	-3101

L= 90 TABLE V-168 W= 50

****

SHELL DATA:

L = 90.0 FT.
W = 50.0 FT.
R = 40.0 FT.
T = 2.5 IN.
PHI= 38.0 DEG.
G = 50.0 LB/FT2

BEAM DATA:

B = 0.75 FT.
D1 = 5.50 FT.
D2 = 4.50 FT.

EXTERNAL BEAM RESULTS:
F = 247665 LB.
M1 = 632442 LB.FT.
M2 = 5895 LB.FT.
S1 = -37199 LB.
S2 = -3040 LB.

INTERNAL BEAM RESULTS:
F = 376733 LB.
M1 = 249866 LB.FT.
S1 = -31722 LB.

EXTERNAL-HALF SHELL RESULTS:

POINT	T1	S	M2	T2
0	-7283	-8645	185	-241
1	-14909	-5818	390	-1717
2	-11719	-2589	3	-2506
3	-4975	-666	-317	-2748
4	-1778	0	-423	-2776

INTERNAL-HALF SHELL RESULTS:

POINT	T1	S	M2	T2
0	10266	-6575	-1164	713
1	-4129	-7161	418	-709
2	-10987	-5257	344	-2007
3	-11553	-2571	-174	-2812
4	-10834	0	-422	-3071

DESIGN TABLES
FOR
SYMMETRICAL CYLINDRICAL SHELL ROOFS
WITHOUT INTERNAL VALLEY BEAMS

L= 40 TABLE F- 1 W= 18

SHELL DATA:

L = 40.0 FT.
W = 18.0 FT.
R = 15.0 FT.
T = 2.5 IN.
PHI= 35.1 DEG.
G = 50.0 LB/FT2

BEAM DATA:

B = 0.75 FT.
D = 2.75 FT.

EXTERNAL BEAM RESULTS:
F = -61 LB.
M1 = 110426 LB.FT.
M2 = 700 LB.FT.
S1 = -8657 LB.
S2 = 41 LB.

EXTERNAL-HALF SHELL RESULTS:

POINT	T1	S	M2	T2
0	24896	4	131	-52
1	7718	-2838	188	-279
2	-3367	-3147	63	-794
3	-9409	-1926	-96	-1235
4	-11304	0	-166	-1403

INTERNAL-HALF SHELL RESULTS:

POINT	T1	S	M2	T2
0	37562	0	-252	759
1	12300	-4371	156	371
2	-4894	-4923	157	-459
3	-14753	-3046	8	-1171
4	-17944	0	-68	-1444

L= 40 TABLE F- 2 W= 20

SHELL DATA:

L = 40.0 FT.
W = 20.0 FT.
R = 15.0 FT.
T = 2.5 IN.
PHI= 39.9 DEG.
G = 50.0 LB/FT2

BEAM DATA:

B = 0.75 FT.
D = 2.75 FT.

EXTERNAL BEAM RESULTS:
F = -3076 LB.
M1 = 103656 LB.FT.
M2 = 1545 LB.FT
S1 = -8454 LB.
S2 = 196 LB.

EXTERNAL-HALF SHELL RESULTS:

POINT	T1	S	M2	T2
0	23782	241	218	-38
1	6677	-2744	248	-260
2	-3190	-2994	55	-813
3	-7851	-1791	-172	-1278
4	-9154	0	-269	-1452

INTERNAL-HALF SHELL RESULTS:

POINT	T1	S	M2	T2
0	28790	0	-331	672
1	9156	-3775	157	311
2	-3885	-4208	155	-483
3	-11075	-2581	-23	-1162
4	-13329	0	-116	-1420

L= 40 TABLE F- 3 W= 22

SHELL DATA:

L = 40.0 FT.
W = 22.0 FT.
R = 17.5 FT.
T = 2.5 IN.
PHI= 37.4 DEG.
G = 50.0 LB/FT2

BEAM DATA:

B = 0.75 FT.
D = 2.75 FT.

EXTERNAL BEAM RESULTS:
F = -5812 LB.
M1 = 105724 LB.FT.
M2 = 1562 LB.FT.
S1 = -8915 LB.
S2 = 281 LB.

EXTERNAL-HALF SHELL RESULTS:

POINT	T1	S	M2	T2
0	23975	456	131	-52
1	6576	-2804	239	-290
2	-3077	-3072	64	-917
3	-7372	-1824	-169	-1440
4	-8504	0	-271	-1634

INTERNAL-HALF SHELL RESULTS:

POINT	T1	S	M2	T2
0	27998	0	-401	756
1	8609	-3967	170	328
2	-3896	-4375	166	-586
3	-10522	-2661	-44	-1359
4	-12532	0	-152	-1651

L= 40 TABLE F- 4 W= 24
****

SHELL DATA:

L = 40.0 FT.
W = 24.0 FT.
R = 17.5 FT.
T = 2.5 IN.
PHI= 41.6 DEG.
G = 50.0 LB/FT2

BEAM DATA:

B = 0.75 FT.
D = 2.75 FT.

EXTERNAL BEAM RESULTS:
F = -9495 LB.
M1 = 102737 LB.FT.
M2 = 2263 LB.FT.
S1 = -9080 LB.
S2 = 444 LB.

EXTERNAL-HALF SHELL RESULTS:

POINT	T1	S	M2	T2
0	23511	745	110	-59
1	5537	-2656	261	-270
2	-3118	-2797	71	-907
3	-5989	-1580	-190	-1414
4	-6475	0	-304	-1595

INTERNAL-HALF SHELL RESULTS:

POINT	T1	S	M2	T2
0	22971	0	-470	679
1	6588	-3553	185	270
2	-3454	-3818	177	-612
3	-8223	-2270	-62	-1343
4	-9512	0	-184	-1614

L= 40 TABLE F- 5 W= 26
****

SHELL DATA:

L = 40.0 FT.
W = 26.0 FT.
R = 20.0 FT.
T = 2.5 IN.
PHI= 39.1 DEG.
G = 50.0 LB/FT2

BEAM DATA:

B = 0.75 FT.
D = 2.75 FT.

EXTERNAL BEAM RESULTS:
F = -11541 LB.
M1 = 105279 LB.FT.
M2 = 2188 LB.FT.
S1 = -9502 LB.
S2 = 501 LB.

EXTERNAL-HALF SHELL RESULTS:

POINT	T1	S	M2	T2
0	23804	906	17	-71
1	5378	-2751	256	-306
2	-3167	-2866	85	-1021
3	-5699	-1590	-182	-1579
4	-6012	0	-300	-1775

INTERNAL-HALF SHELL RESULTS:

POINT	T1	S	M2	T2
0	22960	0	-541	759
1	6219	-3754	206	277
2	-3620	-3956	193	-725
3	-7908	-2312	-76	-1536
4	-8951	0	-213	-1834

L= 40 TABLE F- 6 W= 28
****

SHELL DATA:

L = 40.0 FT.
W = 28.0 FT.
R = 22.5 FT.
T = 2.5 IN.
PHI= 37.3 DEG.
G = 50.0 LB/FT2

BEAM DATA:

B = 0.75 FT.
D = 2.75 FT.

EXTERNAL BEAM RESULTS:
F = -13297 LB.
M1 = 107481 LB.FT.
M2 = 2143 LB.FT.
S1 = -9866 LB.
S2 = 551 LB.

EXTERNAL-HALF SHELL RESULTS:

POINT	T1	S	M2	T2
0	24076	1044	-67	-80
1	5130	-2858	256	-345
2	-3279	-2917	99	-1141
3	-5398	-1576	-177	-1744
4	-5499	0	-299	-1953

INTERNAL-HALF SHELL RESULTS:

POINT	T1	S	M2	T2
0	22890	0	-612	832
1	5781	-3933	228	275
2	-3808	-4047	210	-845
3	-7527	-2312	-90	-1727
4	-8279	0	-240	-2046

L= 42 TABLE F- 7 W= 18

****

SHELL DATA:

L = 42.0 FT.
W = 18.0 FT.
R = 15.0 FT.
T = 2.5 IN.
PHI= 35.1 DEG.
G = 50.0 LB/FT2

BEAM DATA:

B = 0.75 FT.
D = 2.75 FT.

EXTERNAL BEAM RESULTS:

F = 1056 LB.
M1 = 121028 LB.FT.
M2 = 755 LB.FT.
S1 = -8918 LB.
S2 = 8 LB.

EXTERNAL-HALF SHELL RESULTS:

POINT	T1	S	M2	T2
0	27390	-79	185	-46
1	8480	-3056	215	-288
2	-3821	-3372	67	-816
3	-10580	-2063	-110	-1266
4	-12712	0	-186	-1438

INTERNAL-HALF SHELL RESULTS:

POINT	T1	S	M2	T2
0	41255	0	-243	761
1	13611	-4584	167	374
2	-5342	-5175	169	-456
3	-16288	-3207	20	-1170
4	-19848	0	-57	-1443

L= 42 TABLE F- 8 W= 20

****

SHELL DATA:

L = 42.0 FT.
W = 20.0 FT.
R = 15.0 FT.
T = 2.5 IN.
PHI= 39.9 DEG.
G = 50.0 LB/FT2

BEAM DATA:

B = 0.75 FT.
D = 2.75 FT.

EXTERNAL BEAM RESULTS:

F = -1968 LB.
M1 = 112965 LB.FT.
M2 = 1670 LB.FT.
S1 = -8630 LB.
S2 = 161 LB.

EXTERNAL-HALF SHELL RESULTS:

POINT	T1	S	M2	T2
0	26047	147	296	-29
1	7318	-2971	284	-269
2	-3613	-3222	54	-839
3	-8864	-1928	-199	-1316
4	-10357	0	-305	-1495

INTERNAL-HALF SHELL RESULTS:

POINT	T1	S	M2	T2
0	31549	0	-327	673
1	10157	-3956	163	314
2	-4208	-4427	163	-480
3	-12240	-2724	-17	-1161
4	-14785	0	-111	-1421

L= 42 TABLE F- 9 W= 22

****

SHELL DATA:

L = 42.0 FT.
W = 22.0 FT.
R = 17.5 FT.
T = 2.5 IN.
PHI= 37.4 DEG.
G = 50.0 LB/FT2

BEAM DATA:

B = 0.75 FT.
D = 2.75 FT.

EXTERNAL BEAM RESULTS:

F = -5254 LB.
M1 = 115316 LB.FT.
M2 = 1703 LB.FT.
S1 = -9147 LB.
S2 = 260 LB.

EXTERNAL-HALF SHELL RESULTS:

POINT	T1	S	M2	T2
0	26259	393	203	-44
1	7241	-3015	272	-298
2	-3443	-3294	62	-940
3	-8297	-1960	-196	-1476
4	-9606	0	-307	-1675

INTERNAL-HALF SHELL RESULTS:

POINT	T1	S	M2	T2
0	30629	0	-400	758
1	9571	-4155	175	332
2	-4199	-4606	172	-582
3	-11640	-2813	-39	-1359
4	-13935	0	-149	-1653

L= 42 TABLE F- 10 W= 24

****

SHELL DATA:

```
L   = 42.0 FT.
W   = 24.0 FT.
R   = 17.5 FT.
T   =  2.5 IN.
PHI= 41.6 DEG.
G   = 50.0 LB/FT2
```

BEAM DATA:

```
B   = 0.75 FT.
D   = 2.75 FT.
```

EXTERNAL BEAM RESULTS:

```
F  =   -9483 LB.
M1 =  111803 LB.FT.
M2 =    2472 LB.FT.
S1 =   -9321 LB.
S2 =     436 LB.
```

EXTERNAL-HALF SHELL RESULTS:

POINT	T1	S	M2	T2
0	25679	709	187	-50
1	6127	-2843	296	-275
2	-3433	-2999	68	-926
3	-6745	-1705	-221	-1446
4	-7361	0	-345	-1634

INTERNAL-HALF SHELL RESULTS:

POINT	T1	S	M2	T2
0	25009	0	-472	681
1	7369	-3714	190	277
2	-3670	-4029	183	-606
3	-9120	-2414	-59	-1343
4	-10655	0	-184	-1619

L= 42 TABLE F- 11 W= 26

****

SHELL DATA:

```
L   = 42.0 FT.
W   = 26.0 FT.
R   = 20.0 FT.
T   =  2.5 IN.
PHI= 39.1 DEG.
G   = 50.0 LB/FT2
```

BEAM DATA:

```
B   = 0.75 FT.
D   = 2.75 FT.
```

EXTERNAL BEAM RESULTS:

```
F  =  -12008 LB.
M1 =  114717 LB.FT.
M2 =    2403 LB.FT.
S1 =   -9801 LB.
S2 =     505 LB.
```

EXTERNAL-HALF SHELL RESULTS:

POINT	T1	S	M2	T2
0	26011	898	86	-64
1	5990	-2928	287	-310
2	-3450	-3068	82	-1037
3	-6411	-1717	-211	-1609
4	-6848	0	-339	-1812

INTERNAL-HALF SHELL RESULTS:

POINT	T1	S	M2	T2
0	24926	0	-544	762
1	6986	-3919	210	286
2	-3820	-4180	199	-716
3	-8785	-2468	-75	-1538
4	-10078	0	-215	-1841

L= 42 TABLE F- 12 W= 28

****

SHELL DATA:

```
L   = 42.0 FT.
W   = 28.0 FT.
R   = 22.5 FT.
T   =  2.5 IN.
PHI= 37.3 DEG.
G   = 50.0 LB/FT2
```

BEAM DATA:

```
B   = 0.75 FT.
D   = 2.75 FT.
```

EXTERNAL BEAM RESULTS:

```
F  =  -14194 LB.
M1 =  117216 LB.FT.
M2 =    2365 LB.FT.
S1 =  -10215 LB.
S2 =     565 LB.
```

EXTERNAL-HALF SHELL RESULTS:

POINT	T1	S	M2	T2
0	26310	1061	-6	-74
1	5755	-3027	284	-346
2	-3537	-3120	97	-1153
3	-6071	-1708	-203	-1773
4	-6292	0	-336	-1989

INTERNAL-HALF SHELL RESULTS:

POINT	T1	S	M2	T2
0	24767	0	-618	836
1	6532	-4101	233	287
2	-3989	-4283	216	-833
3	-8381	-2481	-90	-1729
4	-9384	0	-244	-2056

L= 44 TABLE F- 13 W= 20

SHELL DATA:

L = 44.0 FT.
W = 20.0 FT.
R = 15.0 FT.
T = 2.5 IN.
PHI= 39.9 DEG.
G = 50.0 LB/FT2

BEAM DATA:

B = 0.75 FT.
D = 3.00 FT.

EXTERNAL BEAM RESULTS:
F = -1055 LB.
M1 = 136582 LB.FT.
M2 = 1420 LB.FT.
S1 = -9842 LB.
S2 = 110 LB.

EXTERNAL-HALF SHELL RESULTS:

POINT	T1	S	M2	T2
0	26247	75	235	-46
1	7668	-2963	259	-274
2	-3630	-3238	63	-815
3	-9348	-1955	-166	-1273
4	-11042	0	-264	-1446

INTERNAL-HALF SHELL RESULTS:

POINT	T1	S	M2	T2
0	34447	0	-322	675
1	11205	-4137	171	317
2	-4550	-4645	171	-477
3	-13461	-2865	-9	-1160
4	-16309	0	-104	-1421

L= 44 TABLE F- 14 W= 22

SHELL DATA:

L = 44.0 FT.
W = 22.0 FT.
R = 17.5 FT.
T = 2.5 IN.
PHI= 37.4 DEG.
G = 50.0 LB/FT2

BEAM DATA:

B = 0.75 FT.
D = 3.00 FT.

EXTERNAL BEAM RESULTS:
F = -4054 LB.
M1 = 138830 LB.FT.
M2 = 1462 LB.FT.
S1 = -10328 LB.
S2 = 198 LB.

EXTERNAL-HALF SHELL RESULTS:

POINT	T1	S	M2	T2
0	26417	289	151	-59
1	7593	-3029	250	-305
2	-3466	-3334	70	-920
3	-8811	-2004	-168	-1438
4	-10334	0	-272	-1633

INTERNAL-HALF SHELL RESULTS:

POINT	T1	S	M2	T2
0	33395	0	-397	760
1	10577	-4344	181	336
2	-4521	-4836	178	-579
3	-12811	-2964	-34	-1358
4	-15401	0	-145	-1654

L= 44 TABLE F- 15 W= 24

SHELL DATA:

L = 44.0 FT.
W = 24.0 FT.
R = 17.5 FT.
T = 2.5 IN.
PHI= 41.6 DEG.
G = 50.0 LB/FT2

BEAM DATA:

B = 0.75 FT.
D = 3.00 FT.

EXTERNAL BEAM RESULTS:
F = -8348 LB.
M1 = 134121 LB.FT.
M2 = 2257 LB.FT.
S1 = -10453 LB.
S2 = 369 LB.

EXTERNAL-HALF SHELL RESULTS:

POINT	T1	S	M2	T2
0	25736	596	152	-63
1	6540	-2867	282	-283
2	-3399	-3075	74	-911
3	-7266	-1783	-203	-1425
4	-8130	0	-323	-1612

INTERNAL-HALF SHELL RESULTS:

POINT	T1	S	M2	T2
0	27158	0	-472	683
1	8183	-3876	194	282
2	-3903	-4237	188	-601
3	-10057	-2555	-57	-1344
4	-11847	0	-183	-1623

L= 44 TABLE F- 16 W= 26

SHELL DATA:

L = 44.0 FT.
W = 26.0 FT.
R = 20.0 FT.
T = 2.5 IN.
PHI= 39.1 DEG.
G = 50.0 LB/FT2

BEAM DATA:

B = 0.75 FT.
D = 3.00 FT.

EXTERNAL BEAM RESULTS:
F = -10811 LB.
M1 = 137159 LB.FT.
M2 = 2193 LB.FT.
S1 = -10936 LB.
S2 = 434 LB.

EXTERNAL-HALF SHELL RESULTS:

POINT	T1	S	M2	T2
0	26023	771	55	-77
1	6415	-2955	275	-318
2	-3401	-3156	88	-1021
3	-6932	-1807	-195	-1589
4	-7626	0	-321	-1794

INTERNAL-HALF SHELL RESULTS:

POINT	T1	S	M2	T2
0	27002	0	-547	765
1	7786	-4087	214	293
2	-4037	-4401	205	-709
3	-9703	-2621	-74	-1539
4	-11251	0	-216	-1847

L= 44 TABLE F- 17 W= 28

SHELL DATA:

L = 44.0 FT.
W = 28.0 FT.
R = 22.5 FT.
T = 2.5 IN.
PHI= 37.3 DEG.
G = 50.0 LB/FT2

BEAM DATA:

B = 0.75 FT.
D = 3.00 FT.

EXTERNAL BEAM RESULTS:
F = -12976 LB.
M1 = 139765 LB.FT.
M2 = 2160 LB.FT.
S1 = -11356 LB.
S2 = 492 LB.

EXTERNAL-HALF SHELL RESULTS:

POINT	T1	S	M2	T2
0	26281	926	-35	-87
1	6197	-3055	274	-353
2	-3469	-3222	103	-1136
3	-6594	-1812	-190	-1755
4	-7081	0	-322	-1975

INTERNAL-HALF SHELL RESULTS:

POINT	T1	S	M2	T2
0	26751	0	-623	840
1	7315	-4271	237	297
2	-4186	-4516	222	-823
3	-9273	-2646	-89	-1731
4	-10534	0	-247	-2065

L= 44 TABLE F- 18 W= 30

SHELL DATA:

L = 44.0 FT.
W = 30.0 FT.
R = 25.0 FT.
T = 2.5 IN.
PHI= 35.8 DEG.
G = 50.0 LB/FT2

BEAM DATA:

B = 0.75 FT.
D = 3.00 FT.

EXTERNAL BEAM RESULTS:
F = -14865 LB.
M1 = 142132 LB.FT.
M2 = 2143 LB.FT.
S1 = -11729 LB.
S2 = 543 LB.

EXTERNAL-HALF SHELL RESULTS:

POINT	T1	S	M2	T2
0	26532	1061	-121	-95
1	5908	-3164	277	-392
2	-3595	-3271	118	-1256
3	-6253	-1796	-187	-1922
4	-6502	0	-325	-2154

INTERNAL-HALF SHELL RESULTS:

POINT	T1	S	M2	T2
0	26513	0	-700	910
1	6797	-4439	261	292
2	-4368	-4590	240	-943
3	-8804	-2634	-104	-1921
4	-9739	0	-277	-2275

L= 46 TABLE F- 19 W= 22

SHELL DATA:

L = 46.0 FT.
W = 22.0 FT.
R = 17.5 FT.
T = 2.5 IN.
PHI= 37.4 DEG.
G = 50.0 LB/FT2

BEAM DATA:

B = 0.75 FT.
D = 3.00 FT.

EXTERNAL BEAM RESULTS:
F = -3182 LB.
M1 = 150479 LB.FT.
M2 = 1580 LB.FT.
S1 = -10581 LB.
S2 = 173 LB.

EXTERNAL-HALF SHELL RESULTS:

POINT	T1	S	M2	T2
0	28742	217	219	-52
1	8280	-3241	281	-313
2	-3860	-3553	69	-942
3	-9812	-2137	-192	-1472
4	-11530	0	-304	-1671

INTERNAL-HALF SHELL RESULTS:

POINT	T1	S	M2	T2
0	36297	0	-394	761
1	11628	-4533	187	339
2	-4861	-5065	185	-575
3	-14035	-3113	-28	-1358
4	-16931	0	-140	-1656

L= 46 TABLE F- 20 W= 24

SHELL DATA:

L = 46.0 FT.
W = 24.0 FT.
R = 17.5 FT.
T = 2.5 IN.
PHI= 41.6 DEG.
G = 50.0 LB/FT2

BEAM DATA:

B = 0.75 FT.
D = 3.00 FT.

EXTERNAL BEAM RESULTS:
F = -7959 LB.
M1 = 144986 LB.FT.
M2 = 2443 LB.FT.
S1 = -10698 LB.
S2 = 354 LB.

EXTERNAL-HALF SHELL RESULTS:

POINT	T1	S	M2	T2
0	27922	543	229	-54
1	7149	-3059	317	-289
2	-3741	-3278	72	-931
3	-8097	-1907	-232	-1456
4	-9110	0	-363	-1649

INTERNAL-HALF SHELL RESULTS:

POINT	T1	S	M2	T2
0	29419	0	-472	685
1	9032	-4039	199	286
2	-4152	-4443	194	-596
3	-11036	-2694	-53	-1344
4	-13086	0	-181	-1626

L= 46 TABLE F- 21 W= 26

SHELL DATA:

L = 46.0 FT.
W = 26.0 FT.
R = 20.0 FT.
T = 2.5 IN.
PHI= 39.1 DEG.
G = 50.0 LB/FT2

BEAM DATA:

B = 0.75 FT.
D = 3.00 FT.

EXTERNAL BEAM RESULTS:
F = -10927 LB.
M1 = 148426 LB.FT.
M2 = 2387 LB.FT.
S1 = -11240 LB.
S2 = 429 LB.

EXTERNAL-HALF SHELL RESULTS:

POINT	T1	S	M2	T2
0	28243	746	124	-69
1	7048	-3137	306	-322
2	-3705	-3357	86	-1038
3	-7712	-1933	-223	-1618
4	-8550	0	-358	-1829

INTERNAL-HALF SHELL RESULTS:

POINT	T1	S	M2	T2
0	29189	0	-549	767
1	8619	-4255	219	300
2	-4271	-4620	210	-703
3	-10660	-2771	-71	-1539
4	-12470	0	-216	-1852

L= 46 TABLE F- 22 W= 28
****

SHELL DATA:

L = 46.0 FT.
W = 28.0 FT.
R = 22.5 FT.
T = 2.5 IN.
PHI= 37.3 DEG.
G = 50.0 LB/FT2

BEAM DATA:

B = 0.75 FT.
D = 3.00 FT.

EXTERNAL BEAM RESULTS:
F = -13543 LB.
M1 = 151351 LB.FT.
M2 = 2362 LB.FT.
S1 = -11709 LB.
S2 = 497 LB.

EXTERNAL-HALF SHELL RESULTS:

POINT	T1	S	M2	T2
0	28524	924	26	-81
1	6845	-3228	303	-356
2	-3744	-3423	101	-1150
3	-7329	-1941	-216	-1782
4	-7956	0	-357	-2009

INTERNAL-HALF SHELL RESULTS:

POINT	T1	S	M2	T2
0	28844	0	-627	843
1	8128	-4443	241	305
2	-4401	-4746	228	-815
3	-10203	-2807	-88	-1732
4	-11728	0	-249	-2072

L= 46 TABLE F- 23 W= 30
****

SHELL DATA:

L = 46.0 FT.
W = 30.0 FT.
R = 25.0 FT.
T = 2.5 IN.
PHI= 35.8 DEG.
G = 50.0 LB/FT2

BEAM DATA:

B = 0.75 FT.
D = 3.00 FT.

EXTERNAL BEAM RESULTS:
F = -15844 LB.
M1 = 153982 LB.FT.
M2 = 2353 LB.FT.
S1 = -12126 LB.
S2 = 557 LB.

EXTERNAL-HALF SHELL RESULTS:

POINT	T1	S	M2	T2
0	28791	1082	-65	-90
1	6564	-3330	304	-392
2	-3846	-3474	116	-1266
3	-6952	-1928	-212	-1947
4	-7335	0	-359	-2187

INTERNAL-HALF SHELL RESULTS:

POINT	T1	S	M2	T2
0	28503	0	-706	914
1	7590	-4612	265	304
2	-4561	-4832	246	-932
3	-9706	-2807	-103	-1924
4	-10906	0	-281	-2285

L= 46 TABLE F- 24 W= 32
****

SHELL DATA:

L = 46.0 FT.
W = 32.0 FT.
R = 25.0 FT.
T = 2.5 IN.
PHI= 38.7 DEG.
G = 50.0 LB/FT2

BEAM DATA:

B = 0.75 FT.
D = 3.00 FT.

EXTERNAL BEAM RESULTS:
F = -18650 LB.
M1 = 152906 LB.FT.
M2 = 2904 LB.FT.
S1 = -12341 LB.
S2 = 666 LB.

EXTERNAL-HALF SHELL RESULTS:

POINT	T1	S	M2	T2
0	28740	1273	-110	-103
1	5397	-3262	317	-401
2	-4191	-3166	123	-1283
3	-5795	-1615	-218	-1915
4	-5462	0	-368	-2123

INTERNAL-HALF SHELL RESULTS:

POINT	T1	S	M2	T2
0	25541	0	-772	843
1	6001	-4328	286	234
2	-4537	-4331	259	-976
3	-8003	-2401	-116	-1902
4	-8458	0	-303	-2228

L= 48 TABLE F- 25 W= 22

SHELL DATA:

L = 48.0 FT.
W = 22.0 FT.
R = 17.5 FT.
T = 2.5 IN.
PHI= 37.4 DEG.
G = 50.0 LB/FT2

BEAM DATA:

B = 0.75 FT.
D = 3.25 FT.

EXTERNAL BEAM RESULTS:
F = -2037 LB.
M1 = 178620 LB.FT.
M2 = 1315 LB.FT.
S1 = -11885 LB.
S2 = 119 LB.

EXTERNAL-HALF SHELL RESULTS:

POINT	T1	S	M2	T2
0	28910	133	154	-67
1	8609	-3238	255	-318
2	-3888	-3570	78	-918
3	-10299	-2164	-158	-1429
4	-12218	0	-261	-1622

INTERNAL-HALF SHELL RESULTS:

POINT	T1	S	M2	T2
0	39333	0	-390	763
1	12723	-4722	194	342
2	-5220	-5294	193	-573
3	-15312	-3261	-21	-1357
4	-18524	0	-134	-1656

L= 48 TABLE F- 26 W= 24

SHELL DATA:

L = 48.0 FT.
W = 24.0 FT.
R = 17.5 FT.
T = 2.5 IN.
PHI= 41.6 DEG.
G = 50.0 LB/FT2

BEAM DATA:

B = 0.75 FT.
D = 3.25 FT.

EXTERNAL BEAM RESULTS:
F = -6674 LB.
M1 = 171618 LB.FT.
M2 = 2207 LB.FT.
S1 = -11923 LB.
S2 = 291 LB.

EXTERNAL-HALF SHELL RESULTS:

POINT	T1	S	M2	T2
0	28019	436	188	-68
1	7533	-3081	300	-296
2	-3735	-3339	78	-916
3	-8622	-1969	-210	-1431
4	-9871	0	-335	-1622

INTERNAL-HALF SHELL RESULTS:

POINT	T1	S	M2	T2
0	31789	0	-471	686
1	9914	-4203	204	290
2	-4417	-4649	200	-593
3	-12057	-2831	-49	-1344
4	-14375	0	-179	-1628

L= 48 TABLE F- 27 W= 26

SHELL DATA:

L = 48.0 FT.
W = 26.0 FT.
R = 20.0 FT.
T = 2.5 IN.
PHI= 39.1 DEG.
G = 50.0 LB/FT2

BEAM DATA:

B = 0.75 FT.
D = 3.25 FT.

EXTERNAL BEAM RESULTS:
F = -9523 LB.
M1 = 175146 LB.FT.
M2 = 2158 LB.FT.
S1 = -12459 LB.
S2 = 361 LB.

EXTERNAL-HALF SHELL RESULTS:

POINT	T1	S	M2	T2
0	28298	623	87	-81
1	7440	-3166	291	-330
2	-3688	-3431	92	-1024
3	-8242	-2007	-204	-1596
4	-9325	0	-335	-1806

INTERNAL-HALF SHELL RESULTS:

POINT	T1	S	M2	T2
0	31484	0	-550	769
1	9484	-4425	223	305
2	-4520	-4838	216	-698
3	-11658	-2918	-69	-1540
4	-13736	0	-215	-1856

L= 48 TABLE F- 28 W= 28

****

SHELL DATA:

L = 48.0 FT.
W = 28.0 FT.
R = 22.5 FT.
T = 2.5 IN.
PHI= 37.3 DEG.
G = 50.0 LB/FT2

BEAM DATA:

B = 0.75 FT.
D = 3.25 FT.

EXTERNAL BEAM RESULTS:
F = -12073 LB.
M1 = 178141 LB.FT.
M2 = 2140 LB.FT.
S1 = -12929 LB.
S2 = 425 LB.

EXTERNAL-HALF SHELL RESULTS:

POINT	T1	S	M2	T2
0	28540	790	-6	-92
1	7252	-3261	289	-364
2	-3711	-3510	107	-1135
3	-7863	-2028	-200	-1762
4	-8745	0	-338	-1990

INTERNAL-HALF SHELL RESULTS:

POINT	T1	S	M2	T2
0	31044	0	-630	846
1	8972	-4616	245	313
2	-4632	-4975	233	-808
3	-11172	-2965	-86	-1733
4	-12966	0	-250	-2078

L= 48 TABLE F- 29 W= 30

****

SHELL DATA:

L = 48.0 FT.
W = 30.0 FT.
R = 25.0 FT.
T = 2.5 IN.
PHI= 35.8 DEG.
G = 50.0 LB/FT2

BEAM DATA:

B = 0.75 FT.
D = 3.25 FT.

EXTERNAL BEAM RESULTS:
F = -14353 LB.
M1 = 180845 LB.FT.
M2 = 2137 LB.FT.
S1 = -13350 LB.
S2 = 483 LB.

EXTERNAL-HALF SHELL RESULTS:

POINT	T1	S	M2	T2
0	28770	939	-95	-101
1	6990	-3364	292	-400
2	-3794	-3575	123	-1251
3	-7487	-2029	-197	-1929
4	-8136	0	-343	-2173

INTERNAL-HALF SHELL RESULTS:

POINT	T1	S	M2	T2
0	30598	0	-711	917
1	8413	-4787	269	314
2	-4771	-5071	252	-923
3	-10645	-2976	-103	-1925
4	-12115	0	-284	-2294

L= 48 TABLE F- 30 W= 32

****

SHELL DATA:

L = 48.0 FT.
W = 32.0 FT.
R = 25.0 FT.
T = 2.5 IN.
PHI= 38.7 DEG.
G = 50.0 LB/FT2

BEAM DATA:

B = 0.75 FT.
D = 3.25 FT.

EXTERNAL BEAM RESULTS:
F = -17534 LB.
M1 = 179174 LB.FT.
M2 = 2698 LB.FT.
S1 = -13579 LB.
S2 = 597 LB.

EXTERNAL-HALF SHELL RESULTS:

POINT	T1	S	M2	T2
0	28618	1147	-137	-114
1	5885	-3279	310	-403
2	-4073	-3286	131	-1263
3	-6311	-1743	-208	-1904
4	-6268	0	-360	-2121

INTERNAL-HALF SHELL RESULTS:

POINT	T1	S	M2	T2
0	27258	0	-782	848
1	6720	-4481	291	248
2	-4686	-4559	267	-962
3	-8812	-2571	-117	-1905
4	-9520	0	-309	-2242

L= 50 TABLE F- 31 W= 22

SHELL DATA:

```
L   = 50.0 FT.
W   = 22.0 FT.
R   = 17.5 FT.
T   =  2.5 IN.
PHI= 37.4 DEG.
G   = 50.0 LB/FT2
```

BEAM DATA:

```
B   = 0.75 FT.
D   = 3.25 FT.
```

EXTERNAL BEAM RESULTS:

```
F  =    -936 LB.
M1 =  192611 LB.FT.
M2 =    1410 LB.FT.
S1 =  -12173 LB.
S2 =      92 LB.
```

EXTERNAL-HALF SHELL RESULTS:

POINT	T1	S	M2	T2
0	31277	58	217	-61
1	9319	-3445	284	-325
2	-4304	-3783	78	-938
3	-11363	-2294	-178	-1458
4	-13493	0	-288	-1654

INTERNAL-HALF SHELL RESULTS:

POINT	T1	S	M2	T2
0	42504	0	-384	764
1	13863	-4912	202	345
2	-5596	-5521	202	-570
3	-16641	-3408	-14	-1356
4	-20182	0	-126	-1657

L= 50 TABLE F- 32 W= 24

SHELL DATA:

```
L   = 50.0 FT.
W   = 24.0 FT.
R   = 17.5 FT.
T   =  2.5 IN.
PHI= 41.6 DEG.
G   = 50.0 LB/FT2
```

BEAM DATA:

```
B   = 0.75 FT.
D   = 3.25 FT.
```

EXTERNAL BEAM RESULTS:

```
F  =   -5928 LB.
M1 =  184519 LB.FT.
M2 =    2370 LB.FT.
S1 =  -12177 LB.
S2 =     269 LB.
```

EXTERNAL-HALF SHELL RESULTS:

POINT	T1	S	M2	T2
0	30229	372	263	-59
1	8160	-3278	334	-302
2	-4105	-3542	76	-935
3	-9526	-2093	-238	-1462
4	-10942	0	-372	-1657

INTERNAL-HALF SHELL RESULTS:

POINT	T1	S	M2	T2
0	34268	0	-469	688
1	10831	-4369	209	294
2	-4698	-4853	206	-589
3	-13119	-2967	-44	-1344
4	-15712	0	-175	-1630

L= 50 TABLE F- 33 W= 26

SHELL DATA:

```
L   = 50.0 FT.
W   = 26.0 FT.
R   = 20.0 FT.
T   =  2.5 IN.
PHI= 39.1 DEG.
G   = 50.0 LB/FT2
```

BEAM DATA:

```
B   = 0.75 FT.
D   = 3.25 FT.
```

EXTERNAL BEAM RESULTS:

```
F  =   -9295 LB.
M1 =  188465 LB.FT.
M2 =    2330 LB.FT.
S1 =  -12772 LB.
S2 =     350 LB.
```

EXTERNAL-HALF SHELL RESULTS:

POINT	T1	S	M2	T2
0	30537	584	155	-75
1	8091	-3352	322	-335
2	-4016	-3631	89	-1041
3	-9087	-2131	-230	-1624
4	-10331	0	-370	-1839

INTERNAL-HALF SHELL RESULTS:

POINT	T1	S	M2	T2
0	33887	0	-549	771
1	10383	-4597	227	309
2	-4786	-5054	221	-694
3	-12696	-3063	-65	-1540
4	-15050	0	-214	-1859

L= 50 TABLE F- 34 W= 28

SHELL DATA:

L = 50.0 FT.
W = 28.0 FT.
R = 22.5 FT.
T = 2.5 IN.
PHI= 37.3 DEG.
G = 50.0 LB/FT2

BEAM DATA:

B = 0.75 FT.
D = 3.25 FT.

EXTERNAL BEAM RESULTS:
F = -12310 LB.
M1 = 191789 LB.FT.
M2 = 2321 LB.FT.
S1 = -13291 LB.
S2 = 423 LB.

EXTERNAL-HALF SHELL RESULTS:

POINT	T1	S	M2	T2
0	30797	773	56	-86
1	7918	-3438	318	-368
2	-4005	-3710	104	-1150
3	-8659	-2154	-225	-1788
4	-9697	0	-372	-2021

INTERNAL-HALF SHELL RESULTS:

POINT	T1	S	M2	T2
0	33350	0	-632	848
1	9848	-4791	249	319
2	-4879	-5202	239	-802
3	-12180	-3120	-84	-1734
4	-14249	0	-251	-2083

L= 50 TABLE F- 35 W= 30

SHELL DATA:

L = 50.0 FT.
W = 30.0 FT.
R = 22.5 FT.
T = 2.5 IN.
PHI= 40.5 DEG.
G = 50.0 LB/FT2

BEAM DATA:

B = 0.75 FT.
D = 3.25 FT.

EXTERNAL BEAM RESULTS:
F = -16485 LB.
M1 = 189100 LB.FT.
M2 = 3021 LB.FT.
S1 = -13549 LB.
S2 = 565 LB.

EXTERNAL-HALF SHELL RESULTS:

POINT	T1	S	M2	T2
0	30495	1035	18	-99
1	6740	-3313	338	-361
2	-4193	-3410	113	-1150
3	-7262	-1872	-240	-1759
4	-7562	0	-396	-1972

INTERNAL-HALF SHELL RESULTS:

POINT	T1	S	M2	T2
0	28960	0	-707	779
1	7892	-4428	270	262
2	-4609	-4666	254	-831
3	-10005	-2720	-101	-1716
4	-11288	0	-283	-2040

L= 50 TABLE F- 36 W= 32

SHELL DATA:

L = 50.0 FT.
W = 32.0 FT.
R = 25.0 FT.
T = 2.5 IN.
PHI= 38.7 DEG.
G = 50.0 LB/FT2

BEAM DATA:

B = 0.75 FT.
D = 3.25 FT.

EXTERNAL BEAM RESULTS:
F = -18696 LB.
M1 = 192820 LB.FT.
M2 = 2938 LB.FT.
S1 = -14010 LB.
S2 = 614 LB.

EXTERNAL-HALF SHELL RESULTS:

POINT	T1	S	M2	T2
0	30836	1174	-81	-109
1	6506	-3433	337	-401
2	-4319	-3475	130	-1270
3	-6964	-1867	-233	-1926
4	-7039	0	-395	-2151

INTERNAL-HALF SHELL RESULTS:

POINT	T1	S	M2	T2
0	29065	0	-789	852
1	7465	-4636	296	260
2	-4849	-4784	274	-950
3	-9652	-2736	-117	-1907
4	-10618	0	-314	-2253

L= 52 TABLE F- 37 W= 22

SHELL DATA:

```
L   = 52.0 FT.
W   = 22.0 FT.
R   = 17.5 FT.
T   =  2.5 IN.
PHI= 37.4 DEG.
G   = 50.0 LB/FT2
```

BEAM DATA:

```
B   = 0.75 FT.
D   = 3.50 FT.
```

EXTERNAL BEAM RESULTS:

```
F  =      20 LB.
M1 =  225788 LB.FT.
M2 =    1124 LB.FT.
S1 =  -13614 LB.
S2 =      48 LB.
```

EXTERNAL-HALF SHELL RESULTS:

POINT	T1	S	M2	T2
0	31435	-1	140	-76
1	9630	-3420	253	-327
2	-4319	-3775	89	-910
3	-11802	-2302	-137	-1409
4	-14117	0	-237	-1598

INTERNAL-HALF SHELL RESULTS:

POINT	T1	S	M2	T2
0	45807	0	-378	766
1	15047	-5103	210	347
2	-5990	-5748	211	-567
3	-18024	-3555	-5	-1356
4	-21903	0	-178	-1657

L= 52 TABLE F- 38 W= 24

SHELL DATA:

```
L   = 52.0 FT.
W   = 24.0 FT.
R   = 17.5 FT.
T   =  2.5 IN.
PHI= 41.6 DEG.
G   = 50.0 LB/FT2
```

BEAM DATA:

```
B   = 0.75 FT.
D   = 3.50 FT.
```

EXTERNAL BEAM RESULTS:

```
F  =   -4565 LB.
M1 =  215914 LB.FT.
M2 =    2113 LB.FT.
S1 =  -13505 LB.
S2 =     211 LB.
```

EXTERNAL-HALF SHELL RESULTS:

POINT	T1	S	M2	T2
0	30355	275	214	-72
1	8519	-3293	314	-309
2	-4118	-3587	84	-920
3	-10046	-2141	-211	-1433
4	-11687	0	-339	-1626

INTERNAL-HALF SHELL RESULTS:

POINT	T1	S	M2	T2
0	36856	0	-467	689
1	11783	-4534	215	297
2	-4994	-5057	213	-586
3	-14223	-3101	-39	-1343
4	-17099	0	-170	-1632

L= 52 TABLE F- 39 W= 26

SHELL DATA:

```
L   = 52.0 FT.
W   = 26.0 FT.
R   = 20.0 FT.
T   =  2.5 IN.
PHI= 39.1 DEG.
G   = 50.0 LB/FT2
```

BEAM DATA:

```
B   = 0.75 FT.
D   = 3.50 FT.
```

EXTERNAL BEAM RESULTS:

```
F  =   -7743 LB.
M1 =  219906 LB.FT.
M2 =    2082 LB.FT.
S1 =  -14085 LB.
S2 =     285 LB.
```

EXTERNAL-HALF SHELL RESULTS:

POINT	T1	S	M2	T2
0	30626	467	112	-86
1	8455	-3378	304	-343
2	-4025	-3691	96	-1027
3	-9619	-2191	-207	-1598
4	-11097	0	-341	-1811

INTERNAL-HALF SHELL RESULTS:

POINT	T1	S	M2	T2
0	36398	0	-549	772
1	11314	-4769	232	313
2	-5068	-5269	227	-690
3	-13774	-3207	-62	-1540
4	-16410	0	-211	-1862

L= 52 TABLE F- 40 W= 28

SHELL DATA:

L = 52.0 FT.
W = 28.0 FT.
R = 22.5 FT.
T = 2.5 IN.
PHI= 37.3 DEG.
G = 50.0 LB/FT2

BEAM DATA:

B = 0.75 FT.
D = 3.50 FT.

EXTERNAL BEAM RESULTS:
F = -10633 LB.
M1 = 223259 LB.FT.
M2 = 2081 LB.FT.
S1 = -14596 LB.
S2 = 354 LB.

EXTERNAL-HALF SHELL RESULTS:

POINT	T1	S	M2	T2
0	30853	642	17	-97
1	8295	-3472	302	-376
2	-4003	-3783	110	-1136
3	-9200	-2226	-205	-1765
4	-10481	0	-347	-1997

INTERNAL-HALF SHELL RESULTS:

POINT	T1	S	M2	T2
0	35762	0	-633	850
1	10755	-4968	253	324
2	-5142	-5428	244	-796
3	-13227	-3273	-82	-1735
4	-15578	0	-250	-2087

L= 52 TABLE F- 41 W= 30

SHELL DATA:

L = 52.0 FT.
W = 30.0 FT.
R = 22.5 FT.
T = 2.5 IN.
PHI= 40.5 DEG.
G = 50.0 LB/FT2

BEAM DATA:

B = 0.75 FT.
D = 3.50 FT.

EXTERNAL BEAM RESULTS:
F = -15102 LB.
M1 = 219565 LB.FT.
M2 = 2797 LB.FT.
S1 = -14846 LB.
S2 = 498 LB.

EXTERNAL-HALF SHELL RESULTS:

POINT	T1	S	M2	T2
0	30450	912	-11	-110
1	7179	-3339	327	-367
2	-4121	-3505	120	-1135
3	-7788	-1970	-226	-1744
4	-8360	0	-381	-1962

INTERNAL-HALF SHELL RESULTS:

POINT	T1	S	M2	T2
0	30923	0	-711	781
1	8670	-4583	274	270
2	-4802	-4877	260	-823
3	-10891	-2870	-100	-1717
4	-12432	0	-285	-2047

L= 52 TABLE F- 42 W= 32

SHELL DATA:

L = 52.0 FT.
W = 32.0 FT.
R = 25.0 FT.
T = 2.5 IN.
PHI= 38.7 DEG.
G = 50.0 LB/FT2

BEAM DATA:

B = 0.75 FT.
D = 3.50 FT.

EXTERNAL BEAM RESULTS:
F = -17288 LB.
M1 = 223483 LB.FT.
M2 = 2719 LB.FT.
S1 = -15316 LB.
S2 = 545 LB.

EXTERNAL-HALF SHELL RESULTS:

POINT	T1	S	M2	T2
0	30760	1044	-109	-119
1	6960	-3459	327	-406
2	-4233	-3581	136	-1254
3	-7491	-1977	-221	-1913
4	-7848	0	-382	-2144

INTERNAL-HALF SHELL RESULTS:

POINT	T1	S	M2	T2
0	30961	0	-796	855
1	8233	-4793	300	271
2	-5026	-5007	280	-939
3	-10523	-2897	-117	-1909
4	-11752	0	-318	-2262

L= 54 TABLE F- 43 W= 24

SHELL DATA:

L = 54.0 FT.
W = 24.0 FT.
R = 17.5 FT.
T = 2.5 IN.
PHI= 41.6 DEG.
G = 50.0 LB/FT2

BEAM DATA:

B = 0.75 FT.
D = 3.50 FT.

EXTERNAL BEAM RESULTS:
F = -3501 LB.
M1 = 231105 LB.FT.
M2 = 2254 LB.FT.
S1 = -13777 LB.
S2 = 186 LB.

EXTERNAL-HALF SHELL RESULTS:

POINT	T1	S	M2	T2
0	32596	203	286	-64
1	9164	-3491	347	-315
2	-4513	-3789	83	-938
3	-11016	-2264	-236	-1462
4	-12840	0	-373	-1658

INTERNAL-HALF SHELL RESULTS:

POINT	T1	S	M2	T2
0	39551	0	-463	691
1	12770	-4701	221	300
2	-5306	-5260	221	-583
3	-15369	-3233	-33	-1343
4	-18537	0	-165	-1633

L= 54 TABLE F- 44 W= 26

SHELL DATA:

L = 54.0 FT.
W = 26.0 FT.
R = 20.0 FT.
T = 2.5 IN.
PHI= 39.1 DEG.
G = 50.0 LB/FT2

BEAM DATA:

B = 0.75 FT.
D = 3.50 FT.

EXTERNAL BEAM RESULTS:
F = -7196 LB.
M1 = 235515 LB.FT.
M2 = 2233 LB.FT.
S1 = -14413 LB.
S2 = 269 LB.

EXTERNAL-HALF SHELL RESULTS:

POINT	T1	S	M2	T2
0	32889	418	179	-80
1	9123	-3566	335	-348
2	-4375	-3889	94	-1043
3	-10527	-2313	-232	-1625
4	-12181	0	-374	-1842

INTERNAL-HALF SHELL RESULTS:

POINT	T1	S	M2	T2
0	39015	0	-547	774
1	12280	-4942	237	317
2	-5365	-5484	233	-686
3	-14894	-3349	-57	-1540
4	-17819	0	-208	-1864

L= 54 TABLE F- 45 W= 28

SHELL DATA:

L = 54.0 FT.
W = 28.0 FT.
R = 22.5 FT.
T = 2.5 IN.
PHI= 37.3 DEG.
G = 50.0 LB/FT2

BEAM DATA:

B = 0.75 FT.
D = 3.50 FT.

EXTERNAL BEAM RESULTS:
F = -10553 LB.
M1 = 239195 LB.FT.
M2 = 2242 LB.FT.
S1 = -14972 LB.
S2 = 346 LB.

EXTERNAL-HALF SHELL RESULTS:

POINT	T1	S	M2	T2
0	33129	613	79	-92
1	8978	-3651	330	-381
2	-4318	-3981	109	-1151
3	-10055	-2349	-229	-1790
4	-11507	0	-379	-2027

INTERNAL-HALF SHELL RESULTS:

POINT	T1	S	M2	T2
0	38279	0	-633	852
1	11695	-5145	258	329
2	-5420	-5652	250	-792
3	-14313	-3423	-79	-1735
4	-16952	0	-249	-2090

L= 54 TABLE F- 46 W= 30
****

SHELL DATA:

L = 54.0 FT.
W = 30.0 FT.
R = 22.5 FT.
T = 2.5 IN.
PHI= 40.5 DEG.
G = 50.0 LB/FT2

BEAM DATA:

B = 0.75 FT.
D = 3.50 FT.

EXTERNAL BEAM RESULTS:
F = -15549 LB.
M1 = 234975 LB.FT.
M2 = 3015 LB.FT.
S1 = -15236 LB.
S2 = 500 LB.

EXTERNAL-HALF SHELL RESULTS:

POINT	T1	S	M2	T2
0	32650	904	52	-104
1	7805	-3502	356	-368
2	-4403	-3690	118	-1147
3	-8522	-2088	-252	-1768
4	-9230	0	-416	-1990

INTERNAL-HALF SHELL RESULTS:

POINT	T1	S	M2	T2
0	32977	0	-715	784
1	9474	-4739	279	277
2	-5009	-5087	266	-816
3	-11809	-3017	-99	-1718
4	-13613	0	-286	-2053

L= 54 TABLE F- 47 W= 32
****

SHELL DATA:

L = 54.0 FT.
W = 32.0 FT.
R = 25.0 FT.
T = 2.5 IN.
PHI= 38.7 DEG.
G = 50.0 LB/FT2

BEAM DATA:

B = 0.75 FT.
D = 3.50 FT.

EXTERNAL BEAM RESULTS:
F = -18152 LB.
M1 = 239278 LB.FT.
M2 = 2941 LB.FT.
S1 = -15753 LB.
S2 = 555 LB.

EXTERNAL-HALF SHELL RESULTS:

POINT	T1	S	M2	T2
0	32981	1056	-52	-114
1	7601	-3616	355	-406
2	-4491	-3769	135	-1263
3	-8195	-2098	-246	-1934
4	-8686	0	-417	-2172

INTERNAL-HALF SHELL RESULTS:

POINT	T1	S	M2	T2
0	32948	0	-801	858
1	9027	-4952	305	280
2	-5217	-5228	287	-930
3	-11425	-3054	-116	-1911
4	-12921	0	-321	-2270

L= 54 TABLE F- 48 W= 34
****

SHELL DATA:

L = 54.0 FT.
W = 34.0 FT.
R = 27.5 FT.
T = 2.5 IN.
PHI= 37.2 DEG.
G = 50.0 LB/FT2

BEAM DATA:

B = 0.75 FT.
D = 3.50 FT.

EXTERNAL BEAM RESULTS:
F = -20450 LB.
M1 = 243185 LB.FT.
M2 = 2892 LB.FT.
S1 = -16216 LB.
S2 = 604 LB.

EXTERNAL-HALF SHELL RESULTS:

POINT	T1	S	M2	T2
0	33300	1189	-151	-123
1	7316	-3738	358	-447
2	-4636	-3831	152	-1384
3	-7854	-2088	-261	-2102
4	-8090	0	-419	-2351

INTERNAL-HALF SHELL RESULTS:

POINT	T1	S	M2	T2
0	32885	0	-888	928
1	8518	-5149	332	276
2	-5452	-5329	308	-1050
3	-10982	-3056	-133	-2101
4	-12125	0	-355	-2481

L= 56 TABLE F- 49 W= 24
****

SHELL DATA:

L = 56.0 FT.
W = 24.0 FT.
R = 17.5 FT.
T = 2.5 IN.
PHI= 41.6 DEG.
G = 50.0 LB/FT2

BEAM DATA:

B = 0.75 FT.
D = 3.75 FT.

EXTERNAL BEAM RESULTS:
F = -2146 LB.
M1 = 267700 LB.FT.
M2 = 1976 LB.FT.
S1 = -15219 LB.
S2 = 134 LB.

EXTERNAL-HALF SHELL RESULTS:

POINT	T1	S	M2	T2
0	32739	120	228	-77
1	9503	-3496	323	-320
2	-4535	-3817	92	-920
3	-11521	-2298	-204	-1430
4	-13557	0	-333	-1622

INTERNAL-HALF SHELL RESULTS:

POINT	T1	S	M2	T2
0	42353	0	-459	692
1	13792	-4868	228	302
2	-5632	-5462	228	-581
3	-16556	-3365	-26	-1342
4	-20024	0	-159	-1633

L= 56 TABLE F- 50 W= 26
****

SHELL DATA:

L = 56.0 FT.
W = 26.0 FT.
R = 20.0 FT.
T = 2.5 IN.
PHI= 39.1 DEG.
G = 50.0 LB/FT2

BEAM DATA:

B = 0.75 FT.
D = 3.75 FT.

EXTERNAL BEAM RESULTS:
F = -5562 LB.
M1 = 272116 LB.FT.
M2 = 1964 LB.FT.
S1 = -15829 LB.
S2 = 210 LB.

EXTERNAL-HALF SHELL RESULTS:

POINT	T1	S	M2	T2
0	33002	312	127	-91
1	9463	-3586	313	-355
2	-4402	-3934	102	-1028
3	-11054	-2360	-204	-1596
4	-12930	0	-340	-1809

INTERNAL-HALF SHELL RESULTS:

POINT	T1	S	M2	T2
0	41738	0	-545	775
1	13279	-5116	243	320
2	-5677	-5697	240	-683
3	-16053	-3489	-52	-1540
4	-19276	0	-204	-1865

L= 56 TABLE F- 51 W= 28
****

SHELL DATA:

L = 56.0 FT.
W = 28.0 FT.
R = 22.5 FT.
T = 2.5 IN.
PHI= 37.3 DEG.
G = 50.0 LB/FT2

BEAM DATA:

B = 0.75 FT.
D = 3.75 FT.

EXTERNAL BEAM RESULTS:
F = -8722 LB.
M1 = 275783 LB.FT.
M2 = 1984 LB.FT.
S1 = -16370 LB.
S2 = 280 LB.

EXTERNAL-HALF SHELL RESULTS:

POINT	T1	S	M2	T2
0	33215	489	34	-102
1	9327	-3683	311	-388
2	-4340	-4041	115	-1137
3	-10597	-2408	-205	-1764
4	-12281	0	-350	-1999

INTERNAL-HALF SHELL RESULTS:

POINT	T1	S	M2	T2
0	40900	0	-632	853
1	12666	-5324	263	333
2	-5714	-5875	255	-788
3	-15438	-3572	-75	-1735
4	-18371	0	-247	-2093

```
L= 56       TABLE F- 52        W= 30
************************************

SHELL DATA:

    L   = 56.0 FT.
    W   = 30.0 FT.
    R   = 22.5 FT.
    T   =  2.5 IN.
    PHI= 40.5 DEG.
    G   = 50.0 LB/FT2

BEAM DATA:

    B   = 0.75 FT.
    D   = 3.75 FT.

EXTERNAL BEAM RESULTS:
    F   = -13929 LB.
    M1  = 270278 LB.FT.
    M2  =   2776 LB.FT.
    S1  = -16610 LB.
    S2  =    434 LB.

EXTERNAL-HALF SHELL RESULTS:

POINT      T1        S       M2      T2

  0     32646      781       19    -113
  1      8214    -3532      343    -375
  2     -4363    -3773      124   -1134
  3     -9057    -2171     -235   -1751
  4    -10026        0     -396   -1975

INTERNAL-HALF SHELL RESULTS:

POINT      T1        S       M2      T2

  0     35120        0     -717     786
  1     10303    -4897      283     283
  2     -5231    -5295      272    -810
  3    -12760    -3161      -97   -1719
  4    -14830        0     -286   -2058
```

```
L= 56       TABLE F- 53        W= 32
************************************

SHELL DATA:

    L   = 56.0 FT.
    W   = 32.0 FT.
    R   = 25.0 FT.
    T   =  2.5 IN.
    PHI= 38.7 DEG.
    G   = 50.0 LB/FT2

BEAM DATA:

    B   = 0.75 FT.
    D   = 3.75 FT.

EXTERNAL BEAM RESULTS:
    F   = -16475 LB.
    M1  = 274756 LB.FT.
    M2  =   2708 LB.FT.
    S1  = -17131 LB.
    S2  =    486 LB.

EXTERNAL-HALF SHELL RESULTS:

POINT      T1        S       M2      T2

  0     32948      924      -83    -124
  1      8023    -3648      343    -413
  2     -4437    -3863      141   -1250
  3     -8733    -2193     -231   -1919
  4     -9494        0     -400   -2160

INTERNAL-HALF SHELL RESULTS:

POINT      T1        S       M2      T2

  0     35023        0     -805     861
  1      9846    -5113      309     288
  2     -5422    -5447      293    -922
  3    -12358    -3209     -115   -1912
  4    -14125        0     -323   -2277
```

```
L= 56       TABLE F- 54        W= 34
************************************

SHELL DATA:

    L   = 56.0 FT.
    W   = 34.0 FT.
    R   = 27.5 FT.
    T   =  2.5 IN.
    PHI= 37.2 DEG.
    G   = 50.0 LB/FT2

BEAM DATA:

    B   = 0.75 FT.
    D   = 3.75 FT.

EXTERNAL BEAM RESULTS:
    F   = -18755 LB.
    M1  = 278830 LB.FT.
    M2  =   2665 LB.FT.
    S1  = -17601 LB.
    S2  =    533 LB.

EXTERNAL-HALF SHELL RESULTS:

POINT      T1        S       M2      T2

  0     33236     1052     -180    -132
  1      7755    -3770      347    -453
  2     -4566    -3936      159   -1370
  3     -8394    -2194     -228   -2088
  4     -8910        0     -406   -2343

INTERNAL-HALF SHELL RESULTS:

POINT      T1        S       M2      T2

  0     34882        0     -895     932
  1      9324    -5311      337     286
  2     -5640    -5559      314   -1040
  3    -11896    -3221     -132   -2103
  4    -13313        0     -359   -2491
```

```
L= 58        TABLE F- 55         W= 26
*************************************

SHELL DATA:

    L   = 58.0 FT.
    W   = 26.0 FT.
    R   = 20.0 FT.
    T   =  2.5 IN.
    PHI= 39.1 DEG.
    G   = 50.0 LB/FT2

BEAM DATA:

    B   = 0.75 FT.
    D   = 3.75 FT.

EXTERNAL BEAM RESULTS:
    F   =  -4735 LB.
    M1  = 290268 LB.FT.
    M2  =   2095 LB.FT.
    S1  = -16180 LB.
    S2  =    190 LB.

EXTERNAL-HALF SHELL RESULTS:

POINT      T1        S      M2      T2

  0     35293      256     191     -85
  1     10147    -3773     343    -360
  2     -4774    -4131     101   -1044
  3    -12018    -2480    -226   -1621
  4    -14085        0    -370   -1838

INTERNAL-HALF SHELL RESULTS:

POINT      T1        S      M2      T2

  0     44567        0    -542     776
  1     14312    -5290     249     323
  2     -6004    -5909     246    -680
  3    -17254    -3629     -47   -1539
  4    -20782        0    -200   -1867
```

```
L= 58        TABLE F- 56         W= 28
*************************************

SHELL DATA:

    L   = 58.0 FT.
    W   = 28.0 FT.
    R   = 22.5 FT.
    T   =  2.5 IN.
    PHI= 37.3 DEG.
    G   = 50.0 LB/FT2

BEAM DATA:

    B   = 0.75 FT.
    D   = 3.75 FT.

EXTERNAL BEAM RESULTS:
    F   =  -8351 LB.
    M1  = 294245 LB.FT.
    M2  =   2125 LB.FT.
    S1  = -16765 LB.
    S2  =    269 LB.

EXTERNAL-HALF SHELL RESULTS:

POINT      T1        S      M2      T2

  0     35514      452      94     -97
  1     10026    -3863     338    -393
  2     -4675    -4236     114   -1151
  3    -11506    -2530    -227   -1788
  4    -13374        0    -379   -2026

INTERNAL-HALF SHELL RESULTS:

POINT      T1        S      M2      T2

  0     43625        0    -631     855
  1     13670    -5503     268     337
  2     -6023    -6097     261    -784
  3    -16603    -3719     -71   -1735
  4    -19837        0    -244   -2095
```

```
L= 58        TABLE F- 57         W= 30
*************************************

SHELL DATA:

    L   = 58.0 FT.
    W   = 30.0 FT.
    R   = 22.5 FT.
    T   =  2.5 IN.
    PHI= 40.5 DEG.
    G   = 50.0 LB/FT2

BEAM DATA:

    B   = 0.75 FT.
    D   = 3.75 FT.

EXTERNAL BEAM RESULTS:
    F   = -14061 LB.
    M1  = 288016 LB.FT.
    M2  =   2974 LB.FT.
    S1  = -17009 LB.
    S2  =    430 LB.

EXTERNAL-HALF SHELL RESULTS:

POINT      T1        S      M2      T2

  0     34856      761      83    -107
  1      8856    -3698     372    -378
  2     -4661    -3958     122   -1147
  3     -9843    -2287    -261   -1773
  4    -10963        0    -431   -2002

INTERNAL-HALF SHELL RESULTS:

POINT      T1        S      M2      T2

  0     37353        0    -718     788
  1     11159    -5055     287     288
  2     -5466    -5502     277    -805
  3    -13742    -3302     -94   -1720
  4    -16085        0    -286   -2062
```

L= 58 TABLE F- 58 W= 32

SHELL DATA:

L = 58.0 FT.
W = 32.0 FT.
R = 25.0 FT.
T = 2.5 IN.
PHI= 38.7 DEG.
G = 50.0 LB/FT2

BEAM DATA:

B = 0.75 FT.
D = 3.75 FT.

EXTERNAL BEAM RESULTS:
F = -17036 LB.
M1 = 292901 LB.FT.
M2 = 2911 LB.FT.
S1 = -17578 LB.
S2 = 490 LB.

EXTERNAL-HALF SHELL RESULTS:

POINT	T1	S	M2	T2
0	35177	922	-25	-119
1	8680	-3807	370	-414
2	-4709	-4049	139	-1260
3	-9486	-2312	-255	-1940
4	-10396	0	-433	-2186

INTERNAL-HALF SHELL RESULTS:

POINT	T1	S	M2	T2
0	37188	0	-808	864
1	10690	-5275	313	295
2	-5642	-5664	298	-916
3	-13323	-3360	-114	-1913
4	-15366	0	-325	-2283

L= 58 TABLE F- 59 W= 34

SHELL DATA:

L = 58.0 FT.
W = 34.0 FT.
R = 27.5 FT.
T = 2.5 IN.
PHI= 37.2 DEG.
G = 50.0 LB/FT2

BEAM DATA:

B = 0.75 FT.
D = 3.75 FT.

EXTERNAL BEAM RESULTS:
F = -19714 LB.
M1 = 297312 LB.FT.
M2 = 2874 LB.FT.
S1 = -18091 LB.
S2 = 544 LB.

EXTERNAL-HALF SHELL RESULTS:

POINT	T1	S	M2	T2
0	35479	1067	-128	-128
1	8423	-3925	373	-452
2	-4817	-4124	158	-1377
3	-9118	-2317	-251	-2108
4	-9781	0	-437	-2368

INTERNAL-HALF SHELL RESULTS:

POINT	T1	S	M2	T2
0	36966	0	-900	935
1	10154	-5476	341	295
2	-5841	-5786	320	-1031
3	-12840	-3383	-132	-2105
4	-14536	0	-362	-2499

L= 58 TABLE F- 60 W= 36

SHELL DATA:

L = 58.0 FT.
W = 36.0 FT.
R = 30.0 FT.
T = 2.5 IN.
PHI= 36.0 DEG.
G = 50.0 LB/FT2

BEAM DATA:

B = 0.75 FT.
D = 3.75 FT.

EXTERNAL BEAM RESULTS:
F = -22101 LB.
M1 = 301440 LB.FT.
M2 = 2851 LB.FT.
S1 = -18558 LB.
S2 = 593 LB.

EXTERNAL-HALF SHELL RESULTS:

POINT	T1	S	M2	T2
0	35778	1197	-226	-136
1	8097	-4049	379	-494
2	-4979	-4183	176	-1499
3	-8743	-2303	-249	-2277
4	-9124	0	-464	-2549

INTERNAL-HALF SHELL RESULTS:

POINT	T1	S	M2	T2
0	36764	0	-992	1002
1	9571	-5663	370	289
2	-6075	-5872	343	-1152
3	-12318	-3372	-149	-2296
4	-13626	0	-398	-2708

L= 60 TABLE F- 61 W= 26

```
SHELL DATA:

    L  = 60.0 FT.
    W  = 26.0 FT.
    R  = 20.0 FT.
    T  =  2.5 IN.
    PHI= 39.1 DEG.
    G  = 50.0 LB/FT2

BEAM DATA:

    B  = 0.75 FT.
    D  = 4.00 FT.

EXTERNAL BEAM RESULTS:
    F  =  -3099 LB.
    M1 = 332454 LB.FT.
    M2 =   1806 LB.FT.
    S1 = -17708 LB.
    S2 =    136 LB.
```

EXTERNAL-HALF SHELL RESULTS:

POINT	T1	S	M2	T2
0	35419	162	131	-97
1	10467	-3784	317	-365
2	-4810	-4160	109	-1026
3	-12531	-2514	-194	-1589
4	-14807	0	-329	-1801

INTERNAL-HALF SHELL RESULTS:

POINT	T1	S	M2	T2
0	47500	0	-538	778
1	15379	-5465	255	326
2	-6345	-6121	254	-677
3	-18495	-3767	-40	-1539
4	-22336	0	-194	-1868

L= 60 TABLE F- 62 W= 28

```
SHELL DATA:

    L  = 60.0 FT.
    W  = 28.0 FT.
    R  = 22.5 FT.
    T  =  2.5 IN.
    PHI= 37.3 DEG.
    G  = 50.0 LB/FT2

BEAM DATA:

    B  = 0.75 FT.
    D  = 4.00 FT.

EXTERNAL BEAM RESULTS:
    F  =  -6428 LB.
    M1 = 336382 LB.FT.
    M2 =   1848 LB.FT.
    S1 = -18265 LB.
    S2 =    207 LB.
```

EXTERNAL-HALF SHELL RESULTS:

POINT	T1	S	M2	T2
0	35620	336	42	-107
1	10352	-3888	316	-399
2	-4714	-4283	121	-1136
3	-12044	-2576	-199	-1759
4	-14133	0	-344	-1994

INTERNAL-HALF SHELL RESULTS:

POINT	T1	S	M2	T2
0	46452	0	-629	856
1	14707	-5683	273	340
2	-6346	-6319	268	-781
3	-17806	-3865	-67	-1735
4	-21350	0	-241	-2097

L= 60 TABLE F- 63 W= 30

```
SHELL DATA:

    L  = 60.0 FT.
    W  = 30.0 FT.
    R  = 22.5 FT.
    T  =  2.5 IN.
    PHI= 40.5 DEG.
    G  = 50.0 LB/FT2

BEAM DATA:

    B  = 0.75 FT.
    D  = 4.00 FT.

EXTERNAL BEAM RESULTS:
    F  = -12241 LB.
    M1 = 328583 LB.FT.
    M2 =   2719 LB.FT.
    S1 = -18467 LB.
    S2 =    366 LB.
```

EXTERNAL-HALF SHELL RESULTS:

POINT	T1	S	M2	T2
0	34887	640	45	-117
1	9237	-3729	357	-385
2	-4649	-4030	128	-1135
3	-10384	-2358	-241	-1755
4	-11754	0	-406	-1983

INTERNAL-HALF SHELL RESULTS:

POINT	T1	S	M2	T2
0	39675	0	-719	789
1	12041	-5215	292	293
2	-5715	-5707	283	-800
3	-14757	-3442	-92	-1720
4	-17377	0	-285	-2066

L= 60 TABLE F- 64 W= 32

SHELL DATA:

```
L   = 60.0 FT.
W   = 32.0 FT.
R   = 25.0 FT.
T   =  2.5 IN.
PHI= 38.7 DEG.
G   = 50.0 LB/FT2
```

BEAM DATA:

```
B   = 0.75 FT.
D   = 4.00 FT.
```

EXTERNAL BEAM RESULTS:

```
F   = -15121 LB.
M1  = 333619 LB.FT.
M2  =   2663 LB.FT.
S1  = -19035 LB.
S2  =    422 LB.
```

EXTERNAL-HALF SHELL RESULTS:

POINT	T1	S	M2	T2
0	35181	791	-59	-128
1	9074	-3843	356	-421
2	-4686	-4131	145	-1248
3	-10032	-2393	-237	-1923
4	-11201	0	-412	-2170

INTERNAL-HALF SHELL RESULTS:

POINT	T1	S	M2	T2
0	39440	0	-811	866
1	11559	-5439	317	301
2	-5875	-5880	304	-910
3	-14320	-3510	-112	-1914
4	-16642	0	-326	-2288

L= 60 TABLE F- 65 W= 34

SHELL DATA:

```
L   = 60.0 FT.
W   = 34.0 FT.
R   = 25.0 FT.
T   =  2.5 IN.
PHI= 41.7 DEG.
G   = 50.0 LB/FT2
```

BEAM DATA:

```
B   = 0.75 FT.
D   = 4.00 FT.
```

EXTERNAL BEAM RESULTS:

```
F   = -19855 LB.
M1  = 329952 LB.FT.
M2  =   3413 LB.FT.
S1  = -19339 LB.
S2  =    554 LB.
```

EXTERNAL-HALF SHELL RESULTS:

POINT	T1	S	M2	T2
0	34887	1039	-96	-143
1	7851	-3729	379	-419
2	-4910	-3839	155	-1253
3	-8593	-2114	-255	-1900
4	-8976	0	-440	-2127

INTERNAL-HALF SHELL RESULTS:

POINT	T1	S	M2	T2
0	34967	0	-894	800
1	9473	-5088	341	246
2	-5659	-5339	321	-939
3	-12022	-3097	-130	-1895
4	-13463	0	-360	-2244

L= 60 TABLE F- 66 W= 36

SHELL DATA:

```
L   = 60.0 FT.
W   = 36.0 FT.
R   = 27.5 FT.
T   =  2.5 IN.
PHI= 39.9 DEG.
G   = 50.0 LB/FT2
```

BEAM DATA:

```
B   = 0.75 FT.
D   = 4.00 FT.
```

EXTERNAL BEAM RESULTS:

```
F   = -21982 LB.
M1  = 335410 LB.FT.
M2  =   3291 LB.FT.
S1  = -19849 LB.
S2  =    591 LB.
```

EXTERNAL-HALF SHELL RESULTS:

POINT	T1	S	M2	T2
0	35240	1151	-201	-152
1	7627	-3866	382	-462
2	-5062	-3923	173	-1376
3	-8331	-2121	-250	-2070
4	-8487	0	-443	-2310

INTERNAL-HALF SHELL RESULTS:

POINT	T1	S	M2	T2
0	35264	0	-986	873
1	9057	-5320	370	244
2	-5955	-5479	343	-1058
3	-11699	-3121	-148	-2086
4	-12799	0	-395	-2456

L= 62 TABLE F- 67 W= 26

SHELL DATA:

L = 62.0 FT.
W = 26.0 FT.
R = 20.0 FT.
T = 2.5 IN.
PHI= 39.1 DEG.
G = 50.0 LB/FT2

BEAM DATA:

B = 0.75 FT.
D = 4.25 FT.

EXTERNAL BEAM RESULTS:
F = -3975 LB.
M1 = 395576 LB.FT.
M2 = 1641 LB.FT.
S1 = -20447 LB.
S2 = 138 LB.

EXTERNAL-HALF SHELL RESULTS:

POINT	T1	S	M2	T2
0	37099	201	97	-103
1	11306	-3838	307	-360
2	-4877	-4268	118	-1012
3	-13304	-2597	-174	-1572
4	-15848	0	-306	-1785

INTERNAL-HALF SHELL RESULTS:

POINT	T1	S	M2	T2
0	53570	0	-571	914
1	17464	-5978	280	428
2	-7106	-6711	279	-640
3	-20960	-4137	-36	-1559
4	-25367	0	-201	-1909

L= 62 TABLE F- 68 W= 28

SHELL DATA:

L = 62.0 FT.
W = 28.0 FT.
R = 22.5 FT.
T = 2.5 IN.
PHI= 37.3 DEG.
G = 50.0 LB/FT2

BEAM DATA:

B = 0.75 FT.
D = 4.25 FT.

EXTERNAL BEAM RESULTS:
F = -7236 LB.
M1 = 399335 LB.FT.
M2 = 1695 LB.FT.
S1 = -20992 LB.
S2 = 206 LB.

EXTERNAL-HALF SHELL RESULTS:

POINT	T1	S	M2	T2
0	37261	366	12	-112
1	11180	-3954	308	-394
2	-4783	-4406	129	-1123
3	-12821	-2670	-183	-1746
4	-15180	0	-325	-1981

INTERNAL-HALF SHELL RESULTS:

POINT	T1	S	M2	T2
0	52185	0	-668	999
1	16666	-6195	296	448
2	-7067	-6909	291	-742
3	-20125	-4236	-66	-1756
4	-24199	0	-252	-2141

L= 62 TABLE F- 69 W= 30

SHELL DATA:

L = 62.0 FT.
W = 30.0 FT.
R = 22.5 FT.
T = 2.5 IN.
PHI= 40.5 DEG.
G = 50.0 LB/FT2

BEAM DATA:

B = 0.75 FT.
D = 4.25 FT.

EXTERNAL BEAM RESULTS:
F = -13270 LB.
M1 = 389578 LB.FT.
M2 = 2627 LB.FT.
S1 = -21149 LB.
S2 = 368 LB.

EXTERNAL-HALF SHELL RESULTS:

POINT	T1	S	M2	T2
0	36453	672	30	-121
1	10065	-3802	335	-379
2	-4691	-4172	135	-1125
3	-11133	-2469	-233	-1750
4	-12776	0	-399	-1982

INTERNAL-HALF SHELL RESULTS:

POINT	T1	S	M2	T2
0	44317	0	-764	921
1	13627	-5660	315	393
2	-6299	-6222	306	-764
3	-16635	-3767	-94	-1740
4	-19683	0	-301	-2108

```
L= 62       TABLE F- 70        W= 32
************************************

SHELL DATA:

    L   = 62.0 FT.
    W   = 32.0 FT.
    R   = 25.0 FT.
    T   =  2.5 IN.
    PHI= 38.7 DEG.
    G   = 50.0 LB/FT2

BEAM DATA:

    B   = 0.75 FT.
    D   = 4.25 FT.

EXTERNAL BEAM RESULTS:
    F   = -16238 LB.
    M1  = 394687 LB.FT.
    M2  =   2575 LB.FT.
    S1  = -21730 LB.
    S2  =    424 LB.
```

EXTERNAL-HALF SHELL RESULTS:

POINT	T1	S	M2	T2
0	36704	822	-74	-131
1	9897	-3918	355	-415
2	-4716	-4282	152	-1237
3	-10770	-2512	-231	-1919
4	-12213	0	-407	-2170

INTERNAL-HALF SHELL RESULTS:

POINT	T1	S	M2	T2
0	43893	0	-860	1003
1	13073	-5886	341	406
2	-6437	-6399	327	-871
3	-16115	-3838	-116	-1935
4	-18845	0	-345	-2333

```
L= 62       TABLE F- 71        W= 34
************************************

SHELL DATA:

    L   = 62.0 FT.
    W   = 34.0 FT.
    R   = 25.0 FT.
    T   =  2.5 IN.
    PHI= 41.7 DEG.
    G   = 50.0 LB/FT2

BEAM DATA:

    B   = 0.75 FT.
    D   = 4.25 FT.

EXTERNAL BEAM RESULTS:
    F   = -21440 LB.
    M1  = 389731 LB.FT.
    M2  =   3373 LB.FT.
    S1  = -22040 LB.
    S2  =    563 LB.
```

EXTERNAL-HALF SHELL RESULTS:

POINT	T1	S	M2	T2
0	36323	1086	-106	-147
1	8680	-3796	382	-409
2	-4899	-4000	162	-1240
3	-9283	-2246	-253	-1901
4	-9942	0	-443	-2136

INTERNAL-HALF SHELL RESULTS:

POINT	T1	S	M2	T2
0	38673	0	-949	928
1	10730	-5481	365	345
2	-6131	-5801	345	-901
3	-13513	-3392	-137	-1915
4	-15290	0	-383	-2287

```
L= 62       TABLE F- 72        W= 36
************************************

SHELL DATA:

    L   = 62.0 FT.
    W   = 36.0 FT.
    R   = 27.5 FT.
    T   =  2.5 IN.
    PHI= 39.9 DEG.
    G   = 50.0 LB/FT2

BEAM DATA:

    B   = 0.75 FT.
    D   = 4.25 FT.

EXTERNAL BEAM RESULTS:
    F   = -23683 LB.
    M1  = 395430 LB.FT.
    M2  =   3248 LB.FT.
    S1  = -22571 LB.
    S2  =    602 LB.
```

EXTERNAL-HALF SHELL RESULTS:

POINT	T1	S	M2	T2
0	36637	1200	-213	-156
1	8454	-3932	385	-451
2	-5041	-4092	181	-1361
3	-9011	-2262	-250	-2071
4	-9445	0	-447	-2321

INTERNAL-HALF SHELL RESULTS:

POINT	T1	S	M2	T2
0	38860	0	-1046	1007
1	10271	-5717	395	368
2	-6415	-5947	369	-1018
3	-13141	-3422	-155	-2107
4	-14563	0	-421	-2502

L= 64 TABLE F- 73 W= 28

SHELL DATA:

L = 64.0 FT.
W = 28.0 FT.
R = 22.5 FT.
T = 2.5 IN.
PHI= 37.3 DEG.
G = 50.0 LB/FT2

BEAM DATA:

B = 0.75 FT.
D = 4.25 FT.

EXTERNAL BEAM RESULTS:
F = -6727 LB.
M1 = 424076 LB.FT.
M2 = 1807 LB.FT.
S1 = -21494 LB.
S2 = 194 LB.

EXTERNAL-HALF SHELL RESULTS:

POINT	T1	S	M2	T2
0	39635	330	63	-108
1	11929	-4127	332	-398
2	-5130	-4594	129	-1135
3	-13799	-2787	-200	-1765
4	-16364	0	-349	-2003

INTERNAL-HALF SHELL RESULTS:

POINT	T1	S	M2	T2
0	55388	0	-664	1000
1	17831	-6387	303	451
2	-7439	-7141	299	-739
3	-21481	-4388	-60	-1756
4	-25897	0	-247	-2143

L= 64 TABLE F- 74 W= 30

SHELL DATA:

L = 64.0 FT.
W = 30.0 FT.
R = 22.5 FT.
T = 2.5 IN.
PHI= 40.5 DEG.
G = 50.0 LB/FT2

BEAM DATA:

B = 0.75 FT.
D = 4.25 FT.

EXTERNAL BEAM RESULTS:
F = -13157 LB.
M1 = 413179 LB.FT.
M2 = 2798 LB.FT.
S1 = -21632 LB.
S2 = 361 LB.

EXTERNAL-HALF SHELL RESULTS:

POINT	T1	S	M2	T2
0	38728	645	90	-115
1	10754	-3968	383	-382
2	-5005	-4355	133	-1136
3	-11996	-2584	-255	-1770
4	-13814	0	-430	-2006

INTERNAL-HALF SHELL RESULTS:

POINT	T1	S	M2	T2
0	46945	0	-764	922
1	14612	-5830	320	396
2	-6589	-6437	312	-760
3	-17772	-3911	-90	-1740
4	-21123	0	-299	-2110

L= 64 TABLE F- 75 W= 32

SHELL DATA:

L = 64.0 FT.
W = 32.0 FT.
R = 25.0 FT.
T = 2.5 IN.
PHI= 38.7 DEG.
G = 50.0 LB/FT2

BEAM DATA:

B = 0.75 FT.
D = 4.25 FT.

EXTERNAL BEAM RESULTS:
F = -16554 LB.
M1 = 418685 LB.FT.
M2 = 2752 LB.FT.
S1 = -22259 LB.
S2 = 424 LB.

EXTERNAL-HALF SHELL RESULTS:

POINT	T1	S	M2	T2
0	38990	812	-18	-127
1	10600	-4078	381	-417
2	-5002	-4465	151	-1247
3	-11596	-2629	-253	-1938
4	-13211	0	-436	-2193

INTERNAL-HALF SHELL RESULTS:

POINT	T1	S	M2	T2
0	46441	0	-861	1005
1	14040	-6060	345	411
2	-6710	-6623	333	-866
3	-17228	-3991	-114	-1935
4	-20262	0	-344	-2337

```
L= 64        TABLE F- 76         W= 34
**************************************

SHELL DATA:

     L   = 64.0 FT.
     W   = 34.0 FT.
     R   = 25.0 FT.
     T   =  2.5 IN.
     PHI= 41.7 DEG.
     G   = 50.0 LB/FT2

BEAM DATA:

     B   = 0.75 FT.
     D   = 4.25 FT.

EXTERNAL BEAM RESULTS:
     F   = -22279 LB.
     M1  = 413113 LB.FT.
     M2  =   3605 LB.FT.
     S1  = -22584 LB.
     S2  =    571 LB.

EXTERNAL-HALF SHELL RESULTS:

POINT      T1        S       M2      T2

  0      38547     1093     -49    -142
  1       9333    -3943     410    -409
  2      -5159    -4174     161   -1248
  3     -10008    -2359    -277   -1919
  4     -10809        0    -475   -2159

INTERNAL-HALF SHELL RESULTS:

POINT      T1        S       M2      T2

  0      40788        0    -954     931
  1      11573    -5636     369     352
  2      -6335    -6013     352    -894
  3     -14472    -3542    -135   -1916
  4     -16532        0    -385   -2293
```

```
L= 64        TABLE F- 77         W= 36
**************************************

SHELL DATA:

     L   = 64.0 FT.
     W   = 36.0 FT.
     R   = 27.5 FT.
     T   =  2.5 IN.
     PHI= 39.9 DEG.
     G   = 50.0 LB/FT2

BEAM DATA:

     B   = 0.75 FT.
     D   = 4.25 FT.

EXTERNAL BEAM RESULTS:
     F   = -24909 LB.
     M1  = 419244 LB.FT.
     M2  =   3481 LB.FT.
     S1  = -23161 LB.
     S2  =    615 LB.

EXTERNAL-HALF SHELL RESULTS:

POINT      T1        S       M2      T2

  0      38876     1222    -162    -152
  1       9122    -4075     411    -449
  2      -5281    -4269     181   -1366
  3      -9715    -2379    -272   -2088
  4     -10292        0    -479   -2343

INTERNAL-HALF SHELL RESULTS:

POINT      T1        S       M2      T2

  0      40918        0   -1052    1010
  1      11108    -5874     400     357
  2      -6604    -6169     375   -1009
  3     -14088    -3582    -155   -2109
  4     -15799        0    -425   -2511
```

```
L= 64        TABLE F- 78         W= 38
**************************************

SHELL DATA:

     L   = 64.0 FT.
     W   = 38.0 FT.
     R   = 30.0 FT.
     T   =  2.5 IN.
     PHI= 38.4 DEG.
     G   = 50.0 LB/FT2

BEAM DATA:

     B   = 0.75 FT.
     D   = 4.25 FT.

EXTERNAL BEAM RESULTS:
     F   = -27254 LB.
     M1  = 424874 LB.FT.
     M2  =   3391 LB.FT.
     S1  = -23683 LB.
     S2  =    656 LB.

EXTERNAL-HALF SHELL RESULTS:

POINT      T1        S       M2      T2

  0      39196     1337    -269    -160
  1       8828    -4214     416    -492
  2      -5457    -4346     200   -1489
  3      -9396    -2377    -269   -2258
  4      -9707        0    -485   -2526

INTERNAL-HALF SHELL RESULTS:

POINT      T1        S       M2      T2

  0      40986        0   -1151    1084
  1      10566    -6095     431     355
  2      -6900    -6285     399   -1129
  3     -13631    -3585    -174   -2300
  4     -14940        0    -463   -2722
```

L= 66 TABLE F- 79 W= 28

SHELL DATA:

L = 66.0 FT.
W = 28.0 FT.
R = 22.5 FT.
T = 2.5 IN.
PHI= 37.3 DEG.
G = 50.0 LB/FT2

BEAM DATA:

B = 0.75 FT.
D = 4.50 FT.

EXTERNAL BEAM RESULTS:
F = -4752 LB.
M1 = 477433 LB.FT.
M2 = 1495 LB.FT.
S1 = -23210 LB.
S2 = 137 LB.

EXTERNAL-HALF SHELL RESULTS:

POINT	T1	S	M2	T2
0	39739	226	0	-119
1	12211	-4140	305	-404
2	-5186	-4616	137	-1118
3	-14307	-2814	-166	-1732
4	-17072	0	-306	-1965

INTERNAL-HALF SHELL RESULTS:

POINT	T1	S	M2	T2
0	58697	0	-660	1001
1	19030	-6578	310	453
2	-7826	-7373	306	-736
3	-22877	-4539	-54	-1755
4	-27644	0	-242	-2144

L= 66 TABLE F- 80 W= 30

SHELL DATA:

L = 66.0 FT.
W = 30.0 FT.
R = 22.5 FT.
T = 2.5 IN.
PHI= 40.5 DEG.
G = 50.0 LB/FT2

BEAM DATA:

B = 0.75 FT.
D = 4.50 FT.

EXTERNAL BEAM RESULTS:
F = -11066 LB.
M1 = 464557 LB.FT.
M2 = 2511 LB.FT.
S1 = -23276 LB.
S2 = 298 LB.

EXTERNAL-HALF SHELL RESULTS:

POINT	T1	S	M2	T2
0	38792	526	46	-124
1	11087	-3998	364	-389
2	-5031	-4409	140	-1125
3	-12535	-2636	-231	-1749
4	-14582	0	-399	-1982

INTERNAL-HALF SHELL RESULTS:

POINT	T1	S	M2	T2
0	49664	0	-762	923
1	15625	-6001	325	400
2	-6891	-6650	319	-757
3	-18944	-4053	-86	-1740
4	-22604	0	-296	-2113

L= 66 TABLE F- 81 W= 32

SHELL DATA:

L = 66.0 FT.
W = 32.0 FT.
R = 25.0 FT.
T = 2.5 IN.
PHI= 38.7 DEG.
G = 50.0 LB/FT2

BEAM DATA:

B = 0.75 FT.
D = 4.50 FT.

EXTERNAL BEAM RESULTS:
F = -14297 LB.
M1 = 470171 LB.FT.
M2 = 2476 LB.FT.
S1 = -23891 LB.
S2 = 357 LB.

EXTERNAL-HALF SHELL RESULTS:

POINT	T1	S	M2	T2
0	39037	680	-57	-135
1	10943	-4116	363	-425
2	-5023	-4530	156	-1237
3	-12145	-2691	-231	-1919
4	-13999	0	-409	-2172

INTERNAL-HALF SHELL RESULTS:

POINT	T1	S	M2	T2
0	49080	0	-862	1007
1	15034	-6235	350	415
2	-6997	-6846	339	-861
3	-18375	-4141	-110	-1936
4	-21718	0	-343	-2340

L= 66 TABLE F- 82 W= 34

SHELL DATA:

L = 66.0 FT.
W = 34.0 FT.
R = 25.0 FT.
T = 2.5 IN.
PHI= 41.7 DEG.
G = 50.0 LB/FT2

BEAM DATA:

B = 0.75 FT.
D = 4.50 FT.

EXTERNAL BEAM RESULTS:
F = -20308 LB.
M1 = 463139 LB.FT.
M2 = 3345 LB.FT.
S1 = -24202 LB.
S2 = 505 LB.

EXTERNAL-HALF SHELL RESULTS:

POINT	T1	S	M2	T2
0	38516	966	-79	-150
1	9734	-3976	397	-416
2	-5117	-4258	167	-1237
3	-10547	-2442	-261	-1905
4	-11614	0	-457	-2147

INTERNAL-HALF SHELL RESULTS:

POINT	T1	S	M2	T2
0	42983	0	-957	933
1	12439	-5792	374	359
2	-6552	-6223	358	-888
3	-15459	-3689	-134	-1917
4	-17805	0	-386	-2299

L= 66 TABLE F= 83 W= 36

SHELL DATA:

L = 66.0 FT.
W = 36.0 FT.
R = 27.5 FT.
T = 2.5 IN.
PHI= 39.9 DEG.
G = 50.0 LB/FT2

BEAM DATA:

B = 0.75 FT.
D = 4.50 FT.

EXTERNAL BEAM RESULTS:
F = -22876 LB.
M1 = 469539 LB.FT.
M2 = 3229 LB.FT.
S1 = -24783 LB.
S2 = 548 LB.

EXTERNAL-HALF SHELL RESULTS:

POINT	T1	S	M2	T2
0	38827	1088	-190	-159
1	9534	-4110	399	-456
2	-5228	-4361	186	-1355
3	-10257	-2471	-258	-2076
4	-11108	0	-463	-2334

INTERNAL-HALF SHELL RESULTS:

POINT	T1	S	M2	T2
0	43057	0	-1057	1013
1	11966	-6033	404	366
2	-6807	-6390	382	-1000
3	-15063	-3739	-154	-2110
4	-17065	0	-428	-2518

L= 66 TABLE F- 84 W= 38

SHELL DATA:

L = 66.0 FT.
W = 38.0 FT.
R = 30.0 FT.
T = 2.5 IN.
PHI= 38.4 DEG.
G = 50.0 LB/FT2

BEAM DATA:

B = 0.75 FT.
D = 4.50 FT.

EXTERNAL BEAM RESULTS:
F = -25195 LB.
M1 = 475421 LB.FT.
M2 = 3146 LB.FT.
S1 = -25313 LB.
S2 = 587 LB.

EXTERNAL-HALF SHELL RESULTS:

POINT	T1	S	M2	T2
0	39127	1199	-295	-167
1	9255	-4249	406	-498
2	-5391	-4448	206	-1477
3	-9941	-2479	-257	-2247
4	-10534	0	-471	-2519

INTERNAL-HALF SHELL RESULTS:

POINT	T1	S	M2	T2
0	43055	0	-1158	1088
1	11415	-6255	436	366
2	-7086	-6515	407	-1118
3	-14591	-3753	-174	-2302
4	-16196	0	-467	-2731

L= 68 TABLE F- 85 W= 30

SHELL DATA:

L = 68.0 FT.
W = 30.0 FT.
R = 22.5 FT.
T = 2.5 IN.
PHI= 40.5 DEG.
G = 50.0 LB/FT2

BEAM DATA:

B = 0.75 FT.
D = 4.50 FT.

EXTERNAL BEAM RESULTS:
F = -10681 LB.
M1 = 491198 LB.FT.
M2 = 2664 LB.FT.
S1 = -23779 LB.
S2 = 287 LB.

EXTERNAL-HALF SHELL RESULTS:

POINT	T1	S	M2	T2
0	41084	493	104	-119
1	11789	-4164	391	-392
2	-5362	-4580	139	-1137
3	-13443	-2748	-252	-1768
4	-15676	0	-428	-2004

INTERNAL-HALF SHELL RESULTS:

POINT	T1	S	M2	T2
0	52475	0	-761	925
1	16667	-6173	331	403
2	-7208	-6863	325	-753
3	-20150	-4194	-81	-1740
4	-24125	0	-292	-2115

L= 68 TABLE F- 86 W= 32

SHELL DATA:

L = 68.0 FT.
W = 32.0 FT.
R = 25.0 FT.
T = 2.5 IN.
PHI= 38.7 DEG.
G = 50.0 LB/FT2

BEAM DATA:

B = 0.75 FT.
D = 4.50 FT.

EXTERNAL BEAM RESULTS:
F = -14336 LB.
M1 = 497207 LB.FT.
M2 = 2635 LB.FT.
S1 = -24440 LB.
S2 = 353 LB.

EXTERNAL-HALF SHELL RESULTS:

POINT	T1	S	M2	T2
0	41339	662	-4	-130
1	11659	-4277	389	-427
2	-5325	-4711	155	-1247
3	-13015	-2805	-251	-1937
4	-15052	0	-437	-2194

INTERNAL-HALF SHELL RESULTS:

POINT	T1	S	M2	T2
0	51809	0	-862	1008
1	16056	-6411	355	419
2	-7297	-7068	345	-857
3	-19555	-4290	-107	-1936
4	-23213	0	-341	-2343

L= 68 TABLE F- 87 W= 34

SHELL DATA:

L = 68.0 FT.
W = 34.0 FT.
R = 25.0 FT.
T = 2.5 IN.
PHI= 41.7 DEG.
G = 50.0 LB/FT2

BEAM DATA:

B = 0.75 FT.
D = 4.50 FT.

EXTERNAL BEAM RESULTS:
F = -20859 LB.
M1 = 489376 LB.FT.
M2 = 3560 LB.FT.
S1 = -24757 LB.
S2 = 508 LB.

EXTERNAL-HALF SHELL RESULTS:

POINT	T1	S	M2	T2
0	40748	963	-22	-145
1	10401	-4125	424	-416
2	-5391	-4431	166	-1246
3	-11313	-2553	-285	-1923
4	-12535	0	-489	-2169

INTERNAL-HALF SHELL RESULTS:

POINT	T1	S	M2	T2
0	45258	0	-959	935
1	13327	-5948	378	364
2	-6782	-6431	364	-882
3	-16474	-3834	-131	-1918
4	-19111	0	-386	-2304

L= 68 TABLE F- 88 W= 36

SHELL DATA:

L = 68.0 FT.
W = 36.0 FT.
R = 27.5 FT.
T = 2.5 IN.
PHI= 39.9 DEG.
G = 50.0 LB/FT2

BEAM DATA:

B = 0.75 FT.
D = 4.50 FT.

EXTERNAL BEAM RESULTS:
F = -23822 LB.
M1 = 496232 LB.FT.
M2 = 3445 LB.FT.
S1 = -25384 LB.
S2 = 557 LB.

EXTERNAL-HALF SHELL RESULTS:

POINT	T1	S	M2	T2
0	41073	1100	-138	-155
1	10215	-4255	425	-455
2	-5480	-4537	186	-1362
3	-11000	-2585	-280	-2093
4	-12007	0	-493	-2356

INTERNAL-HALF SHELL RESULTS:

POINT	T1	S	M2	T2
0	45274	0	-1061	1015
1	12847	-6193	409	373
2	-7022	-6608	388	-993
3	-16066	-3894	-153	-2112
4	-18362	0	-430	-2525

L= 68 TABLE F- 89 W= 38

SHELL DATA:

L = 68.0 FT.
W = 38.0 FT.
R = 30.0 FT.
T = 2.5 IN.
PHI= 38.4 DEG.
G = 50.0 LB/FT2

BEAM DATA:

B = 0.75 FT.
D = 4.50 FT.

EXTERNAL BEAM RESULTS:
F = -26512 LB.
M1 = 502490 LB.FT.
M2 = 3365 LB.FT.
S1 = -25954 LB.
S2 = 601 LB.

EXTERNAL-HALF SHELL RESULTS:

POINT	T1	S	M2	T2
0	41382	1224	-249	-164
1	9946	-4391	430	-496
2	-5624	-4627	206	-1482
3	-10663	-2598	-278	-2263
4	-11412	0	-501	-2541

INTERNAL-HALF SHELL RESULTS:

POINT	T1	S	M2	T2
0	45203	0	-1165	1091
1	12286	-6417	441	375
2	-7285	-6743	414	-1109
3	-15577	-3917	-173	-2304
4	-17482	0	-471	-2740

L= 68 TABLE F- 90 W= 40

SHELL DATA:

L = 68.0 FT.
W = 40.0 FT.
R = 32.5 FT.
T = 2.5 IN.
PHI= 37.1 DEG.
G = 50.0 LB/FT2

BEAM DATA:

B = 0.75 FT.
D = 4.50 FT.

EXTERNAL BEAM RESULTS:
F = -28926 LB.
M1 = 508383 LB.FT.
M2 = 3307 LB.FT.
S1 = -26479 LB.
S2 = 642 LB.

EXTERNAL-HALF SHELL RESULTS:

POINT	T1	S	M2	T2
0	41689	1336	-355	-172
1	9608	-4531	438	-538
2	-5820	-4700	226	-1605
3	-10306	-2591	-277	-2434
4	-10760	0	-510	-2724

INTERNAL-HALF SHELL RESULTS:

POINT	T1	S	M2	T2
0	45125	0	-1268	1162
1	11666	-6628	474	370
2	-7580	-6841	439	-1230
3	-15036	-3906	-193	-2494
4	-16501	0	-511	-2949

L= 70 TABLE F- 91 W= 30

SHELL DATA:

L = 70.0 FT.
W = 30.0 FT.
R = 22.5 FT.
T = 2.5 IN.
PHI= 40.5 DEG.
G = 50.0 LB/FT2

BEAM DATA:

B = 0.75 FT.
D = 4.75 FT.

EXTERNAL BEAM RESULTS:

F = -8509 LB.
M1 = 549033 LB.FT.
M2 = 2358 LB.FT.
S1 = -25523 LB.
S2 = 228 LB.

EXTERNAL-HALF SHELL RESULTS:

POINT	T1	S	M2	T2
0	41162	381	54	-129
1	12102	-4180	369	-399
2	-5403	-4632	146	-1125
3	-13977	-2791	-224	-1744
4	-16430	0	-393	-1978

INTERNAL-HALF SHELL RESULTS:

POINT	T1	S	M2	T2
0	55376	0	-758	926
1	17738	-6345	337	406
2	-7537	-7074	332	-751
3	-21391	-4334	-75	-1740
4	-25687	0	-288	-2116

L= 70 TABLE F- 92 W= 32

SHELL DATA:

L = 70.0 FT.
W = 32.0 FT.
R = 25.0 FT.
T = 2.5 IN.
PHI= 38.7 DEG.
G = 50.0 LB/FT2

BEAM DATA:

B = 0.75 FT.
D = 4.75 FT.

EXTERNAL BEAM RESULTS:

F = -11936 LB.
M1 = 555134 LB.FT.
M2 = 2342 LB.FT.
S1 = -26165 LB.
S2 = 288 LB.

EXTERNAL-HALF SHELL RESULTS:

POINT	T1	S	M2	T2
0	41404	535	-48	-139
1	11979	-4311	369	-435
2	-5365	-4767	162	-1236
3	-13565	-2858	-227	-1915
4	-15830	0	-406	-2170

INTERNAL-HALF SHELL RESULTS:

POINT	T1	S	M2	T2
0	54628	0	-861	1010
1	17105	-6588	360	423
2	-7611	-7289	351	-854
3	-20769	-4438	-103	-1936
4	-24746	0	-338	-2346

L= 70 TABLE F- 93 W= 34

SHELL DATA:

L = 70.0 FT.
W = 34.0 FT.
R = 25.0 FT.
T = 2.5 IN.
PHI= 41.7 DEG.
G = 50.0 LB/FT2

BEAM DATA:

B = 0.75 FT.
D = 4.75 FT.

EXTERNAL BEAM RESULTS:

F = -18668 LB.
M1 = 545564 LB.FT.
M2 = 3285 LB.FT.
S1 = -26455 LB.
S2 = 444 LB.

EXTERNAL-HALF SHELL RESULTS:

POINT	T1	S	M2	T2
0	40745	837	-56	-152
1	10777	-4160	410	-423
2	-5374	-4505	171	-1236
3	-11859	-2626	-266	-1908
4	-13336	0	-466	-2155

INTERNAL-HALF SHELL RESULTS:

POINT	T1	S	M2	T2
0	47611	0	-961	936
1	14238	-6106	383	369
2	-7025	-6639	370	-877
3	-17518	-3977	-129	-1918
4	-20448	0	-386	-2308

L= 70 TABLE F- 94 W= 36
****

SHELL DATA:

L = 70.0 FT.
W = 36.0 FT.
R = 27.5 FT.
T = 2.5 IN.
PHI= 39.9 DEG.
G = 50.0 LB/FT2

BEAM DATA:

B = 0.75 FT.
D = 4.75 FT.

EXTERNAL BEAM RESULTS:
F = -21539 LB.
M1 = 552681 LB.FT.
M2 = 3179 LB.FT.
S1 = -27082 LB.
S2 = 490 LB.

EXTERNAL-HALF SHELL RESULTS:

POINT	T1	S	M2	T2
0	41053	966	-169	-162
1	10602	-4293	411	-462
2	-5454	-4620	191	-1352
3	-11551	-2667	-264	-2079
4	-12821	0	-474	-2344

INTERNAL-HALF SHELL RESULTS:

POINT	T1	S	M2	T2
0	47570	0	-1065	1017
1	13749	-6354	413	379
2	-7251	-6825	395	-987
3	-17096	-4046	-151	-2113
4	-19690	0	-431	-2530

L= 70 TABLE F- 95 W= 38
****

SHELL DATA:

L = 70.0 FT.
W = 38.0 FT.
R = 30.0 FT.
T = 2.5 IN.
PHI= 38.4 DEG.
G = 50.0 LB/FT2

BEAM DATA:

B = 0.75 FT.
D = 4.75 FT.

EXTERNAL BEAM RESULTS:
F = -24176 LB.
M1 = 559173 LB.FT.
M2 = 3108 LB.FT.
S1 = -27656 LB.
S2 = 533 LB.

EXTERNAL-HALF SHELL RESULTS:

POINT	T1	S	M2	T2
0	41345	1085	-277	-171
1	10347	-4430	418	-502
2	-5586	-4720	211	-1471
3	-11217	-2689	-263	-2251
4	-12238	0	-484	-2531

INTERNAL-HALF SHELL RESULTS:

POINT	T1	S	M2	T2
0	47429	0	-1170	1093
1	13178	-6580	445	383
2	-7496	-6969	420	-1101
3	-16591	-4079	-172	-2305
4	-18797	0	-474	-2747

L= 70 TABLE F- 96 W= 40
****

SHELL DATA:

L = 70.0 FT.
W = 40.0 FT.
R = 30.0 FT.
T = 2.5 IN.
PHI= 40.9 DEG.
G = 50.0 LB/FT2

BEAM DATA:

B = 0.75 FT.
D = 4.75 FT.

EXTERNAL BEAM RESULTS:
F = -28990 LB.
M1 = 555531 LB.FT.
M2 = 3831 LB.FT.
S1 = -28006 LB.
S2 = 646 LB.

EXTERNAL-HALF SHELL RESULTS:

POINT	T1	S	M2	T2
0	41133	1301	-321	-188
1	9043	-4342	442	-508
2	-5929	-4420	221	-1485
3	-9799	-2389	-279	-2230
4	-9971	0	-510	-2487

INTERNAL-HALF SHELL RESULTS:

POINT	T1	S	M2	T2
0	43248	0	-1259	1022
1	10980	-6242	473	317
2	-7445	-6389	438	-1140
3	-14226	-3613	-190	-2283
4	-15407	0	-507	-2691

INDEX